VIRTUAL OBSERVATORIES OF THE FUTURE

COVER ILLUSTRATION:

A portion of the plate showing the constellation of Argo, from
Johannes Hevelius, *Firmamentum Sobiescianum Sive Uranographia* (1690).
From *Out of This World,* an on-line exhibit at
http://www.lhl.lib.mo.us/pubserv/hos/stars/welcome.htm

A SERIES OF BOOKS ON RECENT DEVELOPMENTS IN ASTRONOMY AND ASTROPHYSICS

Publisher

THE ASTRONOMICAL SOCIETY OF THE PACIFIC
390 Ashton Avenue, San Francisco, California, USA 94112-1722
Phone: (415) 337-1100   Fax: (415) 337-5205
E-Mail: catalog@aspsky.org   Web Site: www.aspsky.org

ASP CONFERENCE SERIES - EDITORIAL STAFF
Managing Editor: D. H. McNamara   LaTeX-Computer Consultant: T. J. Mahoney
Associate Managing Editor: J. W. Moody   Production Manager: Enid L. Livingston

PO Box 24453, 211 KMB, Brigham Young University, Provo, Utah, 84602-4463
Phone: (801) 378-2111   Fax: (801) 378-4049   E-Mail: pasp@byu.edu

ASP CONFERENCE SERIES PUBLICATION COMMITTEE:
Alexei V. Filippenko   Geoffrey Marcy
Ray Norris   Donald Terndrup
Frank X. Timmes   C. Megan Urry

A listing of all other ASP Conference Series Volumes and IAU Volumes
published by the ASP is cited at the back of this volume

ASTRONOMICAL SOCIETY OF THE PACIFIC
CONFERENCE SERIES

Volume 225

VIRTUAL OBSERVATORIES OF THE FUTURE

Proceedings of a conference held at
California Institute of Technology, Pasadena, California, USA
13-16 June 2000

Edited by

Robert J. Brunner
California Institute of Technology, Pasadena, California, USA

S. George Djorgovski
California Institute of Technology, Pasadena, California, USA

and

Alex S. Szalay
The Johns Hopkins University, Baltimore, Maryland, USA

© 2001 by Astronomical Society of the Pacific. All Rights Reserved

No part of the material protected by this copyright notice may be reproduced or utilized in any form or by any means – graphic, electronic, or mechanical including photocopying, taping, recording or by any information storage and retrieval system, without written permission from the publisher.

Library of Congress Cataloging in Publication Data
Main entry under title

Card Number: 00-112304
ISBN: 1-58381-057-9

ASP Conference Series - First Edition

Printed in United States of America by Sheridan Books, Chelsea, Michigan

This volume is dedicated to the memory and spirit of Fritz Zwicky, a visionary, a pioneer of sky surveys, an explorer of parameter spaces, the discoverer of the existence of dark matter, large scale structure, and much else — may the new generations of virtual sky explorers be as creative and original as he ever was.

Fritz Zwicky, 1889–1974

"... As to Hale's advice of making no mean plans, I felt that the method of morphological research, with whose formulation I was groping at the time, would be a most adequate tool for the following purposes.

(a) Exploring the material contents of the universe, whereby special attention was going to be paid the discovery of objects as yet unknown, but whose existence was predicted from theory, as well as of objects entirely novel in character whose discovery could be hastened by the use of novel detection devices ..."

"... the true age of discovery in astronomy is only just starting ..."

Zwicky picture courtesy of the *Engineering and Science* magazine, California Institute of Technology

Contents

Preface xiii

Organizing Committees xv

Conference Participants xvii

Part 1. New Science with a Virtual Observatory

Precision Cosmology 3
    A.S. Szalay

Panchromatic Studies of Active Galactic Nuclei 13
    C.J. Lonsdale

Gravitational Lensing with the NVO 21
    G.K. Squires, & J.A. Tyson

NVO and the LSB Universe 28
    J.M. Schombert

The New Paradigm: Novel, Virtual Observatory Enabled Science 34
    R.J. Brunner

Precision Galactic Structure 40
    S. Kent, for the SDSS Collaboration

Exploring the Time Domain: Transients 46
    A.H. Diercks

Searches for Rare and New Types of Objects 52
    S.G. Djorgovski, A.A. Mahabal, R.J. Brunner, R.R. Gal, S. Castro, R.R. de Carvalho, & S.C. Odewahn

Exploring the Multi-Wavelength, Low Surface Brightness Universe ... 64
    R.J. Brunner, S.G. Djorgovski, R.R. Gal, A.A. Mahabal, & S.C. Odewahn

The ROSAT Optical X-ray Cluster Survey: A Comparison of X-ray and Optical Cluster Detection Methods and Preliminary Results. ... 69
    M. Donahue, C. Scharf, J. Mack, M. Postman, P. Lee, M. Dickinson, P. Rosati, & J. Stocke

The Luminosity Function of 80 Abell Clusters from the CRoNaRio Catalogs 73
    S. Piranomonte, G. Longo, S. Andreon, E. Puddu, M. Paolillo, R. Scaramella, R. Gal, & S.G. Djorgovski

Group and Cluster Detection in DPOSS Catalogs 80
    E. Puddu, V. Strazzullo, S. Andreon, G. Longo, E. De Filippis, R. Gal, S.G. Djorgovski, & R. Scaramella

The WARPS Galaxy Cluster Survey: A Case Study Using Present-Day
Multi-Wavelength Archives 86
    C.A. Scharf

Scaling and Fluctuations in the Galaxy Distribution: Two Tests to Probe
Large Scale Structure 90
    F. Sylos Labini, & A. Gabrielli

Part 2. Astronomical Approaches to a Virtual Observatory

NASA Mission Archives, Data Centers, and Information Services:
A Foundation for the Virtual Observatory 97
    R.J. Hanisch

An Overview of Existing Ground-Based Wide-Area Surveys 103
    B. McLean

Astronomical Data Centers, Information Systems, and Electronic Libraries 108
    D. Egret

Astro-IT Challenges and Big UK Survey Programs—SuperCOSMOS,
UKIRT WFCAM, and VISTA 114
    A. Lawrence

Solar System Surveys 118
    S.H. Pravdo

SkyView: Experiences Building a Virtual Telescope 125
    T.A. McGlynn, L.M. McDonald, & N.E. White

ISAIA: Interoperable Systems for Archival Information Access 130
    R.J. Hanisch

The Digital Sky Project: Prototyping Virtual Observatory Technologies 135
    *R.J. Brunner, T. Prince, J. Good, T.H. Handley, C. Lonsdale,
    & S.G. Djorgovski*

Architecture of the Infrared Science Archive 142
    *J. Good, G.B. Berriman, N. Chiu, T. Handley, A. Johnson, M. Kong,
    W-P. Lee, C.J. Lonsdale, J. Ma, S. Monkewitz, S.W. Norton, & A. Zhang*

Education & Public Outreach: A View from Research Institutions and
Observatories 148
    C. Christian

Museums, Planetaria and the Virtual Observatory 153
    C.T. Liu

The Virtual Observatory and Education: A View From the Classroom . 159
    J.C. White II

The ISO Data Archive and Links to Other Archives 165
    C. Arviset, J. Hernandez, & T. Prusti

An Overview of the Infrared Science Archive 169
    G.B. Berriman, N. Chiu, J. Good, T. Handley, A. Johnson, M. Kong,
    W-P. Lee, C.J. Lonsdale, J. Ma, S. Monkewitz, S. W. Norton, & A. Zhang

The SOHO Data and Information System 173
    G. Dimitoglou, & L. Sánchez Duarte

The CDS Role in the Virtual Observatory 176
    F. Genova, F. Bonnarel, P. Dubois, D. Egret, P. Fernique, S. Lesteven,
    F. Ochsenbein, M. Wenger, & G. Jasniewicz

Lessons Learned for the Virtual Observatory from the Scientist's Expert
Assistant Project 180
    S. Grosvenor, J. Jones, L. Ruley, M. Fishman, K. Wolf, & A. Koratkar

The Virtual Solar Observatory 184
    F. Hill

3-D Visualizations of Massive Astronomy Datasets with a Digital Dome   188
    C.T. Liu, B. Abbott, C. Emmart, M-M. Mac Low, M. Shara,
    F.J. Summers, & N.D. Tyson

Serving the Sky 192
    A.A. Mahabal, R.J. Brunner, S.G. Djorgovski, R.R. Gal, J. Jacob,
    & S.C. Odewahn

Rapid Cross Identification for the National Virtual Observatory:
The Digital Sky Project 197
    J. Ma, T. Handley, J. Good, A. Johnson, R.J. Brunner, T. Prince,
    R. Rutledge, & R. Williams

GAIA and Virtual Observatories 201
    W. O'Mullane, & X. Luri

Mining the Virtual Sky 205
    P.F. Ortiz, F. Ochsenbein, F. Genova, & A. Wicenec

The Chandra Data Archive 209
    A. Rots

The Space-Time Profile for ISAIA 213
    A. Rots

The Carl Sagan Solar-Stellar Observatory 217
    A. Sanchez-Ibarra, & J. Saucedo-Morales

The Role of Existing Data Archive Centers in the
    International Virtual Observatory: PixelSets and Catalogs 221
    D. Schade

Solar Web: A Web Tool for Searching in Web-Based Solar Databases .. 225
    I. Scholl

Multi-Threaded Decomposition of Queries: The SDSS Model 230
    A.R. Thakar, P.Z. Kunszt, & A.S. Szalay

The ROSAT X-ray Database from All-Sky Survey and Pointed
Observations 234
    W. Voges, T. Boller, J. Englhauser, M. Freyberg, & R. Supper

Part 3. Computer Science & Statistics for a Virtual Observatory

Computer Technology Forecast 241
    J. Gray

Astronomy Image Collections 257
    R.W. Moore

Computational AstroStatistics: Fast Algorithms and Efficient Statistics
for Density Estimation in Large Astronomical Datasets 265
    R.C. Nichol, A.J. Connolly, A.W. Moore, J. Schneider, C. Genovese, & L. Wasserman

Statistical Methodology for the National Virtual Observatory 272
    G.J. Babu, & E.D. Feigelson

Nonparametric Density Estimation: A Brief and Selective Review 279
    C.R. Genovese, L. Wasserman, R.C. Nichol, A.J. Connolly, J. Schneider, & A.W. Moore

Visualization of Large Multi-Dimensional Datasets 284
    J. Welling, & M. Derthick

Large-Scale Visualization of Digital Sky Surveys 291
    J.C. Jacob, & L.E. Husman

High Speed Interconnects and Parallel Software Libraries:
Enabling Technologies for the NVO 297
    J. Kepner, & J. McMahon

Approaches to Federation of Astronomical Data 302
    R. Williams

Distributed Archives Interoperability 316
    C.Y. Cheung

Technologies for Mining Terabytes of Data 323
    A.J. Wicenec

Computer Science Issues in the National Virtual Observatory 329
    M.T. Goodrich

Science User Scenarios for a VO Design Reference Mission 333
    K.D. Borne

A Processing Automaton for Intensive Data 337
    J.D. Scargle, J.P. Crutchfield, C. Glymour, & R. Hewett

Part 4. The US National Virtual Observatory Effort

Summary of USNVO Activities 343
    D.S. De Young

The Design Reference Mission for a Virtual Observatory 347
    T. Boroson

The US NVO White Paper: Toward a National Virtual Observatory:
    Science Goals, Technical Challenges, and Implementation Plan . . 353

Author Index 373

Preface: Why Virtual Observatories?

Astronomy is being deluged by a tsunami of data. New digital sky surveys and archives, with information content measured in multiple Terabytes and detections of billions of sources over many wavelengths, are fundamentally changing observational astronomy and the way we perceive and study the physical universe. Even larger, multi-Petabyte data sets are now on the horizon; the tsunami continues to grow larger. This great richness of data poses substantial challenges, and requires new methods, tools, and even a new style of doing astronomy. Virtual observatories are the conceptual response to these challenges.

The conference "Virtual Observatories of the Future" was held at the California Institute of Technology, in Pasadena, California, on June 13–16, 2000. The conference was motivated in part by a major, community-driven push towards a National Virtual Observatory (NVO), reflected in the recommendation by the Astronomy and Astrophysics Survey Committee of the National Research Council Decadal Report, *Astronomy and Astrophysics in the new Millennium*[1], which listed the NVO as the highest priority item in the small program category.

Our goals were twofold. First, this was the initial large, public forum in which the concept of virtual observatories were discussed by the astronomical community, and the meeting served both as a way of informing the community about ongoing developments, and as a way of gathering some broad input for future directions. Second, we saw a need to clearly define the scientific motivations and needs, and to focus on the technical problems and challenges related to the conception of the NVO and its global equivalents.

The history of the US virtual observatory efforts are detailed in these proceedings, followed by a summary of several specific scientific "use cases" that are examples of the type of demonstration projects that will be needed to drive the design and implementation of the virtual observatories of the future. With the permission of the informal NVO interim steering committee, we have included a draft version of the NVO white paper which was circulated to all conference attendees (and is also available from the conference web-site[2]).

A variety of different scientific topics, including cosmology, star formation, active galactic nuclei, gravitational lensing, low surface brightness sources, Galactic structure, *etc.*, were discussed, with a special emphasis on how current investigations could be enhanced by a virtual observatory. Additional topics, including the study of transient sources, searches for rare and new types of astronomical sources, and novel approaches to utilizing both images and catalogs were also explored. Throughout all of these presentations, the revolutionary nature of virtual observatory science was readily apparent.

These developments start with the solid foundation from the many existing sky surveys and digital archives, and build on their experience. Yet the scope of the changes are so fundamental that it was obvious that novel tech-

---

[1] http://www.nap.edu/books/0309070317/html/

[2] http://www.astro.caltech.edu/nvoconf/

nical approaches are needed. The NVO is indeed technology-enabled, but is science-driven.

The nature of the technical and methodological challenges posed by virtual observatories invites substantive collaborations and partnerships between astronomers, computer scientists, and statisticians, and this point was made clear by a number of excellent presentations at the meeting. While similar challenges are now confronting many other data-intensive fields, astronomy does have some special attributes: our problems are challenging, but doable, and can stimulate progress in applied information sciences; our data sets are interestingly large and complex, and are for the most part freely available. The resulting discussions helped fuel the interdisciplinary synergism that will truly empower the future virtual observer.

The public is fascinated by the night sky and the questions it generates. A virtual observatory can also be a powerful tool for science education and public outreach at all levels, and several presentations addressed the role of the NVO in public outreach efforts, including museums and planetaria, and scientific education in the classroom.

The NVO will hopefully grow into a Global Virtual Observatory, serving as the fundamental information infrastructure for astronomy and astrophysics in the next century. It was thus imperative to have as broad of an international involvement during this conference as possible. The conference participants hailed from ten different countries, representing the majority of astronomical data centers. In addition, the conference was attended by a substantial number of non-astronomers, including computer scientists, statisticians, and educators, who all provided interesting insights into defining a virtual observatory.

We are very grateful to acknowledge financial assistance from the National Science Foundation, the National Aeronautics and Space Administration, Sun Microsystems, and Microsoft Research. Considerable thanks are also due to the scientific organizing committee for planning a lively and interesting conference. We would also like to thank the members of the local organizing committee for their tireless work in planning and running this conference. We are especially grateful to Diane Fujitani, Gina Armas, and Mary Ellen Barba for devoting such a considerable fraction of their time, without which we would not have had such a smoothly run and well presented meeting. Finally, we would also like to thank the legion of unpaid volunteers and Caltech students who helped out with a wide variety of requests, all in the interest of learning more about the virtual side of Astronomy.

We hope that this was just the first of many meetings to come, as this new approach to astronomy grows and expands.

> Robert J. Brunner, S. George Djorgovski
> California Institute of Technology

> Alex S. Szalay
> The Johns Hopkins University

December 15, 2000

Organizing Committees

Scientific Organizing Committee

M. Albrecht
C. Alcock
R.J. Brunner
C. Cheung
J. Condon
T. Cornwel
D. Curkendall
R. de Carvalho
D. Deyoung
S.G. Djorgovski (Co-Chair)
R. Doyle
D. Durand
D. Egret
G. Fabbiano
U. Fayyad
B. Hanisch
G. Helou
M. Kurtz
G. Lake
G. Longo
C. Lonsdale
B. Madore
T. McGlynn
B. McLean
R. Moore
F. Murtagh
R. Nichol
S. Pravdo
T. Prince
D. Slutz
S. Strom
A. Szalay (Co-Chair)
N. White
R. Williams

Local Organizing Committee

R.J. Brunner (Chair)
M. E. Barba
S. Castro
S.G. Djorgovski
G. Helou
J. Jacob
C. Lonsdale
B. Madore
A. Mahabal
R. Williams

Conference Participants

Charles Alcock, Lawrence Livermore National Laboratory, Livermore, CA ⟨alcock1@llnl.gov⟩

Gina Armas, Caltech, Pasadena, CA ⟨gina@its.caltech.edu⟩

Christophe Arviset, ESA/ISO Data Centre, Madrid, Spain ⟨carviset@iso.vilspa.esa.es⟩

Jogesh Babu, Penn State University, University Park, PA ⟨babu@stat.psu.edu⟩

Mary Ellen Barba, Caltech, Pasadena, CA ⟨meb@ipac.caltech.edu⟩

Kevin Barron, CENIC/ITP, UCSB, Santa Barbara, CA ⟨kevin@itp.ucsb.edu⟩

Robert Becker, UC-Davis/IGPP-LLNL, Livermore, CA ⟨bob@igpp.ucllnl.org⟩

Bruce Berriman, IPAC, Caltech, Pasadena, CA ⟨gbb@ipac.caltech.edu⟩

Brett Blacker, STScI, Baltimore, MD ⟨blacker@stsci.edu⟩

Kirk Borne, Raytheon ITSS, GSFC, Greenbelt, MD ⟨Kirk.D.Borne.1@gsfc.nasa.gov⟩

Todd Boroson, NOAO, Tucson, AZ ⟨tyb@noao.edu⟩

Joe Bredekamp, NASA Office of Space Science, Washington, DC ⟨joe.bredekamp@hq.nasa.gov⟩

Roger Brissenden, CfA/SAO, Cambridge, MA ⟨brissenden@cfa.harvard.edu⟩

Robert Brunner, Caltech, Pasadena, CA ⟨rb@astro.caltech.edu⟩

Bob Carswell, Institute of Astronomy, Cambridge, Cambridge, England ⟨rfc@ast.cam.ac.uk⟩

Sandra Castro, Caltech, Pasadena, CA ⟨smc@astro.caltech.edu⟩

Cynthia Cheung, NASA Goddard Space Flight Center, Greenbelt, MD ⟨cynthia.cheung@gsfc.nasa.gov⟩

Carol Christian, STScI, Baltimore, MD ⟨carolc@stsci.edu⟩

Yaoquan Chu, Univ. of Sci. & Tech. of China, Hefei, Anhui, China ⟨yqchu@ustc.edu.cn⟩

James Condon, NRAO, Charlottesville, VA ⟨jcondon@nrao.edu⟩

Kem Cook, Lawrence Livermore National Laboratory, Livermore, CA ⟨kcook@llnl.gov⟩

Jeff Cooke, University of California, San Diego, San Diego, CA ⟨cooke@physics.ucsd.edu⟩

Michael Corbin, The University of Arizona, Tucson, AZ ⟨mcorbin@as.arizona.edu⟩

Tim Cornwell, NRAO, Socorro, New Mexico ⟨tcornwel@nrao.edu⟩

Ingrid Cotoros, Caltech, Pasadena, CA ⟨ingrid@its.caltech.edu⟩

Emory Creel, National Center for Data Mining, UIC, Chicago, IL ⟨emory@lac.uic.edu⟩

Roc Cutri, Caltech, Pasadena, CA ⟨roc@ipac.caltech.edu⟩

Conference Participants

David S. De Young, NOAO, Tucson, AZ ⟨deyoung@noao.edu⟩
Alan Diercks, Caltech, Pasadena, CA ⟨ad@astro.caltech.edu⟩
George Dimitoglou, SOHO ESA/NASA PS Team - SM&A, Greenbelt, MD ⟨george@esa.nascom.nasa.gov⟩
George Djorgovski, Caltech, Pasadena, CA ⟨george@astro.caltech.edu⟩
Elaine Dobinson, JPL, Pasadena, CA ⟨elaine.dobinson@jpl.nasa.gov⟩
Megan Donahue, STScI, Baltimore, MD ⟨donahue@stsci.edu⟩
Daniel Durand, National Research Council Canada / CADC, Victoria, BC, Canada ⟨daniel.durand@nrc.ca⟩
Daniel Egret, CDS, Observatoire de Strasbourg, Strasbourg, France ⟨Daniel.Egret@astro.u-strasbg.fr⟩
Eileen Friel, National Science Foundation, Arlington, VA ⟨efriel@nsf.gov⟩
Diane Fujitani, Caltech, Pasadena, CA ⟨dsf@astro.caltech.edu⟩
Niall Gaffney, STScI, Baltimore MD ⟨gaffney@stsci.edu⟩
Roy Gal, Caltech, Pasadena, CA ⟨rrg@astro.caltech.edu⟩
Severin Gaudet, CADC, National Research Council Canada, Victoria, B.C., Canada ⟨severin.gaudet@nrc.ca⟩
Francoise Genova, CDS, Observatoire de Strasbourg, Strasbourg, France ⟨genova@astro.u-strasbg.fr⟩
Christopher Genovese, Dept. of Statistics, Carnegie Mellon University, Pittsburgh ⟨genovese@stat.cmu.edu⟩
John Good, Caltech, Pasadena, CA ⟨jcg@ipac.caltech.edu⟩
Michael Goodrich, Computer Science, Johns Hopkins Univ., Baltimore, MD ⟨goodrich@cs.jhu.edu⟩
Bonnie Gordon, Astronomy Magazine, Waukesha, WI ⟨bgordon@astronomy.com⟩
Krzysztof Gorski, European Southern Observatory, Garching bei Muenchen, Germany ⟨kgorski@cso.org⟩
Robert Granat, JPL, Pasadena, CA ⟨Robert.A.Granat@jpl.nasa.gov⟩
Jim Gray, Microsoft Research, San Francisco, CA ⟨gray@microsoft.com⟩
William Green, Caltech, Pasadena, CA ⟨bgreen@ipac.caltech.edu⟩
Gretchen Greene, STScI, Baltimore, MD ⟨greene@stsci.edu⟩
R. Elizabeth Griffin, Oxford University, Oxford, UK ⟨remg@astro.ox.ac.uk⟩
Sandy Grosvenor, Booz-Allen & Hamilton, Greenbelt, MD ⟨sandy.grosvenor@gsfc.nasa.gov⟩
Tom Handley, Caltech, Pasadena, CA ⟨thh@ipac.caltech.edu⟩
Robert Hanisch, STScI, Baltimore, MD ⟨hanisch@stsci.edu⟩
George Helou, Caltech, Pasadena, CA ⟨gxh@ipac.caltech.edu⟩
Frank Hill, National Solar Observatory, Tucson, AZ ⟨fhill@noao.edu⟩
Tin Ho, Bell Labs, Lucent Technologies, Murray Hill, NJ ⟨tkh@bell-labs.com⟩

J. Steven Hughes, JPL, Pasadena, CA ⟨steve.hughes@jpl.nasa.gov⟩
Matthew Hunt, Caltech, Pasadena, CA ⟨mph@astro.caltech.edu⟩
Laura Husman, JPL, Pasadena, CA ⟨laura@waltz.jpl.nasa.gov⟩
Joseph Jacob, JPL, Pasadena, CA ⟨Joseph.Jacob@jpl.nasa.gov⟩
Bhuvnesh Jain, Johns Hopkins University, Baltimore, MD
⟨bjain@pha.jhu.edu⟩
Inger Jorgensen, Gemini Observatory, Hilo, HI ⟨ijorgensen@gemini.edu⟩
Konstantinos Kalpakis, UMBC, Baltimore, MD ⟨kalpakis@csee.umbc.edu⟩
Stephen Kent, Fermilab, Batavia, IL ⟨skent@fnal.gov⟩
Jeremy Kepner, MIT Lincoln Lab, Lexington, MA
⟨jvkepner@astro.princeton.edu⟩
Martin Kessler, ISO Data Centre, ESA, Madrid, Spain
⟨mkessler@iso.vilspa.esa.es⟩
Peter Kunszt, Johns Hopkins University, Baltimore, MD
⟨kunszt@pha.jhu.edu⟩
Michael Kurtz, Harvard-Smithsonian Center for Astrophysics, Cambridge, MA
⟨kurtz@cfa.harvard.edu⟩
George Lake, University of Washington, Seattle, WA
⟨lake@astro.washington.edu⟩
Andy Lawrence, Royal Observatory Edinburgh, Edinburgh, UK ⟨al@roe.ac.uk⟩
James Lewis, Institute of Astronomy, Cambridge, Cambridge, UK
⟨jrl@ast.cam.ac.uk⟩
Patricia Liggett, JPL, Pasadena, CA ⟨patricia.k.liggett@jpl.nasa.gov⟩
Don Lindler, Advanced Computer Concepts, Inc., Potomac, MD
⟨lindler@rockit.gsfc.nasa.gov⟩
Charles Liu, Deptartment of Astrophysics, American Museum of Natural
History, New York, NY ⟨cliu@amnh.org⟩
Giuseppe Longo, Osservatorio Astronomico di Capodimonte, Napoli, Italy
⟨longo@na.astro.it⟩
Carol Lonsdale, Caltech, Pasadena, CA ⟨cjl@ipac.caltech.edu⟩
Thomas Lutterbie, STScI, Baltimore, MD ⟨lutterbi@stsci.edu⟩
Jin Ma, Caltech, Pasadena, CA ⟨ma@ipac.caltech.edu⟩
Barry Madore, NASA/IPAC Extragalactic Database, Pasadena, CA
⟨barry@ipac.caltech.edu⟩
Ashish Mahabal, Caltech, Pasadena, CA ⟨aam@astro.caltech.edu⟩
Alfredo Manrquez, Sonora, Mexico ⟨alman61@hotmail.com⟩
Janet A. Mattei, AAVSO, Cambridge, MA ⟨jmattei@aavso.org⟩
Thomas McGlynn, NASA/GSFC/HEASARC, Greenbelt, MD
⟨tam@lheapop.gsfc.nasa.gov⟩
Brian McLean, STScI, Baltimore, MD ⟨mclean@stsci.edu⟩

Conference Participants

Aronne Merrelli, Caltech, Pasadena, CA ⟨ajm@astro.caltech.edu⟩
Dave Monet, USNO Flagstaff Station, Flagstaff, AZ ⟨dgm@nofs.navy.mil⟩
Reagan Moore, San Diego Supercomputer Center, La Jolla, CA
 ⟨moore@sdsc.edu⟩
Harvey Newman, Caltech, Pasadena, CA ⟨newman@hep.caltech.edu⟩
Robert Nichol, Carnegie Mellon, Pittsburgh, PA ⟨nichol@cmu.edu⟩
Maria Nieto-Santisteban, STScI, Baltimore, MD ⟨nieto@stsci.edu⟩
Francois Ochsenbein, CDS, Strasbourg, France ⟨francois@astro.u-strasbg.fr⟩
William O'Mullane, GAIA, Noordwijk, Netherlands
 ⟨womullan@astro.estec.esa.nl⟩
Earl O'Neil, Steward Observatory, Tucson, AZ ⟨eoneil@as.arizona.edu⟩
Patricio Ortiz, CDS Strasbourg, Strasbourg, France
 ⟨portiz@astro.u-strasbg.fr⟩
Clive Page, University of Leicester, Leicester, U.K. ⟨cgp@star.le.ac.uk⟩
William Pence, NASA/GSFC, Greenbelt, MD ⟨pence@tetra.gsfc.nasa.gov⟩
Ryszard Pisarski, NASA/GSFC, Greenbelt, MD ⟨rlp@ros5.gsfc.nasa.gov⟩
Raymond Plante, NCSA, Urbana, IL ⟨rplante@ncsa.uiuc.edu⟩
Marc Postman, STScI, Baltimore, MD ⟨postman@stsci.edu⟩
Steve Pravdo, JPL, Pasadena, CA ⟨spravdo@jpl.nasa.goc⟩
Tom Prince, Caltech, Pasadena, CA ⟨prince@caltech.edu⟩
Timo Prusti, ISO Data Centre, ESA, Madrid, Spain ⟨tprusti@iso.vilspa.esa.es⟩
Peter Quinn, European Southern Observatory, Garching, Germany
 ⟨pquinn@eso.org⟩
George Reinhart, University of Illinois, Chicago, Chicago, IL
 ⟨georg@lac.uic.edu⟩
John Rice, University of California, Berkeley, Berkeley, CA
 ⟨rice@stat.berkeley.edu⟩
Doug Roberts, Adler Planetarium, Chicago, IL
 ⟨doug-roberts@northwestern.edu⟩
Joseph Roden, JPL, Pasadena, CA ⟨Joseph.C.Roden@jpl.nasa.gov⟩
Arnold Rots, CfA/CXC, Cambridge, MA ⟨arots@head-cfa⟩
Robert Rutledge, Caltech, Pasadena, CA ⟨rutledge@srl.caltech.edu⟩
Marco Salvati, Osservatorio Astrofisico di Arcetri, Firenze, Italy
 ⟨msalvati@arcetri.astro.it⟩
Antonio Sanchez-Ibarra, Area de Astronomia/UNISON, Sonora, Mexico
 ⟨asanchez@cosmos.cifus.uson.mx⟩
Jeff Scargle, NASA-Ames Research Center, Moffett Field, CA
 ⟨jeffrey@sunshine.arc.nasa.gov⟩
David Schade, Canadian Astronomy Data Centre, Victoria, B.C., Canada
 ⟨David.Schade@hia.nrc.ca⟩

Caleb Scharf, STScI, Baltimore, MD ⟨scharf@stsci.edu⟩
Marion Schmitz, Caltech, Pasadena, CA ⟨zb4ms@ipac.caltech.edu⟩
Jeff Schneider, Carnegie Mellon University, Pittsburgh, PA
  ⟨Jeff.Schneider@cs.cmu.edu⟩
Isabelle Scholl, IAS-CNRS, Orsay, France ⟨scholl@medoc-ias.u-psud.fr⟩
James Schombert, University of Oregon, Eugene, OR ⟨js@abyss.uoregon.edu⟩
Ethan Schreier, STScI, Baltimore, MD ⟨ejs@stsci.edu⟩
Greg Schwarz, AAS Journals Editorial Scientist, Tucson, AZ
  ⟨gschwarz@as.arizona.edu⟩
Steve Scott, Caltech/OVRO, Big Pine, CA ⟨scott@ovro.caltech.edu⟩
Robert Seaman, NOAO, Tucson, AZ ⟨seaman@noao.edu⟩
Ed Shaya, ADC/NASA/ITSS, Greenbelt, MD ⟨shaya@xfiles.gsfc.nasa.gov⟩
Patrick Shopbell, Caltech, Pasadena, CA ⟨pls@astro.caltech.edu⟩
Ralph Shuping, SOFIA, Boulder, CO ⟨shuping@casa.colorado.edu⟩
Donald Slutz, Microsoft Research, San Francisco, CA ⟨dslutz@microsoft.com⟩
Gordon Squires, Caltech, Pasadena, CA ⟨gks@astro.caltech.edu⟩
Brian Stalder, Caltech, Pasadena, CA ⟨stal@its.caltech.edu⟩
Paul Stolorz, JPL, Pasadena, CA ⟨pauls@aig.jpl.nasa.gov⟩
Francesco Sylos Labini, University de Geneve, Sauverny, Switzerland
  ⟨sylos@amorgos.unige.ch⟩
Alexander Szalay, The Johns Hopkins University, Baltimore, MD
  ⟨szalay@jhu.edu⟩
Ani Thakar, The Johns Hopkins University, Baltimore, MD
  ⟨thakar@pha.jhu.edu⟩
Doug Tody, NOAO, Tucson, AZ ⟨tody@noao.edu⟩
Wolfgang Voges, MPE-Garching, Garching, Germany ⟨wvoges@mpe.mpg.de⟩
Boyd Waters, NRAO, Socorro, NM ⟨bwaters@aoc.nrao.edu⟩
Joel Welling, Pittsburgh Supercomputing Center, Pittsburgh, PA
  ⟨welling@psc.edu⟩
Martin White, Harvard-Smithsonian CfA, Cambridge, MA
  ⟨mwhite@cfa.harvard.edu⟩
James White, Astronomical Society of the Pacific, San Francisco, CA
  ⟨jwhite@aspsky.org⟩
Nicholas White, HEASARC, Greenbelt, MD ⟨nwhite@lheapop.gsfc.nasa.gov⟩
Andreas Wicenec, European Southern Observatory, Garching, Germany
  ⟨awicenec@eso.org⟩
Roy Williams, Caltech, Pasadena, CA ⟨roy@cacr.caltech.edu⟩

Part 1
New Science with a Virtual Observatory

Precision Cosmology

A.S. Szalay

Department of Physics and Astronomy, The Johns Hopkins University, Baltimore, MD 21218

Abstract. Ongoing large scale surveys of the galaxy distribution are fundamentally changing cosmology. Multicolor catalogs are approaching a billion galaxies, redshift surveys under way will soon have a million objects. The errors in our cosmological measurements are becoming more and more accurate — we are approaching high-precision cosmology! These large surveys are the first, where the dominant source of noise is systematic errors, requiring novel techniques in their statistical analyses. The multicolor data of these surveys can be used as a low resolution spectrograph: we can recover both redshifts and spectral types for the galaxies, at least in a statistical fashion. This open new windows, we can select samples of distant galaxies based on rest-frame criteria. However, much of the necessary data will reside in separate surveys, one will need to combine UV, optical and IR data to complete such analyses. Without a Virtual Observatory such a task cannot easily be accomplished.

1. Introduction

The study of large scale structure is one of the most dynamically evolving areas of astrophysics today. Cosmology and large scale structure is growing into an accurate science and requires correspondingly more sophisticated methods of analysis. Twenty years ago the estimates of the fluctuation amplitude were about 10^{-3}, almost a factor of 100 off of today's measurements. Ten years ago we could only hope for high precision measurements of large scale structure, there were less than 5000 redshifts measured, and only a handful of normal galaxies with $z > 1$ were known. Computer models of structure formation had just begun to consider non-power-law spectra based on physical models like hot/cold dark matter. As a consequence there was considerable freedom in adjusting parameters in the various galaxy formation scenarios. In contrast, many of today's debates are about factors of 2 and soon we will be arguing about 10% differences. The shape of the primordial fluctuation spectrum, first derived from philosophical arguments (Harrison 1970, Zeldovich 1972), can now be quantified from detections of fluctuations in the CBR made by COBE (Smoot et al. 1991). The number of available redshifts is beyond 50,000, and soon we will have redshift surveys surpassing 1 million galaxies. N-body simulations are becoming more sophisticated, of higher resolution, and incorporating complex gas dynamics. The unprecedented number of new observations currently under way give us hope that over the next decade we will gain a clear understanding of the shape and evolution of the primordial fluctuation spectrum, understand from first principles how galaxies were formed, and make quantitative comparisons and tests to differentiate among the various galaxy formation scenarios. The different forms of evolution (luminosity, clustering and cosmological) are hard to separate, we need more accurate measurements, and larger samples.

2. Quantifying Large Scale Structure — Key Questions

Structure in the universe evolves from the initially small primordial fluctuations. These fluctuations can arise during an inflationary expansion or come from topological defects later. They grow in amplitude, due to gravitational instability, and the shape of the fluctuation spectrum is altered by different physical processes. The nature of the dark matter, whether hot or cold, believed to dominate the mass density of the universe, determines the shape of the power spectrum on small (< 100 Mpc) scales. On the other hand, the shape of the large scale part of the fluctuations (> 200 Mpc) remains remarkably unchanged, because no scale in the evolutionary process becomes this large.

· What are the most important measurements we can make in order to differentiate between proposed cosmologies? Overlap between scales probed by CBR experiments and redshift surveys in the 'local' universe would place strong constraints on the power spectrum. Measurements of galaxy clustering on scales of 200-500 Mpc from redshift surveys would tell us whether the gravity wave/tilted model is correct, measure the bias factor, and determine the shape of the spectrum on scales where most of today's models differ but which are too small for COBE and beyond the scale of current galaxy measurements. For the same reason, many CBR experiments are probing 1-2 degree scales, corresponding to a comoving scale of about ≈ 120 Mpc. In the next section we outline how novel statistical techniques under development will bring this goal within reach, using galaxy redshift surveys.

3. Structure Beyond 100 Mpc — the Power Spectrum

To estimate the power spectrum from a galaxy redshift survey, we must take into account the sampling density (determined by the magnitude limit) and geometry of the survey (determined by the angular coverage and depth). The sampling process and the fact, that only integer numbers of galaxies can be counted, introduces shot noise into the power spectrum (the noise per mode is constant and thus easily subtracted, but contributes to the uncertainty). The observed power spectrum is a convolution of the true power with the Fourier transform of the spatial window function of the survey ($W(\mathbf{x}) = 1$ inside the survey and 0 outside), $P_{obs}(\mathbf{k}) = \int P_{true}(\mathbf{k'})|W(\mathbf{k}-\mathbf{k'})|^2 d^3k'$. One can attempt to deconvolve the true power spectrum or compare to convolved theoretical spectra, but in either case the survey geometry limits both the resolution and the largest wavelength for which an accurate measurement can be obtained.

The standard methods for power spectrum estimation work reasonably well for data in a large, contiguous, three-dimensional volume, with homogeneous sampling of the galaxy distribution. The weighting scheme is optimized for shot-noise dominated errors. Using these techniques, nearby wide-angle redshift surveys (CfA, SSRS, IRAS 1.2, QDOT) yield strong constraints on the power spectrum on scales up to 100 h^{-1} Mpc. Because the uncertainty in the power spectrum depends on the number of independent cells of a given wavelength which we sample, constraints on larger scales require deeper surveys. Due to the difficulty of obtaining redshifts for fainter galaxies and limited telescope time, deep redshift surveys typically have complex geometry, e.g., deep pencil beams or slices.

However, the standard methods are not efficient when applied to data in oddly-shaped and/or disjoint volumes, or when the sampling density of galaxies varies greatly over these regions. Systematic effects, like extinction or calibration zeropoints can substantially contribute to the errors. Convolution of the true power with the complex window function causes power in different modes to be highly coupled. In other words, plane waves do not form an optimal eigenbasis for expansion of the the galaxy density field sampled by redshift surveys. We desire methods for power spectrum estimation that optimally weight the data in each region of the survey, taking into account our prior knowledge of the nature of the noise and clustering in the galaxy distribution. A detailed comparison of all available power spectrum estimation methods is given in Tegmark et al. (1998). There are several contaminating effects, like aliasing from non-spherical survey geometries, extinction, redshift distortions, effects non-linear fluctuation growth. In the section below we outline how to create an analysis tool that goes considerably beyond the present state of the art, and can take all these effects into consideration.

4. The Karhunen-Loève Transform

One can find an optimal set of spatial filters to probe the density fluctuations. Rather than directly compute the Fourier transform of the distribution of objects, we expand the observed density field in the natural orthonormal basis determined for each survey from our prior knowledge of the survey geometry, selection function, and clustering of galaxies, and find the most likely power spectrum model in a Bayesian fashion (Vogeley and Szalay 1996). Expansion of the observed density field in this basis is known as the Karhunen-Loève transform (e.g., Therrien 1992), hereafter the KL transform. Dividing the survey volume into cells V_i, we compute the correlation matrix of expected counts

$$C_{ij} = \langle N_i N_j \rangle = \langle N_i \rangle \langle N_j \rangle \left(1 + \langle \xi_{ij} \rangle\right) + \delta_{ij} \langle N_i \rangle + \eta_{ij}, \qquad (1)$$

where $\delta_{ij} = 0$ for $i \neq j$, N_i is the galaxy count in the ith cell, η_{ij} is additional noise arising from systematic effects and

$$\langle \xi_{ij} \rangle = \frac{1}{V_i V_j} \int \xi(\mathbf{x_i} - \mathbf{x_j}) dV_i dV_j. \qquad (2)$$

We compute ξ from a model, which is our null hypothesis. The eigenvectors $\mathbf{\Psi}_j$ which diagonalize the correlation matrix are the signal-to-noise eigenfunctions of the density field of the survey (solving the equation $\mathbf{C} \cdot \mathbf{\Psi}_j = \lambda_j \mathbf{\Psi}_j$). The eigenvectors have a very simple physical meaning: they contain the optimal weight of a given cell associated with each mode. This weight — via the matrix diagonalization — automatically considers all the different sources of errors, incorporated in the shot-noise term, and η, and the asymmetric geometry of the survey, then computes the optimal weight for each cell and each mode. The eigenvalues represent the statistical information content of the given mode. One can also see, that ranking the modes by decreasing eigenvalues, the list begins with the modes containing large scale power. The eigenvectors of larger rank mostly describe shot-noise.

We expand the observed counts in this orthonormal basis: $N_i = B^j \Psi_{ij}$, which defines the transform $B^j = \Psi^{ij} N_i$. Sorting these functions by decreasing eigenvalue λ yields the set of eigenfunctions in order of decreasing signal to noise. Because the B^j are statistically orthogonal and because we can easily compute the expectation value and variance of the power per eigenmode for any power spectrum model

$$\langle B_j^2 \rangle = \mathbf{\Psi}_j^{-1} \cdot \mathbf{C}^{model} \cdot \mathbf{\Psi}_j, \qquad (3)$$

hypothesis testing is a straightforward process. Note that this method requires an initial guess at the power spectrum, but the form of the eigenfunctions does not depend sensitively on this assumption, and we can easily iterate the process. Summarizing the main features: the KL transform automatically determines the 'correlation eigenmodes' of a complex survey geometry, *each optimally weighted* to measure power on a certain scale. The expansion of the density field in terms of these modes still contains phase information, and the modes are orthogonal, independent, thus statistical hypothesis testing is quite easy. *What are the problems, where are further improvements necessary?* These will be discussed in the next sections, and we will outline the way these aspects can be improved.

4.1. Adaptive Pixelization

The method in its present form requires a pixelization, which was assumed to be given. By changing to smaller cells, the resolution is increasing, but at the cost of a bigger matrix. At the same time, the largest eigenvalues carry most of the clustering signal, most of the higher ranked eigenvectors deal with the representation of the shot-noise. Let us start out with a coarse grid, and compute the eigenvalues and eigenvectors. Next, we subdivide each cell into two halves. This can be considered as a perturbation on the eigensystem, just like the level splitting of the H-atom in an external magnetic field. The resulting eigenvalues and eigenvectors can be computed from perturbation theory. If we are only interested in the first few thousand eigenvectors, we can set an accuracy threshold, beyond which we do not subdivide the cells any further. This threshold should be set on sum of the first M eigenvalues. This technique will automatically guarantee the coarsest pixelization which is still within our required error bounds. The sensitivity of the eigensystem with respect to splitting individual cells can also be computed this way.

4.2. Diagonalizing Large Matrices

Since the standard techniques of matrix diagonalization (SVD, Jacobi, Gauss-Seidel) are typically proportional to N^3, where the matrix is $N \times N$, compute times start to become prohibitive beyond matrix sizes of $N > 8000$. On the other hand, there is an algorithm, developed by C. Lanczos in the 1920's, that is widely used in various areas of computational physics, like nuclear physics and QCD, to diagonalize matrices several million in size. This technique can compute an approximation to the first M eigenvalues and eigenvectors of large matrices, with a savings of over a factor of a 100 in CPU time. A variant of this technique can also be used at the likelihood computation stage, where we calculate directly the (approximate) scalar product of the data vectors with a hypothesis matrix.

4.3. Redshift Space Distortions

The pixelization will occur in redshift space, thus the computation of the correlation matrix has to be done in redshift space. On large scales the effects from the thermal motion of galaxies ('fingers of God') are negligible, but the linear infall effects can be considerable. At the same time, the distribution of angles between lines of sight depends on the survey geometry, redshift distortions are greatest in pencil beams, smaller in slices, and even smaller, but non-negligible in wide angle surveys. Most of the simple results for redshift distortions have been computed using the plane-parallel approximation (Kaiser 1986, Hamilton 1992), where the lines of sight to the two galaxies are close to each other. In a general wide angle survey this is not the case, and explicit expressions are needed for the redshift space correlations. A numerical computation has been first done by Zaroubi and Hoffman (1996) and recently a simple analytic expression has been obtained by Szalay, Matsubara and Landy (1998), for cells with an arbitrary angle between the lines of sight, just what is needed here. We will use this expression to compute the KL correlation matrix. Our prior model will incorporate the usual parameter $\beta = \Omega^{0.6}/b$, connecting density perturbations to peculiar velocities. This also means that in our parameter estimation not only the shape of the real-space power spectrum, but also β are simultaneously recovered.

4.4. Including Systematic Effects

There are various systematic effects, which can easily become the dominant source of error for the next generation surveys. These include zero-point errors in the photometric calibrations, which are typically done over fields several degrees in size, all the way to every fourth of the 6 degree plates in the APM survey. A zero-point error causes a correlated shift in every magnitude, thus in certain areas of the sky the survey goes deeper. A similar large angular scale error can be caused by the galactic extinction, which is quite clumpy and can be several tenths of a magnitude. As the selection function starts to fall steeply, these effects modulate the outer edges of the survey — resulting in mock large scale features in the power spectrum. On scales beyond 200 h^{-1} Mpc this is a huge effect (Vogeley and Connolly 1997). One can compensate by correcting with an extinction map, but still the errors in the galaxy counts in cells at the same part of the sky will be correlated. This can be taken into account by an additional variance to the affected cells in the KL correlation matrix, which will effectively down-weight these cells.

Similarly, constraints from a fixed number of fibers in a given patch of the sky (like in the LCRS) can be considered by using an increased multivariate variance instead of Poisson, for the cells along the same line of sight, since the sum of the galaxies must add up to the total number of fibers! Data from different surveys with differing systematics can also be easily combined into a common analysis. These and many other effects can be taken into account in a simple fashion in the Karhunen-Loève framework, impossible to incorporate in any other algorithm currently available.

5. Photometric Redshifts

The idea of photometric redshifts has been around for a long time, starting with Baum (1962), and Koo (1981). Since the technique requires accurate multicolor photometry, the emergence of CCD detectors with a broad wavelength sensitivity has made the necessary imaging really possible. In particular, with the large format mosaic detectors with close to 100 million pixels, to cover a large area of the sky in multiple wavelength requires only a few nights rather than a multiyear team effort. The accessible wavelength range is also widening: new large IR detectors and arrays are becoming common place, and deep J,H and K band surveys cover large areas of the sky.

A galaxy can be characterized by its luminosity, spectral type and redshift. Its light is passing through the IGM, and is also reddened by the dust in our Galaxy. The resulting broadband fluxes can easily be estimated. Ideally, one would like to have a full inversion: from the observed fluxes determine the luminosity, SED type and redshift. Given enough photometric information, it is possible to reconstruct all the primary physical parameters, not just the redshift. The broadband fluxes represent a really coarse resolution spectrograph. At the end of the inversion process, the parameters will be correlated, but we can recover an idealized SED type parameter and the luminosity as well.

5.1. Clustering Evolution with Photometric Redshifts

The angular correlation function has long been used to study the clustering properties of galaxies. It was especially popular in the time, when there were no redshift surveys available (Peebles 1980). The amplitude of the angular two-point correlation function is tied to the spatial correlation function through Limber's equation, an integral transformation containing the radial distribution of the sample. Since $w(\theta)$ measures the projected galaxy distribution, there will be many galaxy pairs which are random projection on the sky, diluting the clustering signal. The broader the radial distribution, the weaker the angular clustering. Traditionally it has been very difficult to disentangle the different evolutionary effects, due to substantial degeneracies. Luminosity evolution can increase the depth of the sample, decreasing the amplitude. Clustering evolution also clearly affects the correlation strengths. Changes in the angular diameter distance relation, due to cosmological curvature are also important. Also, a sample selected on a single bandpass is not homogeneous, due to the K-correction as a function of redshift. These combined effects have made the interpretation of deep angular correlation functions in terms of physical parameters extremely difficult.

The generalized photometric redshifts provide a very nice way out. Not only can we control the sample selection in terms of rest-frame quantities, but we can define coarse redshift ranges, drastically decreasing the effects of the projection (Connolly et al. 1998). A detailed discussion of this problem is given by Brunner et al. (1999). The first results indicate, that the amplitude of the intrinsic correlations are varying much slower than the expected linear growth of mass correlations. Combined with the strong observed clustering of the Ly-break galaxies at $z > 3$, the implication is that the evolution of the bias is the cause of this effect.

Another approach to angular clustering is the inversion of the large scale power spectrum, pioneered by Baugh and Efstathiou (1994). This is the best measure of the large scale power spectrum to date. The statistical noise in a large scale angular catalog with millions of objects is considerably smaller than the rather sparse redshift surveys today. Even though redshift surveys will get inevitably larger (2dF and SDSS), multicolor angular data will soon be available over 10,000 square degrees, though the imaging part of the SDSS survey. With multicolor photometry one can create thick slices in the radial direction, using photometric redshifts. The use of such slices would provide several benefits: (a) the effects of projection are minimized, (b) the different redshift bins are separated. Given enough galaxies, one can select the same rest-frame population in each of the slices, and perform the angular inversion. Preliminary tests of this idea (Jain et al. 1999) using simulated data look extremely promising.

5.2. Target Selection

Photometric redshifts are already revolutionizing spectroscopy! With the advent of 8 to 10 meter telescopes it is very important to identify from the millions of possible targets which are the objects of interest. The key to a successful spectroscopic program is efficiency — one needs to have a target selection algorithm based on a priori assumptions. Using the generalized photometric redshifts, it is possible to identify potential targets restricted to a given redshift range, and to a reasonably narrow class of SED type.

- *Finding high redshift galaxies* through colors using the Ly-break is another application of photometric redshifts. Preselecting the U-band dropouts as the potential targets for spectroscopic followup has been extremely successful. We have now a better determination of the galaxy correlation function and luminosity function at a redshift 3, than at $z = 1.5$. It is rather interesting, that the high redshift galaxies are clustered rather strongly, probably due to a strong bias.

- *Cluster finding* is another perfect application. Since most cluster galaxies are early types, one can already filter the galaxies based on their SED types, and subdivide the sample into the thick redshift bins. The projection effects are minimized, and the peaks become much more obvious. A more statistically correct approach is to create a matched filter based not only on the surface density profile and luminosity but include colors as well. The mean redshift of the cluster members should give the usual $\sqrt{N}$ gain in accuracy.

- *Intermediate redshift spectroscopy* is becoming increasingly important. The redshift range $0.7 < z < 1.2$ is clearly accessible for ground-based spectroscopy, but there is much projection. A large fraction of a magnitude limited sample would be foreground or background. Using photometric redshifts for target selection can overcome this problem, and assure a high efficiency for identifying proper targets. If the redshift range for spectroscopy is restricted this way a priori, one can optimize the settings of the grating to maximize resolution in the given redshift range.

- *Gravitational lensing* as strongly affected by contamination from foreground objects. Photometric redshifts can be well used for such a segregation. Simpler versions have already been considered: galaxies redder than the E/S0 sequence of the lensing cluster must be in the background. One can refine this considerably. When solving for the surface potential of a lens it helps to know the redshifts of the sources. The pre-factor in the lens-equation has a weak dependence on the source redshift, thus even if one can get a coarse constraint on the individual source redshifts, their inclusion will help to increase the accuracy of the mass model. For weak lensing an important parameter is the dN/dz distribution of the background sources. Photometric redshifts will help to make the estimates more precise.

6. Summary

Over the next few years several new large scale surveys will start producing data, like the Sloan Digital Sky Survey (SDSS) and the 2dF. The analysis techniques outlined here, based on the Karhunen-Loève transform, combined with the new dataset, could result in major new developments in understanding the nature of the fluctuations on scales over 100 Mpc. They can measure the shape of the fluctuation spectrum in an overlap region with COBE. The method is capable of including systematic effects, redshift distortions, incompleteness in the data, representing considerable improvements over current state of the art techniques.

There are several large multicolor surveys under way, which will each cover a large fraction of the sky, and overlap with one another. SDSS will observe 10,000 square degrees in five optical bands, u', g', r', i', z', and will detect over 100 million galaxies, to a limit of $r' = 23$, with a pixel scale $0.4''$. The GALEX satellite will observe the whole sky in u_{1500}, u_{2500} UV bands, and detect over 10 million galaxies. 2MASS is already well under way, and will find the brightest one million galaxies in the J,H and K bands. In terms of deep IR imaging, the field is still in its infancy. With the new large format detectors and mosaics built from them deep all sky IR surveys should soon be possible.

The KPNO deep survey, the NTT, SOAR are all planning to cover tens to hundreds of square degrees to a greater depth. The two Keck telescopes, and the DEIMOS spectrograph, the four VLTs, the two Gemini, will observe deep redshifts, and 2dF and SDSS will collect over a million bright spectra.

A little farther in the future, VISTA will cover most of the southern sky both in the optical and the near IR bands. The proposed Dark Matter Telescope (DMT) will cover the sky in a few night's observing, and build up deep multicolor photometry at a rate of a Petabyte/year. The SWIFT spectroscopic survey telescope is proposing to do ultra deep multiobject spectroscopy.

From the high redshift galaxies we will soon be able to measure the IGM absorption much more accurately, providing an even stronger constraint on the star formation rate as a function of redshift. From the precise multicolor photometry of a large fraction of galaxies at intermediate redshift on can build up direct maps of extinction due to the Milky Way.

Using photometric redshifts to the 100 million galaxies in the SDSS photometric catalog, we will be able to measure the evolution in galaxy clustering as a function of redshift and type. With the help of the orthogonal templates, and the subspace spanned by them, we will be able to see quantitative differences between low and high redshift galaxy populations. Determining templates from the hybrid method we will be able to compare to the stellar population synthesis models and modify them accordingly. Photometric redshifts are well on their way to open an entirely new window to study the statistical properties of the galaxy population in the universe.

6.1. Impact of the Virtual Observatory

These all underline the importance of a Virtual Observatory. Many of these surveys will only cover some of the bands necessary for good photometric redshifts — we need to fuse data from several. In order to merge the data sets, we need to cross-identify the objects, we need to compensate for the different systematics between surveys and we need to have absolute calibrations. Once these hurdles are passed, we will easily select rest-frame subsamples at various redshifts, and perform meaningful statistical comparisons between the low and high-redshift universe.

New algorithms will emerge to deal with the large number of objects, we already heard several attempts for creating an $N \log N$ algorithm for 2-point correlations. Similar breakthroughs are expected for the power spectrum determinations. Grid-aware databases will interface with these statistical tools, and they will return power spectra and correlation functions as 'Virtual Data' from the queries. These will be computed over the computational grid, using hundreds of CPUs, allocated on demand. Monte-Carlo and numerical simulations of the galaxy distribution will contain close to trillion particles, only possible with these heavy duty tools. In this new world it will be hard to imagine, that the first measurements of the galaxy correlation function only contained a few thousand objects.

Acknowledgments. It is exciting to live in this age when such a paradigm shift happens. This meeting happened at a moment, when the number of people getting interested in a Virtual Observatory suddenly has grown from tens to hundreds. I would like to thank Robert Brunner and George Djorgovski for organizing this conference, Jim Gray for introducing me to the science of large databases, my colleagues of the SDSS Project, which started me on the whole process of thinking about large datasets, and Tom Prince whose ideas were essential in formulating what a VO could be.

References

Baum, W. A. 1962, IAU Symposium No. 15, 390

Baugh, C.M. and Efstathiou, G.P. 1994, MNRAS, 267, 323
Brunner, R.J, Connolly, A.J., & Szalay, A.S. 1999, ApJ, 516, 563
Connolly, A.J., Szalay, A.S., & Brunner, R.J. 1998, ApJ, 499, L125
Hamilton, A.J.S. 1992, ApJ, 385, L5
Harrison, E.R. 1970, Phys.Rev.D, 1, 2726
Jain, B. *et al.* 1999, in preparation
Kaiser, N. 1986, MNRAS, 219, 785
Koo, D.C. 1981, ApJ, 252, L75
Peebles, P.J.E. 1980, *The Large Scale Structure of the Universe*, Princeton University Press
Smoot, G.F., *et al.* 1992, ApJ, 396, L1
Szalay, A.S., Matsubara, T. & Landy, S.A., ApJ, 498, 1, 1998
Tegmark, M., Hamilton, A.J.S., Strauss, M., Vogeley, M.S., & Szalay, A.S. 1998, ApJ, 499, 555
Therrien, C. W. 1992, Discrete Random Signals and Statistical Signal Processing, (New Jersey: Prentice-Hall)
Vogeley, M.S. & Connolly, A.J., private communication, 1998
Vogeley, M.S. & Szalay, A.S. 1996, ApJ, 465, 34
Zaroubi, S., Hoffmann, Y. 1996, ApJ, 462, 25
Zel'dovich, Ya.B. 1972, MNRAS, 160, 1P

Panchromatic Studies of Active Galactic Nuclei

Carol J. Lonsdale

IPAC, Caltech, Pasadena, CA 91125

Abstract. Infrared, radio and x-ray studies suggest that our census of AGN may be substantially incomplete, especially with respect to Type 2 objects. Panchromatic studies of AGN populations will provide a powerful tool for completing the census, testing Unified AGN models, determining the role of AGN activity in galaxy formation and evolution and assessing the importance of accretion-power to the Universal energy budget.

1. AGN Populations and the Case of the Missing Type 2 QSOs

Unification models have been quite successful in unifying certain subsets of the AGN population, most notably using orientation to the line-of-sight as the unifying factor. In such models (Figure 1) the Seyfert 1s are observed close to the pole of a dusty, optically-thick torus so that the high velocity motions in the nuclear broad line clouds can be seen directly, while the Seyfert Type 2s are seen closer to edge-on through the torus which blocks the broad line region from view in the UV and optical (UVO), except by reflection. This scenario has successfully explained most of the major observed properties of Seyfert galaxies (see Antonnuci 1993 for a review; and, *e.g.*, Malkan *et al.* 1998, for contrasting evidence and viewpoints). Other aspects of AGN populations can also be unified by orientation, in particular core-dominated and lobe-dominated radio galaxies via orientation of the line-of-sight with respect to the radio jets (Urry and Padovani 1995).

A serious question remains regarding the more luminous QSOs (defined in terms of power, QSOs are usually defined as AGNs with absolute blue magnitude greater than -23 or bolometric luminosity $>10^{12}$ $L_\odot$). Almost all known UVO-selected QSOs are of Type 1 (broad emission line (BEL) objects). This is the so-called "missing" Type 2 QSO problem; until recently, virtually none were known (Padovani 1997; Becker *et al.* 2000; Djorgovski *et al.* 1999), although the popular unification geometries would predict that there should be several times more Type 2 AGN than Type 1s. Following Padovani (1997) the redshift distribution of all AGN identified by type in the Veron-Cetty & Veron (1996) catalog is illustrated schematically in Figure 1; Type 1 and Type 2 objects differ dramatically in redshift distribution with Type 2 objects being virtually absent beyond $z = 0.2$.

Padovani (1997) also clearly demonstrated that our catalogs of known AGN are strongly biased by wavelength: the detection rate of AGN in large sky surveys in a particular wavelength band is roughly inversely proportional to the best

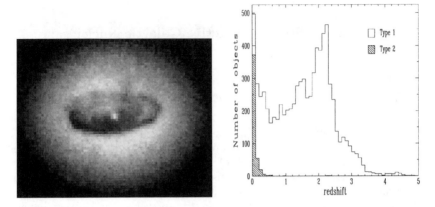

Figure 1.   NGC 4261 central regions (Ferrarese, Ford, and Jaffe 1996) (left) redshift distributions of Type 1 vs. Type 2 AGN (right). The absence of Type 2 AGN beyond $z \sim 0.2$ suggests that our census of AGN may be substantially incomplete, possibly due to obscuration from a central dusty, molecular torus, as suggested by Unified AGN models (Padovani 1997).

survey sensitivity available in that band. It follows, then, that the deficiency of Type 2 QSOs is most likely due to lack of sensitivity in large area AGN surveys in the IR, hard X-ray, and radio (for radio-quiet AGN), relative to the UVO, radio (for radio loud objects) and soft X-ray. This conclusion is broadly consistent with the torus orientation scenario for the lower luminosity Seyfert galaxies: the Type 2 QSOs would be expected to be more heavily obscured and therefore more easily detected in the wavelength bands with the lowest optical depths. It just so happens that these which are very bands with poorer large area survey sensitivity: the hard X-ray, far-infrared and radio (for radio quiet objects).

Therefore it is clear that our current catalogs of AGN — dominated by objects discovered in the UV, optical and soft X-rays — are likely very incomplete, and that a true inventory of all AGN in the Universe will require deep surveys at all wavelengths and extensive datamining within archives to discover them.

2.   Heavily Obscured AGN Populations and the Cosmic Backgrounds

X-ray extragalactic surveys are dominated by AGN and hot intracluster gas in galaxy clusters. The sensitive ROSAT all sky survey resulted in a dramatic increase in our AGN catalogs, dominated by AGN with soft X-ray excesses. Meanwhile the X-ray Background (XRB) has been a puzzle for decades because its spectral energy distribution could not be reconciled with the combined SEDs of the known component populations of (mostly) AGN that were expected to make it up.

With the advent of the sensitive hard X-ray satellites Ginga, ASCA, BeppoSax, RXTE, Chandra and XMM, it is becoming clear that a large fraction of AGN may be highly obscured throughout the soft x-ray to far-infrared, and many are completely Compton thick (Maiolino et al. 1998; Matt et al. 2000; Risaliti, Maiolino and Salvati 1999), and that Einstein and ROSAT AGN populations are biased (unsurprisingly) towards the lower column, softer x-ray-bright subset of the AGN population. Models can reproduce the spectral energy distribution of the XRB if a high fraction (80% or more) of the total AGN energy is absorbed by large columns (Comastri et al. 1995).

The attenuated energy in turn must be re-radiated in the FIR/Submm by warm dust, and may be responsible for a (strongly dust temperature and redshift model-dependent) 10-30% fraction of the Cosmic IR Background (Gunn and Shanks 2000). The CIB in turn is now known to account for 40%, or perhaps significantly more, of the entire integrated energy density of the Universe.

However the fact that Chandra has resolved most of the hard XRB (Mushotzky et al. 2000) and that Fabian et al. (2000) and Hornschemeier et al. (2000) find little or no overlap between the dozen or so Chandra and SCUBA sources in the HDF and 2 Abell cluster fields, is only marginally consistent with these absorbed-AGN XRB and CIB models.

3. AGN, ULIRG and Starburst Connections: Unification by Time

Although it follows from the above line of reasoning that more sensitive IR and radio surveys may be expected to be highly successful at discovering large numbers of heavily obscured radio-quiet AGN, including Type 2 QSOs (*cf.* Becker et al. 2000), it is not clear whether the Seyfert galaxy unification by orientation scheme can be extended to the more luminous QSO population.

The IRAS Bright Galaxy Sample (Soifer et al. 1989), which is complete to $z \sim 0.1$ for an IR luminosity of 10^{12}, is dominated at these QSO-scale luminosities by the ULIRGs (Ultra Luminous IR Galaxies), thus if a population of IR-luminous Type 2 QSOs exists locally, *it can only be found* within the ULIRG population. However ULIRGs are typically found in advanced merger systems with disturbed structure, and frequently have strong circumnuclear starbursts which may heavily contribute to the high IR luminosity. Therefore, since many ULIRGs are clearly in transitory, chaotic states of interaction or merger, it is important to consider not only unification of Type 1 and Type 2 objects by orientation, but also unification by time: the ULIRGs may be QSO 1s in the making, triggered by a merger event (Sanders et al. 1988).

Another consideration concerning the idea of unifying QSO 1s and 2s by orientation of a dusty torus alone, like Seyferts, is that if models in which 80% or more of AGNs responsible for the XRB are obscured at some significant level, then the average covering factor of absorbing material in the "QSO 2s" would have to be much larger than observed in the Seyfert 2 population.

It has recently been claimed that ISO spectroscopy has for the most part solved the question of whether starbursts or AGN dominate the FIR/Submm luminosity of local ULIRGs, by demonstrating that most local ULIRGs do not show the high ionization lines characteristic of AGN even in the mid-infrared ($5 < \lambda < 30 \mu m$) (Genzel & Cesarsky 2000). On the other hand, recent hard

X-ray detections from the very dusty starburst ULIRGs NGC 6240 and NGC 1614 have clearly demonstrated that powerful AGN can be completely hidden by column densities, $N_H > 10^{24}$ (Iwasawa 1999, Risaliti et al. 2000), even in the mid-IR frequency range surveyed by ISO LWS. For Compton thick sources, even hard X-ray imaging will not detect AGN directly; they will be visible only in reflected/scattered softer X-rays.

This X-ray picture is consistent with the extremely high line-of-sight optical extinctions indicated by radio imaging of ULIRGs in CO, which can greatly exceed $A_v =100$ mag. (Sanders & Mirabel 1996). Consistent with these results is the evidence from spectroscopy and polarimetry (e.g., Hines et al. 1995, 1999) that a large fraction of the local ULIRGs and the "hyperluminous" IR galaxies (infrared-warm, IRAS-selected galaxies with L$> 10^{13}$ L$_\odot$; Cutri et al. 1994) are chiefly powered by obscured QSOs.

Sensitive VLBI radio imaging at high frequencies, where the free-free optical depth is minimized, is a viable alternative method of detecting AGN within heavily obscured systems. Our decade-long VLBI campaign to determine the far-infrared power sources of ULIRGs by studying the radio power on parsec scales, and where possible, by imaging the high brightness temperature radio emission, has been highly successful (Lonsdale, Smith & Lonsdale 1993). The importance of imaging has been dramatically demonstrated, both by the unexpected detection of a cluster of radio supernovae in Arp 220 (Smith et al. 1998), and the unambiguously AGN-related structure of the ULIRG UGC 5101 and the well-known IR luminous LoBal QSO Mrk 231 (see Figure 2).

The VLBI radio observations provide direct information about the existence of an AGN in a ULIRG, from the brightness temperature of the emission and from the structure in imaging data (see Figure 2). In addition it can potentially help tie down the power of the AGN, since the ratio $L_{radio-core}/L_{bol}$ for radio-quiet quasars (RQQs) has small scatter (Lonsdale, Smith & Lonsdale 1995), so for RQQs the radio core flux is a good tracer of bolometric AGN power. This may also be the case for obscured QSOs in ULIRGs.

Another consideration may also be important here, namely that the true population of QSOs with non-detectable BELs due to orientation-dependent absorption may be much smaller than for the lower luminosity Seyferts because once the QSO turns on, the higher radiation pressure in these more powerful objects may dispel the dusty material extremely quickly. Perhaps the unusual QSO UNJ1025-0040 (Brotherton et al. 2000) — with a bright, UV-strong, QSO nucleus and apparently low extinction, and a highly luminous post-starburst population of A stars — is one of these rare transition objects between dusty starburst and UV-bright QSO.

Support also comes from the new results from the FIRST survey that BAL QSOs are more frequent than previously believed, and can be radio-strong. Becker et al. (2000) favor a model in which the BAL QSOs are not normal QSOs with line-of-sight skimming the torus (e.g., Hines and Wills, 1995), but newly-formed QSOs dispelling their dusty cocoons. In this regard it is notable that BAL QSOs are often FIR-luminous; in particular the LoBal Mrk 231 is a ULIRG in our sample which we have shown to possess a parsec scale double radio source from the newly re-activated central engine, plowing into a heavily free-free absorbed medium (Smith et al. 1999; see also Carilli & Taylor 2000).

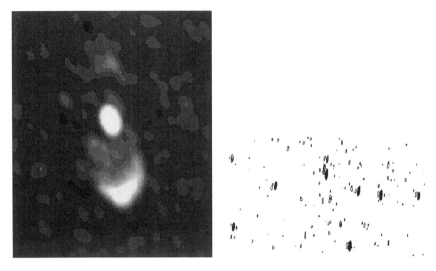

Figure 2. Left: 18cm VLBI images of the central regions of the ULIRGs Mkn 231, which exhibits an AGN core (left), and Arp 220, showing about a dozen luminous radio supernovae—evidence for Starburst excitation (right) (Smith et al. 1999).

Other dimensions besides orientation and time are certainly relevant; in particular black hole mass and AGN fueling rate will affect the characteristics of high vs. low luminosity AGN. We do not address these issues directly here, but note that QSOs and ULIRGs seem much more likely to be linked to strong merger events than Seyferts, which can inhabit more normal galaxies and can be stimulated by tidal interactions, satellite infall or bar instabilities. Thus the time dimension during a rapid and chaotic merger event is much more likely to dominate the appearance of the high luminosity QSOs than the lower luminosity Seyferts.

4. Evolution with Cosmic Time

These issues have important implications not only for AGN inventories but also for the confrontation of galaxy formation models with observations. Recently there has been a substantial theoretical effort addressing the connection between QSO and Galaxy formation based on (1) the strong observational evidence for causal and/or evolutionary connections between nuclear starbursts and AGN in the local Universe (Sanders 1999); (2) the similarity in shape of the number density of QSOs with redshift and the integrated Star Formation History (Madau et al. 1996; Steidel et al. 1999); and (3) the correlation between black hole mass and spheroid mass in the local Universe (Magorian et al. 1998), implying a strong link in the astrophysics of spheroid and black hole growth.

5. AGN Discovery in Panchromatic Archives: The Importance of a Virtual Observatory

To make progress in understanding the formation and history of AGN, their evolutionary connection to starburst events, their importance to galaxy formation and evolution, and their energetic contribution to the energy budgets of the Universe, it is necessary to (1) have a full census of all AGN as a function of cosmic time, and (2) to understand in depth the connections between AGN episodes within galaxies and overall galaxy evolution, including starburst episodes.

Djorgovski et al. (1999) have used an optical color-selection technique in the Digitized Palomar Sky Survey (DPOSS) archive to search for a population of Type 2 Quasars by their strong [OIII] equivalent width on the F plates at redshifts of $3.1<z<3.7$; i.e., as outliers in the (r-i) vs. (g-r) color-color diagram. The success rate for finding Type 2 QSOs is high, and the implied surface density is consistent with being the same, or up to a factor of 10 times higher than that of Type 1 QSOs (depending on the magnitude limit assumed for the QSO Type 1 population). These numbers are therefore consistent with the XRB models, and represent a strong confirmation of QSO unification models, whether by orientation or time. This result is a strikingly successful example of the importance of cluster analysis in a Virtual Observatory.

The Two Micron All-Sky Survey, 2MASS, provides another striking example of the power of simple cluster analysis in a large database. Cutri et al. (2000) have obtained an extremely high success rate in discovering new, red QSOs as outliers in (J-K) color. In this case, however, most of the new QSOs are of Type 1, therefore it is likely that the survey has uncovered a highly incomplete sampling of the reddened tail of the known local QSO Type 1 population. These first results from mining the enormously rich 2MASS database represent only the tip of the iceberg: the brightest and reddest objects which are easily isolated in color-color space from red stars and galaxies.

The FIRST radio survey is also turning up significant samples of Type 2 red objects with intermediate radio strength (White et al. 2000).

Lonsdale et al. (2001) address a subset of the AGN population that is bright in both the mid- to far-infrared and the soft (1-2.5 kev) X-rays (see Figure 3). Compared to UVO-selection, such a population will include a much higher fraction of objects in which the engine is not viewed directly, since optical depths in both the soft X-rays and the infrared are far less than in the UVO. On the other hand the selection criteria may not be very sensitive to the most highly extinguished populations, since the highest columns so far observed, $\sim 10^{24}$, are large enough to completely block 1-2.5 kev X-rays, which would be detectable by ROSAT, therefore, only in reflection, if at all. Such columns also imply dust optical depths high enough to strongly suppress mid-infrared emission, as noted above. Thus the soft X-ray/FIR ratio and the 25/60 micron color are expected to be good discriminators of optical depth effects in moderately highly obscured AGN populations (Risaliti et al. 2000). Lonsdale et al. (2001) have therefore sought to construct the most complete local sample possible, from the all sky ROSAT and IRAS surveys, of soft X-ray and mid- to far-IR bright AGN and starbursts. They indeed find a strong correlation between the 60/25 micron color and the X-ray/Far-IR luminosity, for a sample of ~ 600 IRAS-ROSAT candidate AGNs, >200 of them previously unidentified.

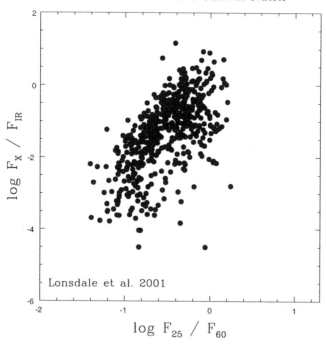

Figure 3. Correlation between 25/60 and X-ray/IR for ROSAT-IRAS AGN (Lonsdale *et al.* 2001)

6. SIRTF AGN Surveys

One of the six newly-announced SIRTF Legacy programs is SWIRE, the SIRTF Wide-area InfraRed Extragalactic Survey (C. Lonsdale P.I.). Surveying 50-100 square degrees, SWIRE will easily detect IR-loud (dust-obscured) QSOs to z>2.5 with MIPS, whether obscured by edge-on molecular tori or by large columns associated with merger-induced starbursts. SWIRE's IRAC survey will be much more sensitive than 2MASS to the heavily reddened population (but not those with extremely high N_H). Most low and moderate extinction AGN will be revealed through warm dust in the IRAC bands and at 24μm, though strong PAH features can confuse interpretation of broadband colors (Clavel *et al.* 2000; Laurent *et al.* 2000). The biggest difficulty will be in identifying the most highly obscured AGN, even in the mid-IR, as is the case in the local Universe. Hard X-rays are uniquely capable of penetrating all but the highest columns, so one of the SWIRE fields is coincident with the deep XMM-LSS survey of M. Pierre (1998).

References

Antonucci, R. 1993, ARA&A, 31 473
Becker, R., *et al.* 2000, ApJ, 538, 72

Brotherton, M. S., *et al.* 2000, ApJ, in press
Carilli, C., & Taylor 2000, ApJ, 532, 95
Clavel, J., *et al.* 2000, A&A, 357, 839
Comastri, *et al.* 1995, A&A, 296, 1
Cutri, R., *et al.* 1994, ApJ, 424, 65
Djorgovski, S. G., *et al.* 1999, B.A.A.S., 195, #65.02
Fabian, A., *et al.* 2000, MNRAS, 315, L8
Ferrarese, L., Ford, H. C., & Jaffe, W., 1996,ApJ, 470, 444
Genzel, R. & Cesarsky, C. 2000, ARA&A, 38, in press
Gunn, K., & Shanks, T. 2000, MNRAS, in press (Astroph/9909089)
Hines, D., & Wills, B. 1995, ApJ, 448, L69
Hines, D., Schmidt, G., Smith, P., Cutri, R. & Low, F. 1995, ApJ, 450, L1
Hines, D., Schmidt, G., Wills, B., Smith, P., & Sowinski, L. 1999, ApJ, 512, 145
Hornschemeier, A., *et al.* 2000, ApJ, 54, 49
Iwasawa 1999, MNRAS, 302, 961
Laurent, O., *et al.* 2000, A&A, 359, 887
Lonsdale C. J., Smith, H. E., & Lonsdale C. J. 1993, ApJ, 405, L9
Lonsdale, C. J., Smith, H. E., & Lonsdale, C. J. 1995, ApJ, 438, 632
Lonsdale, C. J., Voges, W. H., Boller, Th., & Wolstencroft 2001, in preparation
Magorrian, J., *et al.* 1998, AJ, 115, 2285
Maiolino, R., *et al.* 1998, A&A, 338, 781
Madau, P., *et al.* 1996, MNRAS, 283, 1388
Malkan, M., Gorjian, V., & Tam, R. 1998, ApJS, 117, 25
Matt, *et al.* 2000, MNRAS, 318, 173
Mushotzky, R., Cowie, L., Barger, A. & Amaud, K. 2000, *Nature*, 404, 459
Padovani, P. 1997, In B. McLean, editors, *The Proceedings of the 179th IAU on New Horizons from Multi-Wavelength Sky Surveys*, IAU Symposium No. 179., 424
Pierre, M. 1998, in *The Young Universe*, ASP Conference Series No. 146, 405
Risaliti, G., Maiolino, R., & Salvati, M. 1999, ApJ, 552, 157
Risoliti, G., Gilli, R., Maiolino, R., & Salvati, M. 2000, A&A, 357, 13
Sanders, D. 1999, ApJS, 266, 331
Sanders, D., & Mirabel, I. F. 1996, ARA&A34, 749
Sanders, D., *et al.* 1988, ApJ, 325, 74
Smith, H. E., Lonsdale, C., Lonsdale, C. & Diamond, P. 1998, ApJ, 493, L17
Smith, H. E., Lonsdale, C., Lonsdale, C., & Diamond, P. 1999, ApJS, 266, 125
Soifer, B. T., *et al.* 1989, AJ, 98 766
Steidel, C., *et al.* 1999, ApJ, 519, 1
Urry, M., & Padovani, P. 1995, PASP, 107, 803.
Veron-Cetty, M.-P. & Veron, P. 1996, ESO Scientific Report, 7th ed.
White, *et al.* 2000, ApJS, 126, 333

Gravitational Lensing with the NVO

Gordon K. Squires

California Institute of Technology, M/C 105-24, 1200 E. California Blvd., Pasadena, CA 91125

J. Anthony Tyson

Bell Labs, Lucent Technologies, 1D-432, 700 Mountain Ave, Murray Hill, NJ 07974

Abstract. Weak lensing probes of the dark matter distribution on all scales, from galaxy-galaxy lensing (on scales of $\simeq$ few h^{-1}kpc) to cosmic shear (probing scales of $\simeq$ 100 h^{-1}Mpc), all share common data requirements: large area coverage, moderate depth, multiwavelength information, and good image quality (in terms of seeing, resolution and psf stability). As such, the NVO will provide a useful feeder for weak gravitational lensing studies, and simultaneously will benefit by archiving the types of observations obtained for lensing programs. We present two examples of lensing observations, one ongoing and one proposed, as prototypes for the type of data lensing surveys will provide for the NVO.

1. Introduction

In the broadest of terms, programs designed to detect weak gravitational lensing signals typically are performed on moderate to large aperture telescopes ($\gtrsim$ 4m), obtaining moderately deep integrations ($t_{exp} \gtrsim$ 1 hr) in good seeing conditions (FWHM$\lesssim$ 1″.0), in two or more passbands, and over as large a field of view as possible (ranging from $O(2' \times 2')$ in space, to $O(30' \times 30')$ with ground-based CCD arrays). One can imagine that the resulting data sets would be useful for a wide range of studies including transient detection, identification of Kuiper belt objects, studying the Galactic halo stellar population, and in fields containing clusters of galaxies: studies of the Fundamental Plane, the cluster galaxy color-magnitude relation, and the Butcher-Oemler effect. As such, it would clearly be of great use to a large number of researchers to have access to a central repository of data obtained from weak gravitational lensing observations.

In this short contribution, we outline the data requirements for weak lensing analyses, and suggest what format the data should be provided to enable a wide range of applications. We also describe two projects, one ongoing and one planned, that would provide useful feeders to the NVO.

2. Data Requirements

Weak gravitational lensing imaging studies are relatively straightforward in their data requirements: obtain large area coverage with moderate depth, with a stable and mappable psf.

Currently, the best example of a successful weak lensing study coming from the type of data one might access with the NVO is the detection of galaxy-galaxy lensing in the Sloan Digital Sky Survey (SDSS) commissioning data (Fischer et al. 2000). With only 2 nights on a 2.5m telescope and seeing of $\gtrsim 1''.25$, the SDSS collaboration was able to detect the lensing signal of foreground galaxies on their faint background counterparts, and constrain the mean halo velocity dispersion to be $\sigma = 150 - 190$ km/s and the halo scale length to be $r_h > 260$ h^{-1}kpc (at the 95% confidence level). The surprising aspect of this study is not that a signal was detected, but rather that is was detected on such a modest data set in terms of depth and seeing conditions. In making this detection possible, the key features were the relatively large area covered by the observations ($225\square°$) and the fact that the psf was sufficiently stable to enable mapping and correction.

Heretofore, signals extracted from all of the weak gravitational lensing observations, from small scales with galaxy-galaxy lensing (Fischer et al. 2000), through cluster lensing (e.g., Hoekstra, Franx & Kuijken 2000), to the largest scales and cosmic shear (Van Waerbeke et al. 2000; Wittman et al. 2000; Kaiser, Wilson & Luppino 2000), have been limited in signal-to-noise by the intrinsic ellipticities of the faint, lensed galaxies. Conversely, systematic shape distortions, at least at the $\gtrsim 1\%$ level, are well-corrected (e.g., Kaiser, Squires & Broadhurst 1995; Luppino & Kaiser 1997), and likely well below this level as well (e.g., Kuijken 1999; Kaiser 2000).

As the primary source of noise in weak gravitational lensing studies is statistical, the requirement for increasing the signal-to-noise is simple: increase the number of background galaxies. This is done either with deep integrations (and preferably with a large aperture telescope) or covering a larger area of sky.

As an example of this, consider the signal-to-noise in extracting cluster masses. The mass in the lensing cluster is typically extracted via the aperture mass statistic

$$\zeta(\theta_1, \theta_2) = 2 \left(1 - \frac{\theta_1^2}{\theta_2^2}\right)^{-1} \int_{\theta_1}^{\theta_2} d\ln\theta \langle \gamma_t(\theta) \rangle \qquad (4)$$

which measures the mean surface density interior to radius θ_1, relative to the mean in an aperture with $\theta_1 \leq \theta \leq \theta_2$, where $\gamma_t(\theta)$ is the tangential component of the shear. For an isothermal cluster with velocity dispersion σ_{1000} (measured in units of 1000 km/s), the signal-to-noise in the aperture mass statistic is

$$S/N = 4.1 \sigma_{1000}^2 \beta_{0.5} \left(\frac{\bar{n}}{20\square'}\right) \left(\frac{\langle \gamma^2 \rangle}{0.3}\right)^{-1/2} \left(\frac{a-1}{a+1}\right)^{1/2} \qquad (5)$$

where $\theta_2 = a\,\theta_1$, and $\bar{n}$ is the surface density of lensed galaxies.

The signal-to-noise in extracting cluster masses (and similar arguments apply to galaxy-galaxy lensing as well as cosmic shear) is maximized either by deeper exposures (i.e., increasing $\bar{n}$) or by increasing the outer radius of the control aperture, and thus requires the largest possible field of view. The latter is usually much more efficient, and has the added benefit of probing outside the cluster mass distribution and thereby removing the systematic mass underestimation made via this technique. Similarly, for detecting cosmic shear and constraining the form of the underlying matter power spectrum $P(k)$, the large survey area is critical to probe a physically interesting range of scales.

While the requirement for lots of (good) data is intuitive, it is of equal importance to be able to correct for systematic, non-gravitational effects which also perturb the observed shapes of the faint galaxies. Determining reliable and optimal methods for correcting psf anisotropy has been a subject of intense scrutiny (Kaiser, Squires & Broadhurst 1995; Luppino & Kaiser 1997; Hoekstra et al. 1998; Kuijken 1999; Kaiser 2000; Wittman et al. 2000). The important aspect of all of the correction algorithms, is that the calculations are done in the image plane (i.e., access to each individual image is required).

3. Deep Lens Survey

A good example of a lensing feeder to the NVO is an ongoing effort to detect cosmic shear. This consists of an ultra-deep multiband optical survey of $7 \times 4\square°$ fields, using the Mosaic CCD imagers at the Blanco and Mayall 4m telescopes. The team of some twenty five Co-Is worldwide will take five years to complete the "Deep Lens Survey" in four bands: B, V, R, z' to 29/29/29/28 mag per square arcsecond in surface brightness. The deep combined data and catalogs for sub-fields are being released to the community as they are completed. In addition, optical transient events and supernova candidates are released in real time. Moving object lists (asteroids, KBOs, comets) are listed separately as they are found. Finally, the individual flat-fielded exposures will also be publicly available for those who wish to analyze the images themselves.

The main goal of the survey is to produce unbiased maps of the large-scale structure of the mass distribution beyond the local universe, via very deep multicolor imaging of $7 \times 2°$ fields and color-redshifts. The shear of distant galaxies induced by the mass of foreground structures will be measured. These weak-lensing observations are sensitive to all forms of clumped mass and will yield unbiased mass maps with resolution of one arcmin in the plane of the sky (about $120 \, h^{-1}$kpc at z = 0.2). These maps will measure for the first time the change in large scale structure from z=1 to the present epoch, and test the current theories of structure formation, which predict that mass in the low-redshift universe has a particular filamentary/sheetlike structure. These observations will directly constrain the clustering properties of matter, most notably Ω_m and Ω_Λ, and, when compared with the results from microwave background anisotropy missions, will test the basic theory of structure formation via gravitational instability.

While this is the main goal of the survey, a wide-field imaging survey has a myriad of other uses. In addition, the group is acquiring the data in a way which makes it possible to detect variable objects on scales of hours to months, by spreading observations of individual subfields over 4 runs over two years. These transient events are released onto the website in real-time.

The goal behind releasing transient events so fast is to allow for spectroscopic followup by the community while the events are still bright enough to be captured. The March and April 2000 dark runs at CTIO produced transient event listings in the 10h and 14h equatorial fields, and the KPNO dark runs in November and December 2000 will target the two northern fields. During the April 2000 CTIO observing run, a supernova was detected in real time, and the community notified via an IAU circular (IAUC 7398). Transients shorter than a day or so may be very interesting: this is our first glimpse of transient phenomena at the faint 20–25 mag attainable with this 4m survey.

The individual exposures are 600–900s each, depending on the filter; taken in blocks consisting of five pointings dithered by up to $8'$ to shift all low surface brightness features off themselves (to provide adequate sky-flatfielding) and cover $42' \times 42'$ in one subfield (9 subfields = one $2° \times 2°$ field). This dither is repeated (with about a one arcminute offset) in each filter – and then off to another subfield within one of two fields. In the first two survey runs, the survey team observed five subfields in the 0053+12 field and four sub-fields in 0919+30 to roughly half of the final depth in B, R and z'. The survey observations go deepest in R (to a surface brightness limit of 29 mag/$\square''$) and concentrate on R in good seeing in order to obtain the best lensing signal possible. The other bands are imaged less deep (approximately half of the exposure time of R, in order to provide photometric redshift estimates.)

The dataset from the completed survey will be large by today's standards. The images will comprise of order 200 Gb + catalogs. Updates are available at the project website[1].

4. A Future Feeder to the NVO: The Dark Matter Telescope

Gravitational lensing observations are optimized, as with most extragalactic imaging programs, via a balance of, on one hand, a large area of the sky covered, and, on the other hand, superlative image quality on large aperture telescopes. Current generation telescopes and instrumentation typically attempt to optimize one of these characteristics: in space, the Hubble Space Telescope offers exquisite image quality, but over a relatively small field of view (FOV); the Wide Field Planetary Camera subtends only a few square arcminutes, and even the Advanced Camera for Surveys increases this only to a $202'' \times 202''$ FOV. On the other hand, ground based efforts have been channeled towards ever increasing light gathering power in mirror size, but with image quality limited by atmospheric blurring. Strategies employing adaptive optics improve image quality drastically, of course, but again are currently operational over an extremely small area. The design of the Dark Matter Telescope (DMT/LSST; Tyson, Wittman, & Angel 2000) has been created to provide a balance between light gathering capability, and the delivery of pristine image quality over a large field of view.

To maximize the figure of merit for an imaging telescope, it should be located at a site with excellent seeing and the telescope focal length must be chosen so the detector pixels will adequately sample seeing limited images. Typically, good telescopes at the best sites (e.g., in Hawaii or Chile) will deliver images of $0.''5$ on occasion, more frequently in the near infrared. The pixel sampling

---

[1] http://dls.bell-labs.com

should thus be no worse than 0".25 (the Nyquist sampling criterion) to avoid further significant image degradation, thus each square degree on the sky must be sampled by about 200 million pixels in the detector mosaic.

The DMT/LSST is designed to cover the $0.3 - 2.4\mu m$ spectral range where the atmosphere is largely transparent and has relatively low emissivity. This will require two detector array types, most likely silicon CCDs below $1\mu m$ wavelength and HgCdTe arrays above. Both types have quantum efficiencies $\eta \simeq 1$. Individual 2048×2048 arrays of HgCdTe will be practical in the near future with $15\mu m$ pixels, but not much smaller. Thus the optimum focal length is $10 - 12m$ and hence requires a relatively fast focus of $f/1.25$.

To achieve this, the telescope mirror design is a three mirror system, as explored by Paul (1935). He envisaged a parabolic primary, convex spherical secondary and a concave spherical tertiary of equal but opposite curvature. In this model, the image is formed midway between secondary and tertiary, with good correction over a wide field. Indeed, this design was utilized by McGraw et al. (1982), using a 1.8m parabolic primary at f/2.2 to yield a telescope with a one degree field. The central obscuration was 22% by area, and the design gave images no more than 0".2 rms diameter at the edge of the field.

In the DMT/LSST design, the mirror placement has been changed somewhat: the mirrors are arranged so the light from the secondary passes through a hole in the primary to a near-spherical tertiary behind, and the light comes to a focus near the primary vertex. Obscuration by the detector is minimized by making the primary and secondary together afocal. The telescope is very compact, little longer than the primary mirror diameter, and centrally balanced. It thus requires an enclosure much smaller and less expensive than for standard 8m designs. Other conventional designs, including Schmidt telescopes and other corrected systems based on one or two mirrors, are incapable of wide fields at so fast a focus.

The detectors are required to have sensitivity in the optical through infrared. To achieve this over a large field of view, circular mosaics of imaging detectors 55cm in diameter are required, with CCDs for wavelengths $0.3 - 1\mu m$ and HgCdTe devices for $1 - 2.4\mu m$. To operate both in the optical and infrared, two interchangeable cameras are envisaged, each having fused silica windows 1.3m diameter and 79mm thick. Interference filters 60cm diameter will be used to select bands as narrow as $\delta\lambda/\lambda = 3$ %.

The DMT/LSST is proposed to operate in two modes: a "Deep Probe" mode is proposed to concentrate on ten 100 □° fields in the sky, opening the window on the developing mass structure formation from 7 billion years ago to the present, several thousand very distant supernovae (to chart any acceleration in the expansion of the universe), and ultra-faint transient events associated with energetic phenomena in the distant universe. At least five wavelength filter bands will be required to get redshift distance information, creating a 3-dimensional color image of the distant universe in luminosity. This enables a motion picture of developing mass structures from half the age of the universe to the present.

The all-sky mode can survey all the sky (about 20,000 □°) visible from the telescope site to 24th magnitude in a few nights. This could be repeated several times during the same month. This huge database of 20s exposures (through a broad red filter) would create a motion picture of the dynamic sky sufficiently faint to detect 90% of the near Earth objects larger than 300m in diameter after a decade of operation. In addition, several hundred thousand nearby supernovae would be discovered and their luminosity charted.

The data rate expected from this telescope is immense. In $\simeq$ 4 nights the fast-slewing telescope could be used to survey the entire visible sky (20,000 $\square°$) to a (5σ) limiting magnitude of V=24, I=23, or Ks=19, yielding $\simeq$ 1 Tb/night of imaging data. Alternatively, in 50 hours of observation an 8 color composite image of a single 7$\square°$ field could be made to a 5σ limiting magnitudes U = 26.7, B = 27.8, V = 27.9, R = 27.6, I = 26.8, J =2 4.8, H = 23.5 and Ks = 22.8.

In the telescope design, it is expected that all of the data (except the 1TB/night all-sky mode imaging) could be on spinning disks by 2005, and all data by 2010, extrapolating current trends in storage capability. For the DMT/LSST, as with the NVO, the challenge will be to develop efficient database search tools.

5. Conclusions

Current generation weak gravitational lensing studies probing the dark matter on all scales have the common feature that the dominant source of noise is statistical. As such, the NVO can provide the key ingredients for lensing analyses that has been thus far lacking: large sky coverage. Similarly, the type of imaging data that will be obtained by lensing studies are likely to be useful to a wide range of other scientific pursuits such as transient detection (including supernovae), identification of Kuiper belt objects, studying the Galactic halo stellar population, studies of the Fundamental Plane, the evolution of the cluster galaxy color-magnitude relation, and the Butcher-Oemler effect.

Weak lensing observations are already producing data at a prodigious rate. Over the course of four years of observations, the Deep Lens Survey will yield $\simeq$ 200 Gb of imaging data covering 7 $\times$ 4$\square°$ in B, V, R, z' to 29/29/29/28 mag/$\square''$ in surface brightness. The final dataset from the SDSS will be enormous (40 Tb covering $10^4\square°$ in five wavebands). Similarly, the proposed Dark Matter Telescope will produce data at an incredible rate: in full-sky survey mode $\simeq$ 1 Tb/night. These types of observations provide a challenge to the NVO: correction for non-gravitational systematic image distortion requires access to the original images, and hence a search and delivery mechanism allowing users to access data in the image-plane will be required.

References

Fischer, P., & the SDSS collaboration 2000, astro-ph/9912119
Hoekstra, H., Franx, M., & Kuijken, K. 2000, ApJ, 532, 88
Hoekstra, H., Kuijken, H., Franx, M., & Squires, G. 1998, ApJ, 504, 636
Kaiser, N. 2000, ApJ, 537, 555
Kaiser, N., Wilson, G., & Luppino, G. 2000, astro-ph/0003338
Kaiser, N., Squires, G., & Broadhurst, T. 1995, ApJ, 449, 460

Kuijken, K. 1999, A&A, 352, 355

Luppino, G.A., & Kaiser, N. 1997, ApJ, 475, 20

McGraw, J. T., Stockman, H. S., Angel, J. R. P., Epps, H., & Williams, J. T. 1982, Proc. SPIE, 331, 137

Paul, M. 1935, Rev. d'Optique, 14, 169

Tyson, J. A., Wittman, D., & Angel, J. R. P. 2000, in Sources and Detection of Dark Matter in the Universe 2000, ed. D. Cline (Marina Del Rey: Springer), in press. See also: astro-ph/0005381

Van Waerbeke, L., et al. 2000, A&A, 358, 77

Wittman, D. M., Tyson, J. A., Kirkman, D. Dell'Antonio, I., & Bernstein, G. 2000, Nature, 405, 143

NVO and the LSB Universe

James M. Schombert

Department of Physics, University of Oregon, Eugene, OR

Abstract. There is tremendous scientific potential in a National Virtual Observatory, particularly for projects that need to mine large databases for rare or unusual objects. However, the NVO will also make an impact on any project, large or small, the requires a mixture of datasets to explore a wide range of astrophysical phenomenon. In this article I discuss the influence of the NVO on research into the formation and evolution of low surface brightness (LSB) galaxies. In particular, I present the preliminary results from an NVO-style project that combines the DPOSS and 2MASS datasets to search for giant disk galaxies.

1. The Dataset Revolution

Astrophysical problems seem to increase in complexity with each successive generation. Observationally, new wavelengths and new flux limits bring about new phenomenon that demands more telescope time and begs theoretical interpretation. In the last 15 years we have seen an explosion in the amount, wavelength coverage and diversity of our datasets that have lead to numerous discoveries, but have also buried us in the sheer quantity of information.

Our community has also grown parallel to our data growth, but most of the high powered observational tools still lie in the possession of a few institutions. This disparity in big telescope resources has been offset, in a large part, by the formation of national data centers and the distribution of analysis software. Now an astronomer, regardless of the size of their home institution, can have access to high quality data and produce cutting edge science. With the addition of small grant programs (*i.e.*, NASA's ADP program), an astronomical industry developed during the last two decades and discovery has moved from the hands of the few to the hands of the many. One only need to compare the impact of HST science on the astronomical community to that of Keck to see how the existence of non-proprietary datasets can push forward science.

As datasets have grown larger, there has been a strong emphasis on data mining and computational power. However, intelligent and cleverly designed projects depend more on access to direct tools rather then sophisticated algorithms. Thus, many astronomical projects today involve teams of scientists who bring together the diverse talents needed to attack the details of massive amounts of data. This is the goal of NVO, to provide the infrastructure and intellectual support for astronomical programs that compliment our already existing observational and theoretical foundations.

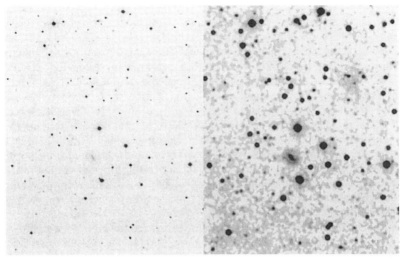

Figure 1. A subset of DPOSS normal (left panel) and smooth with a 5 pixel Gaussian filter (right panel). A LSB disk galaxy is barely visible in the normal frame. A second LSB dwarf galaxy is discovered in the upper right-hand section of the smoothed frame. Both objects where detected in H I at Arecibo, the LSB disk has a velocity of 5,550 km/s, the dwarf is at 2,500 km/s.

Within the NVO concept there lies the goal of removing the division between the type of science and the type of data. Indeed, the greatest benefit the NVO may bring to us is to eliminate the specialization that divides our fields of research (stellar versus extragalactic) and between the wavelength regions (divided mostly due to the technology used to observe within the spectral regions of interest). As an example, one area of astronomy that could derive enormous benefit from the NVO is research into low surface brightness (LSB) galaxies.

2. LSB Universe

One area that is particularly challenging to the observational world is the universe of LSB objects. We can only study what we know to exist, and with respect to galaxies that means the object must reside in some catalog. While there has always been a push to find the faintest objects (meaning the lowest in mass as stellar luminosity maps into baryonic mass) or the most distant (meaning the closest to the galaxy formation epoch), it has only been recently that there has been much concern for objects with low luminosity density.

The LSB realm is interesting to many astrophysical problems. For example, star formation is normally a phenomenon associated with high gas density environments. Yet, LSB galaxies display many characteristics that indicates a history of recent star formation. Thus, studies into their past will probe star formation in new parameter space. The range of galaxy properties requires an examination of the LSB universe because galaxies at the extreme ends of the mass spectrum (dwarf and giant) tend to be LSB in nature. LSB galaxies also differ from their brighter cousins in that their gas masses often exceed their stellar masses. Thus, baryon counts of the Universe at high redshift will be underestimated without some knowledge of the distribution of LSB galaxies.

In some sense, the lack of pursuit of LSB research is technical in nature. To find the faintest, or most distant, galaxies becomes a simple process of building larger and larger collecting surfaces. However, achieving fainter levels of surface brightness to explore the LSB universe battles against the natural glow of the night sky and is not overcome with larger pieces of glass. Space imaging has the immediate advantage of getting above the sky glow, but the emphasis in space has been on small and faint, so pixel sizes to take advantage of high resolution images work against LSB objects by reducing the number of counts per pixel. Even in the non-optical portions of the spectrum the emphasis is always first on the detection threshold of point sources rather than design for sky brightness (see, for example, the 2MASS survey).

The first expeditions into the LSB universe were taken, of course, by Zwicky who hypothesized on the existence of 'hidden' galaxies as a counter to Hubble's notion that the galaxy luminosity function is Gaussian in shape. By the 1960's, numerous galaxy catalogs by Arp, Sandage, van den Bergh (DDO) and de Vaucouleurs (RC2) had defined the Hubble classification system. These catalogs were primarily defined by HSB galaxies, but there were always appendices or notes concerning 'diffuse' objects, usually assumed to be nearby dwarfs.

Disney (1976) was the first to place the concept of galaxy visibility in an analytic form and to demonstrate that the mean central surface brightness of our galaxy catalogs was, in fact, a function of the natural sky brightness and not imposed by astrophysics. While the importance of this work is recognized today, galaxy evolution was a relatively new field at the time and the study of LSB galaxies remained in the background (no pun intended) until the mid-1980's.

3. LSB Detection

There was very little that the typical observational astronomer could do about the night sky problem until the mid-1980's with the evident of the Second Palomar Sky Survey. While the sky brightness had only degraded since the first Sky Survey, the finer emulsions and deeper plate material allowed for, at least, a cursory examination of the difference between the angular limited UGC and the new plate material. This resulted in the POSS-II LSB catalogs (Schombert & Bothun 1988, Schombert et al. 1992, Schombert et al. 1997) which were visual surveys, but demonstrated an increase of one magnitude per square arcsecond to the old catalogs. More importantly, it provided a jumpstart to the LSB field by simply providing a new list of objects in which to study.

The first visual surveys were extremely crude but spurred a more exacting search for LSB's using CCDs (O'Neil, Bothun & Cornell 1997, Dalcanton et al. 1997). Flattening is always a key parameter for LSB galaxy detection. Most CCDs on 1-meter class telescopes are sky limited in a few minutes of exposure time. The critical component in finding LSB galaxies, and measuring fluxes, is how well you know the sky value and how flat (on large and small scales) you can make your data. Transit CCD surveys offer the best method for sky flattening, since they allow the sky to pass through each pixel which is then summed for

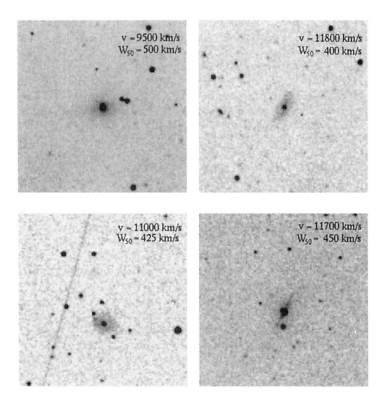

Figure 2. Four AGN Malin objects discovered from a combined search of DPOSS and 2MASS datasets. The near-IR 2MASS catalogs provide the low luminosity AGN sources (due to their anomalous colors) typical to the Malin class. The optical catalog is fast filtered for the presence of an LSB disk.

the entire scan. Long scans can suffer from temporal variations, but these can at least be quantified.

LSB detection is functionally a very difficult problem. The standard procedure is to detect all the HSB objects, mask them out, smooth and run the detection scheme again. Unfortunately, this will usually remove any LSB galaxies with bright bulges. They will be in the original catalog, but their LSB nature may not be known from whatever parameters are stored in the catalog.

One example of a smooth and search algorithm is shown in Figure 1. Here a sample of the sky from a J plate of DPOSS is shown in its raw form. A single LSB disk galaxy is obvious near the center of the frame. Smoothing with a Gaussian filter enhances the detectable of the LSB disk and also discovers a second LSB galaxy in the top right portion of the scan. Both objects were detected in H I at Arecibo with velocities of 5,550 km/s and 2,500 km/s respectfully.

The smoothed image demonstrates the two major difficulties in automating LSB detection algorithms. The first is that the number of pixels distorted by stellar sources is extremely high at the surface brightness levels of interest. It is practically impossible to detect, in an automatic fashion, LSB objects near stars (although the human brain seems to carry out the task fairly well). Second, the uncorrelated background noise varies substantially over even this small piece of sky. This makes a systematic survey, to specific threshold levels, a technical challenge (what NASA would call completeness and reliability).

4. AGN Malin Search

One of the most intriguing classes of LSB galaxies is the supergiant disk systems, the Malin class. The prototype to this class, F568-6, was discovered from a visual search of the POSS-II (Bothun et al. 1990). While low in central surface brightness ($\mu_o = 23.4B$ magnitude per square arcsecond), F568-6 is by no means low in luminosity ($M_B = -21.1$) nor low in total or HI mass. The Malin class contains the largest galaxies in the Universe, yet are notorious difficult to find and catalog (Sprayberry, Impey & Irwin 1996).

One of their properties provides a promising avenue for the cataloging of a significant number. Most Malin class galaxies have a weak AGN in their core. The current theory is that the copious gas supply in the disk provides the fuel for a central engine, even if at a low intensity (Schombert 1998). While the weak AGN appears as a point source, its near-IR colors would distinguish it from a stellar SED. Unfortunately, a survey of the sky in the near-IR will not, by itself, identify the Malin objects since the sky brightness at 2.2 μm is 2000 times higher than in the optical making their disk regions invisible. Thus, this project requires a 'virtual observatory', in this case the combination of two existing databases, DPOSS (optical) and 2MASS (near-IR).

The procedure is straight-forward, first isolate all the objects in the 2MASS catalog with non-thermal colors (i.e., outside some boundary defined by normal stars). Second, search the near-IR source positions on the blue plates of DPOSS with a fast area scan. A series of circular apertures are placed around the point source then tested against the local sky. LSB galaxy detection is effective if only a particular region is being tested against the background since varying diameters are checked which maximize the signal from the LSB disk versus the signal from sky.

A preliminary search was undertaken last winter using eight plates from DPOSS that contained some fraction of 2MASS coverage (about 3 square degrees). Forty candidates were produced of which ten of these objects were searched with the new, upgraded Arecibo telescope. The Gregorian system at Arecibo has a much wider velocity range, a critical element since the large Malin objects tend to be at velocities greater than 8,000 km/s. Eight of the ten candidates were detected at 21-cm.

Four of the detection's DPOSS images are shown in Figure 2. All have the characteristics AGN nucleus surrounded by a LSB disk. All eight also have HI widths in excess of 350 km/s (the typical spiral has a rotation width of 250 km/s). Assuming these galaxies follow the baryonic TF relation (McGaugh et al. 2000), then their masses will exceed $10^{12} M_\odot$.

5. Conclusions

It is universally recognized that the NVO would be a powerful tool for the astronomical community. A particular emphasis will be placed on the need for the NVO to make the most efficient use of the large datasets in our present holdings and to future projects. However, perhaps one of the most common uses of the NVO will be of the one human/one workstation type project.

One such project I have described herein, the search for AGN Malin galaxies typifies my vision of how small projects will use the NVO. The merger of multi-wavelength datasets and simple tools, combined with a researchers experience in the astrophysical phenomenon, was used here to achieve a catalog of a new, and exciting, type of galaxy.

While each of our individual research interests may reap the benefits of the NVO, there is no doubt that the sum of the contributions of many small projects will also service to build the framework of the NVO. A system that we hope will be wavelength and research field independent.

Acknowledgments. I wish to thank my collaborators in the LSB universe (G. Bothun, J. Eder, S. McGaugh and K. O'Neil) for their insights. Many different telescopes have contributed to the LSB searches, but a special thanks is given to the Arecibo scientists and staff for their speedy upgrade which has increased LSB's visibility by many orders of magnitude. I also wish to thank the organizers of the conference for inviting me and the NASA grant (NAG5-6109) which has supported my LSB programs.

References

Bothun, G., Schombert, J., Impey, C. & Schneider, S. 1990, ApJ, 360, 427

Dalcanton, J., Spergel, D., Gunn, J., Schmidt, M & Schneider, D. 1997, AJ, 114, 635

Disney, M. 1976, *Nature*, 263, 573

O'Neil, K., Bothun, G. & Cornell, M. 1997, AJ, 113, 1212

Schombert, J & Bothun, G. 1988, AJ, 95, 1389

Schombert, J, Bothun, G., Schneider, S. & McGaugh, S. 1992 1988, AJ, 103, 1107

Schombert, J., Pildis, R. & Eder, J. 1997, ApJ, 481, 157

Schombert, J. 1998, AJ, 116, 1650

Sprayberry, D., Impey, C. & Irwin, M. 1996, ApJ, 463, 535

The New Paradigm: Novel, Virtual Observatory Enabled Science

R.J. Brunner

Department of Astronomy, California Institute of Technology, Pasadena, CA, 91125

Abstract. A virtual observatory will not only enhance many current scientific investigations, but it will also enable entirely new scientific explorations due to both the federation of vast amounts of multiwavelength data and the new archival services which will, as a necessity, be developed. The detailing of specific science use cases is important in order to properly facilitate the development of the necessary infrastructure of a virtual observatory. The understanding of high velocity clouds is presented as an example science use case, demonstrating the future synergy between the data (either catalog or images), and the desired analysis in the new paradigm of a virtual observatory.

1. Introduction

A diverse and exciting array of scientific possibilities, whose exploration are enhanced by the existence of a Virtual Observatory, are detailed elsewhere in this volume. Certain lines of scientific inquiry, however, are not just enhanced by a virtual observatory, but are actually enabled by it. For example, a panchromatic study of active galactic nuclei (see, *e.g.*, Boroson, these proceedings), studies of the low surface brightness universe (see, *e.g.*, Schombert, these proceedings), a study of Galactic structure (see, *e.g.*, Kent, these proceedings), or a panchromatic study of galaxy clusters, are all extremely interesting projects that are facilitated by a virtual observatory.

In this article, I will discuss some specific technical challenges which must be overcome in order to fully enable this new type of scientific inquiry. This is not as difficult as it may first appear, as many of these challenges are already being tackled, as is evidenced by the prototype services which are currently available at many of the leading data centers. In order to truly make revolutionary, and not merely evolutionary, leaps forward in our ability to answer the important scientific questions of our time, we need to "think outside the box", not just in the design and implementation of a virtual observatory, but in the actual scientific methodology we wish to employ (see, *e.g.*, Figure 1, which demonstrates this concept by combining large image viewing with the ability to selectively mark objects in the image based on their statistical properties).

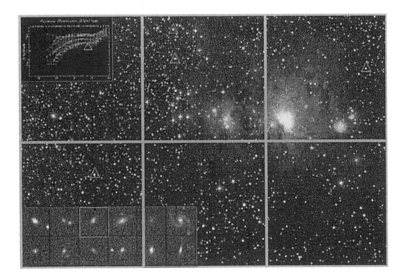

Figure 1. A prototype of the visualization services which would empower scientists to not only tackle current scientific challenges, but also to actually aid in the exploration of the, as yet unknown, challenges of the future. Note the intelligent combination of image and catalog visualizations to aid the scientist in exploring parameter space. Figure courtesy of Joe Jacob, JPL.

2. Technological Challenges

While many of the technical challenges are rather self-evident upon a cursory examination, such as the federation of existing archival centers, other challenges are considerably more difficult to elucidate. This effect is primarily a result of the difficulty in designing scientific programs for the, as yet unavailable, virtual observatory. This is exactly the time where "thinking outside the box" applies, as one needs to ask not *"what can I do right now?"*, but *"what would I like to be able to do?"*.

The first step in this process is to consider, in its entirety, all of the data which might be available for ingestion into a virtual observatory. This includes the obligatory data catalogs, which are the most often used derivative of survey programs, and perhaps more importantly, the original imaging data and any associated metadata (that is, data which describes the data). Similar extensions likewise apply to other types of astronomical data, including spectral and temporal.

After taking this revolutionary leap, we can now consider querying not just catalogs, but also the data from which the catalogs were extracted. This would allow for new techniques to be applied, which might, for example, perform source extraction using multiple wavelength images simultaneously (*e.g.*, χ^2 detection, Szalay *et al.* 1999), or perhaps to extract flux limits for objects detected in other wavelengths, or, finally, to extract matched parameters (*e.g.*, matched aperture photometry).

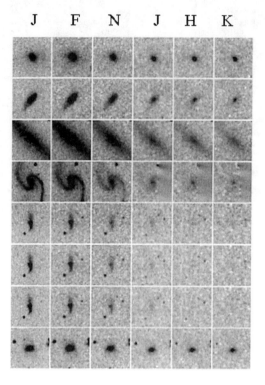

Figure 2. The multiwavelength nature of nearby galaxies, constructed from the optical data of the DPOSS survey (J, F, and N) and the near-infrared data of the 2MASS survey (J, H, and K). Figure courtesy of Steve Odewahn, Arizona State University.

This is demonstrated in Figure 2, where the multiwavelength nature of nearby galaxies is explored, from the optical, extracted from the DPOSS survey (Djorgovski *et al.* 1998), to the near-infrared, extracted form the 2MASS survey (Skrutskie *et al.* 1997). As this example demonstrates, multiwavelength image processing is a pressing need, since objects bright in one wavelength are often much fainter, if even detected at all, at other wavelengths.

Another example of the need for image reprocessing is shown in Figure 3, where the detection of a nearby, low surface brightness dwarf spheroidal galaxy is demonstrated. The vast majority of survey pipelines are designed to detect the dominant source population, namely high surface brightness point-type sources. As a result, an implicit surface-brightness selection effect exists in nearly all catalogs (see, *e.g.*, Schombert, these proceedings for a more detailed account). In the future, one would ideally like to be able to reprocess survey data in an effort to find objects at varying spatial scales and surface brightnesses.

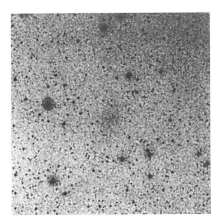 

Figure 3. A demonstration of the detection process which must be used to detect the low surface brightness sources missed by most survey pipelines. On the left is the original DPOSS plate image which contains the dwarf spheroidal galaxy Andromeda V (Armandroff *et al.* 1998), which is visible with a large image stretch. On the right is a reprocessed version of the original image which is designed to emphasize subtle background variations which can be caused by low surface brightness sources (see, *e.g.*, Brunner *et al.* 2001).

3. Science Use Case: Understanding High Velocity Clouds

As a demonstration of how a virtual observatory can enable new science, consider the specific science use case of understanding high velocity clouds (HVCs). HVCs are defined as systems consisting of neutral Hydrogen which have velocities that are incompatible with simple models of Galactic rotation (Wakker and van Woerden 1997). Their origin, however, remains uncertain, with various arguments being made in support of a wide range of hypothesis, including that they are Galactic constituents, that they are the remnants of galaxy interactions, or that they are fragments from the hierarchical formation of our local group of galaxies.

In an effort to truly understand these systems, we also would like to understand their composition. Although these systems are, by definition, found in neutral Hydrogen surveys, we can perform either follow-up observations at other wavelengths, or else correlate the HI data with existing surveys at other wavelengths in order to learn more about them (see, *e.g.*, Figure 4 for a demonstration of multiwavelength image correlation for a known HVC). This process can often require the construction of large image mosaics involving multiple POSS-II photographic plates in order to map structures that span several tens of square degrees. This service should clearly be one of the principal design requirements for a virtual observatory.

The most powerful method for understanding the composition of HVCs, however, is to study their absorption effects on the spectra of background sources, most notably quasars. In order to find suitable targets, we need to be able to dynamically correlate the HVC images with published quasar catalogs in order to determine the optimal line-of-sights for quantifying the composition of the intervening HVCs with follow-up spectral observations.

 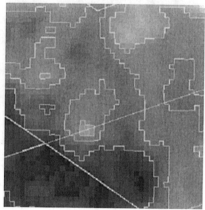

Figure 4. A demonstration of the image correlation process for a known high velocity cloud complex. The figure on the left is a DPOSS F (approximately R) plate image, selected since it would contain any H_α emission. The HVC is detected by the low surface brightness emission near the center of the image. The figure on the right is the correlated IRAS 100μm image, which demonstrates a remarkable correlation with the diffuse emission seen in the figure on the right (see Brunner et al. 2001).

Finally, we also would like to understand the evolution of these systems, which has obvious implications for understanding their origin. This can optimally be done by comparing the predictions of theoretical models to our correlated multiwavelength observations. This implies a need for a virtual observatory to allow seamless access to not only astronomical data but also the results of dynamical analysis, either through persisted calculations or a real-time process.

4. The New Paradigm

To accomplish these ambitious scientific goals, we need powerful tools, which should be implemented as part of a virtual observatory. First, we need the ability to process and visualize large amounts of imaging data. This should be done in both a manner which is suitable for public consumption (*i.e.*, the virtualsky.org project) and also a manner which preserves scientific calibrations. These services will also need to provide coordinate transformations, overlays and arbitrary re-pixelizations. Ideally, these operations occur as part of a service which can also accept user-defined functions to further process the data, minimizing the size of the data stream which must be established with the end-user.

Next, we need the ability to federate an arbitrary collection of catalogs, selected from geographically diverse archives, a prime computational grid application. To completely enable the discovery process, we also need intelligent display mechanisms to explore the high-dimensionality spaces which will result from this federation process. We also will need to allow the user to post-process these federations using user-defined tools or functions (*e.g.*, statistical analysis) as well as combine these processes with image operations and visualizations.

Finally, a complete census and subsequent description of science use cases (*e.g.*, the previous section, see also, Boroson, these proceedings), inevitably leads one to the formulation of a new paradigm for doing astronomy with a virtual observatory. In the future, anyone, anywhere, will be able to do cutting edge science, as researchers will only be limited by their creativity and energy, not their access to restricted observations or telescopes. Not only will this revolutionize the scientific output of our community, but it will also have an important effect on the sociology of our field as well, since students will need to be trained in these new tools and techniques.

Acknowledgments. This work was made possible in part through the NPACI sponsored Digital Sky project and a generous equipment grant from SUN Microsystems. RJB would like to acknowledge the generous support of the Fullam Award for facilitating this project. Access to the DPOSS image data stored on the HPSS, located at the California Institute of Technology, was provided by the Center for Advanced Computing Research. The processing of the DPOSS data was supported by a generous gift from the Norris foundation, and by other private donors.

References

Armandroff, T.E., Davies, J.E., & Jacoby, G.H. 1998, AJ, 116, 2287

Brunner, R.J., Roth, N., Hokada, J., Gal, R., Mahabal, A.A., Odewahn, S.C., & Djorgovski, S.G. 2001, AJ, in preparation

Djorgovski, S., de Carvalho, R.R., Gal, R.R., Pahre, M.A., Scaramella, R., and Longo, G. 1998, In B. McLean, editors, *The Proceedings of the 179th IAU on New Horizons from Multi-Wavelength Sky Surveys*, IAU Symposium No. 179, 424

Skrutskie, M.F., *et al.* 1997, In F. Garzon *et al.* , editors, *The Impact of Large Scale Near-IR Sky Surveys*, Kluwer, 25

Szalay, A.S., Connolly, A.J., & Szokoly, G.P. 1999, AJ, 117, 68

Wakker, B.P., & van Woerden, H. 1997, ARA&A, 35, 217

Precision Galactic Structure

Stephen Kent

Fermilab, P. O. Box 500, Batavia, IL 60510

For the SDSS Collaboration

Abstract. Optical and IR surveys in progress or in the planning stages will lead to substantial improvements in our picture of the Milky Way as a consequence of their providing large volumes of data with much improved photometric and positional measurements compared with existing datasets.

1. Introduction

The structure of the Milky Way can be described to first order by a canonical model that is the sum of a circularly symmetric disk (or two disks, one thick, one thin), a bulge (either an oblate spheroid or a triaxial bar), and an oblate spheroidal halo; each component has characteristic shape parameters, stellar populations, and kinematic properties. While the major features of each component are reasonably well known, gaps remain (*e.g.*, the mass function for low mass stars); further, in detail the Milky Way has a more complex structure than that of the canonical model. Existing and future optical and near IR surveys such as SDSS and 2MASS will provide a wealth of new data on the Milky Way, and it would seem that it should be possible to improve our models of the Milky Way immensely. However, interpretation of the data is not necessarily straightforward, and this paper will discuss how precision Galactic Structure requires data of high quality, not just large quantity.

The data sets provided by the photometric surveys typically consist of star counts as a function of position, apparent magnitude, and mulitple colors. The first problem one encounters is that interpretation of these data requires that one model the galaxy in equal detail; a study of Galactic structure ends up being as much an exercise in modeling stellar populations as it is in modeling the structure itself. If one wishes to probe the structure of the disk at large distances from the sun, one must additionally deal with problems of confusion due to high star densities and extinction by dust. Fortunately the dust seems to follow the total gas content of the Galaxy, so maps can be contructed based on HI and CO surveys (Kent, Dame and Fazio 1991). Finally, to probe the kinematic properties of the Galaxy, one would like to combine radial velocity and proper motion surveys to obtain 3-dimensional velocities of large samples of stars.

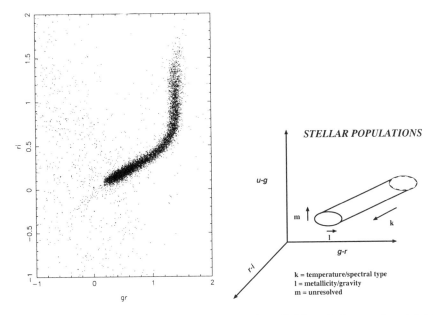

Figure 1. Stellar locus in two SDSS colors (left) and three colors (right)

2. Stellar Populations

The observable properties of a population of stars can be expressed as a multidimensional function of luminosity and color. Stars do not uniformly fill this space, but are concentrated in a narrow locus.

Figure 1 shows a typical projection of the stellar locus in two SDSS colors. On the right is a schematic representation of a section of the locus in a 3 dimensional color space. Newberg et al. (1997) have shown that the locus is remarkably well behaved. If one defines principal axes k, l, m as shown, then the k axis primarily measures temperature, l measures a mix of metallicity and gravity, while the m dimension is essentially unresolved. The widths in the l and m direction are .07 mag and .03 mag respectively (FWHM). Newberg et al. have shown that it is possible to measure metallicities of F and G main sequence stars with an accuracy of 0.13 dex.

In a typical magnitude-limited sample of stars, stars with the same apparent magnitude but different colors can span a wide range of intrinsic luminosity, making it difficult to analyze such samples as a single set. Often it is possible to select a subset of stars based on color alone that have a narrow range in luminosity compared with the ensemble of all stars, easing the task of making detailed structure maps. Two features that will be used in the next section include the main sequence turnoff, which is readily identified in any histogram of stars vs. color, and the blue horizontal branch (BHB), which can be picked out using u, g, r colors or their equivalent. Color alone cannot distinguish blue horizontal branch stars from blue stragglers, however.

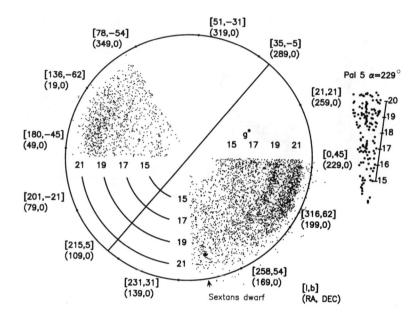

Figure 2. Distribution of horizontal branch stars on the celestial equator.

3. Halo

Properties of the halo that are not yet well known include:

1. Radial density profile: $\rho(r)$.

2. Metallicity vs. radius.

3. Triaxiality: $a : b : c$, which may be a signature of the underlying dark matter distribution.

4. Substructure.

Figure 2 is a plot from Yanny et al. (2000) of the distribution of BHB stars along the equator in early SDSS data. The most notable feature is the presence of bands of stars between 19th and 21st mag (g) in both the Northern and Southern Galactic hemispheres. The double band seen in the North (lower right part of the figure) is thought to be a single structure seen in BHB's at 19th mag and blue stragglers at 21st mag. The bands are thought to be the tidally disrupted remnant of companion galaxies of the Milky Way captured at some previous time. The masses are a few million solar masses.

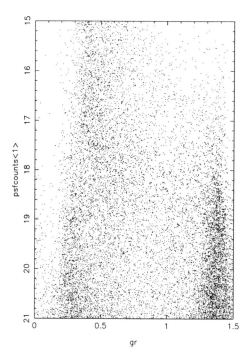

Figure 3. g-r color vs. apparent g magnitude for a sample of stars near the SGP

4. Disk/Halo Transition

Figure 3 is a color/apparent-magnitude diagram for stars from the SDSS commissioning data (adapted from Chen et al. 2000) covering approximately 8 square degrees of sky in the South Galactic Hemisphere. This line-of-sight is close to perpendicular to the Galactic disk and thus samples primarily the old disk and the nearby halo. The near-vertical band on the left side of the figure arises from stars at the main sequence turnoff; the band fainter than $g = 18$ on the right side of the diagram arises from K and M dwarfs in the Galactic disk. The turnoff stars show an abrupt change in color from $g - r = 0.4$ to $g - r = 0.25$ at around $g = 18$. This change reflects the transition from the metal-deficient thick disk to the even more metal-poor halo (Chen et al. 2000). The vertical extent of the thick disk has been notoriously tricky to measure; with accurate photometry, its separation from the halo is straightforward.

5. Precision Dynamics

Measurement of dynamical quantities such as the local mass density in the solar neighborhood proves to be perniciously difficult when one attempts to attains accuracies of better than a factor 2. In principle the problem is well defined: one only needs to measure the density and velocity dispersions, including their gradients, for a homogeneous subpopulation of stars, and then compute mass densities using the combined Boltzmann and Poisson equations. In practice, the amount of data required to do a complete job is prohibitive, and virtually all analyses introduce simplifying assumptions, at the cost of possibly introducing biases in the results (*e.g.*, Kuijken and Gilmore 1989). Limiting factors include the lack of combined radial velocity/proper motion data for large samples of stars that are needed to construct 3-dimensional velocity ellipsoids. Future proper motion surveys (*e.g.*, as might be constructed by combining scans of plates from the two Palomar Sky Surveys) could have a major impact.

A classic problem is measurement of the tilt of the velocity ellipsoid for a population of stars as one moves out of the Galactic plane (Kent and De Zeeuw 1991). Kuijken and Gilmore (1989) showed that this term makes a 20% contribution in the calculation of the local surface mass density. For stars in the plane of the disk, the velocity dispersion in the radial direction is about twice that perpendicular to the disk. If the mass of the Galaxy is spherically symmetric, then as one moves out of the plane, the major axis of the velocity distribution function rotates so as to continue pointing to the Galactic center, but if the mass is primarily in the disk, the major axis will remain parallel to the disk. A measurement of this direction gives information on the ratio of disk to (dark) halo mass. Halo stars, for example, have a characteristic velocity dispersion of 150 km s^{-1}, so measurement of the tilt term at a height of 10 kpc above the plane requires accurate radial velocities combined with proper motions that have errors of order 3 milliarcseconds per year.

6. Conclusions

The field of Galactic Structure will benefit greatly from the new generation of surveys, both because of their completeness and their accuracy. The problems and possibilities discussed above are just a fraction of the total that one can imagine. Full utilization of these surveys will require joint analyses of combined data sets.

Acknowledgments. The Sloan Digital Sky Survey (SDSS) is a joint project of The University of Chicago, Fermilab, the Institute for Advanced Study, the Japan Participation Group, The Johns Hopkins University, the Max-Planck-Institute for Astronomy, Princeton University, the United States Naval Observatory, and the University of Washington. Apache Point Observatory, site of the SDSS, is operated by the Astrophysical Research Consortium. Funding for the project has been provided by the Alfred P. Sloan Foundation, the SDSS member institutions, the National Aeronautics and Space Administration, the National Science Foundation, the U.S. Department of Energy, and Monbusho. The SDSS Web site is http://www.sdss.org/.

References

Chen, B. Stoughton, C., Smith, J. A., Uomoto, A., Pier, J. R., Yanny, B., & Ivezic, Z. 2000, in preparation

Kent, S. M., Dame, T., & Fazio, G. 1991, ApJ, 378, 131

Kent, S. M., & de Zeeuw, T. 1991, AJ, 102, 1994

Kuijken, K, & Gilmore, G. 1989, MNRAS, 239, 605

Newberg, H. J., & Yanny, B. 1997, ApJS, 113, 89

Yanny, B., Newberg, H. J., Kent, S., Laurent-Muehleisen, S. A., Pier, J. R., Richards, G. T., Stoughton, C., Anderson, J. E., Annis, J., Brinkmann, J., Chen, B., Csabai, I., Doi, M., Fukugita, M., Hennessy, G. S., Ivezic, Z., Knapp, G. R., Lupton, R., Munn, J. A., Nash, T., Rockosi, C. M., Schneider, D. P., Smith, J. A., & York, D. G. 2000, ApJ, in press

Virtual Observatories of the Future
ASP Conference Series, Vol. 225, 2001
R.J. Brunner, S.G. Djorgovski, and A.S. Szalay, eds.

Exploring the Time Domain: Transients

Alan H. Diercks

California Institute of Technology, Department of Astronomy, Pasadena, CA 91125

Abstract. Although the National Virtual Observatory (NVO) is often conceived of as a tool for in depth analysis of static, archival data, it also has the potential to contribute significantly to astronomy in the time domain. I focus here on a few applications to the study of transient phenomena including the use of the distributed facilities of the NVO to perform real-time searches for rare events.

1. Introduction

Transient astronomical phenomena share several characteristics: 1) they are typically rare and often detected only through dedicated observing programs, 2) they are often "one time" events (*e.g.*, supernovae, Gamma-ray bursts) and therefore time is of the essence and data are precious, 3) many are non-thermal, often explosive events which emit photos over a wide range of wavelengths and are best studied with a multi-wavelength approach.

Although the detection of transients often requires special instrumentation, follow-up observations generally employ conventional techniques, and the challenges are primarily operational. By their very nature, these events are unscheduled and often brief, making them difficult to accommodate in traditional block-scheduled observing programs and putting significant pressure on observers who must scramble to collect data as quickly as possible.

Even when transients are not localized in real-time, statistical studies of properties such as position, flux, duration, variability, and spectral shape can provide important clues to their nature. Even before the first Gamma-ray burst (GRB) afterglow was localized in the optical and the distance scale determined from spectroscopy, the isotropic distribution of error-boxes provided strong evidence that the events were not concentrated in the Galactic plane or towards M31 and hence likely to be either at cosmological distances or within the solar system. By necessity, transient observers have become adept at making due with scant information.

Although studies of variable stars have a rich history, and some types of transients such as novae, supernovae, and X-ray bursters are well characterized, several new satellite missions are poised to significantly boost the detection rates and narrow the error boxes of high-energy transients. In addition, large scale ground-based surveys using wide-field instruments at a variety of wavelengths are monitoring an increasing fraction of the night sky, providing hope that new types of phenomena discovered at high energies will be localized and studied at longer wavelengths.

The announcement at the 5th Huntsville Gamma-ray Burst Symposium of the discovery of a class of short, intense X-ray flashes with properties similar to GRBs but lacking gamma-ray emission (Heise 1999) is just one example of a new transient phenomena that have yet to be detected at any other wavelength. It is likely that the combination of space-based detection and wide-field, ground-based imaging will be required to localize and study these events.

I discuss below three areas where the NVO could contribute significantly to the study of transients, with an emphasis on high-energy events: rapid localizations, searches in archival data, and real-time discovery.

2. Rapid Localizations of High-Energy Transients

For the foreseeable future, short duration transients detected at Gamma and X-ray wavelengths will continue to have localizations which exceed an optical seeing disk. The predicted error radii for localizations from future missions such as HETE-II (Ricker 1997) and SWIFT (Gherles 1999) vary from a few arcminutes to several arcseconds. Although these error boxes are comfortably accommodated by ground-based optical imaging instruments on large telescopes, spectroscopic observations require localizing the optical counterpart to sub-arcsecond precision.

The challenges associated with studying GRB afterglows illustrate the importance of rapid localization. Because they fade rapidly, the opportunity for high-signal to noise spectroscopic observations is short-lived. Currently, localization requires either comparison with an existing survey such as DPOSS (Djorgovski et al. 1998) or two epochs of observation to identify the fading afterglow. The first option is limited to rather bright events which fall above the magnitude limits of existing surveys, while the second requires two epochs of observation sufficiently separated in time for the afterglow to have faded sufficiently to be detectable against the non-variable background.

If we assume that fading by 0.2 magnitudes is sufficient for robust detection, the typical afterglow power-law decay index of -1.2 means that the second epoch of observation must occur later than $1.16t_1$ where t_1 is the time of the first observation. Although this is not too costly for afterglows which are observed within hours of the Gamma-ray event, continuous observations are not possible for all sky positions and the delay in identification of the transient can extend to several days.

Immediate access to the full resolution pixel data from the new, deeper sky surveys such as Sloan (York et al. 2000) at a variety of wavelengths through a single, uniform interface such as the NVO would greatly assist in rapidly identifying new afterglows from a single epoch of data, allowing spectroscopic follow-up to occur while the events are still bright. The scientific payoff includes detailed measurement of the spectral energy distribution of the afterglow, measurement of the host environment through its imprint on the spectrum, and possibly, information about the absorption in the inter-galactic medium along an unbiased line of sight to high redshift (Lamb & Reichart 2000). All of these measurements require bright, and hence rapidly localized, afterglows.

3. Transient Discovery in Archival Data

Many large-scale surveys which were not undertaken with variable sources in mind, nevertheless contain implicit temporal information. For example, approximately 50% of the DPOSS survey region is covered twice due to field overlap (R. Gal private communication) allowing for the detection of sources which vary on a wide variety of time-scales. This large-scale, global search would be best done through the mechanism of the NVO where the scientific payoff would be considerably leveraged by ready access to data from substantially different wavelengths both for statistical studies of large populations and to identify the most interesting objects for follow-up with large telescopes.

In addition to the large, archival data sets which the NVO aims to federate, there are a significant number of smaller projects which are directed at specific science goals but produce data which could be invaluable for other purposes. Examples include near-Earth asteroid searches such as LINEAR (Viggh et al. 1998) and LONEOS (Koehn 1999), micro-lensing surveys such as MACHO (Alcock et al. 2000), OGLE (Udalski, Kubiak & Szymanski 1997), and EROS (Renault et al. 1998), supernova searches such as the Supernova Cosmology Project (Perlmutter et al. 1998) and the High-z Team (Schmidt et al. 1998), and studies of weak gravitational lensing such as Deep Lens Survey (Tyson 2000). Many of these projects are explicitly temporal in nature and some even focus on a particular type of transient (e.g., supernovae), however the full power of the data sets is generally not exploited for several reasons.

The critical path item in many of these projects is the development of custom software. Little time or resources are generally available for exploration of the data beyond the primary science goals. The NVO could fill this gap by providing a set of data standards and general purpose software tools to make the data from these smaller, more focussed studies available to the community.

One example of a "small" program which would benefit greatly from the NVO is the NOAO Deep Lens Survey (Tyson 2000) conducted at both KPNO and CTIO (see Squires & Tyson, this volume for a more detailed description). The primary science goal of this project is to measure the cosmic mass distribution in the field through weak gravitational lensing. The project targets seven 2-degree fields, which will be imaged in the B, V, R, and z' passbands to flux limits near 29 magnitude per square arcsecond. The survey strategy has been designed to spread total integration on each field over five years, so that variable objects can be detected as the sub-fields are repeatedly imaged on a variety of time-scales over the course of the project.

As each frame is acquired, it is compared with a pre-determined set of comparison images selected from previous observations of the same field. Processing an entire 8k × 8k mosaic image including flat-fielding, alignment, image subtraction, and object detection takes approximately 900 seconds on a 4 processor 550 MHz Pentium III computer. This strategy has so far allowed searches for variable objects over a one month baseline, and is yielding interesting transients at the $R \sim 21$–23 magnitude level. A few examples of very fast fading, as yet unidentified optical transients are shown in Figure 1. While this effort has been successful at regularly detecting a potentially new class of objects, a much more thorough exploration of the data, in both the image and catalog domains, would be possible through access to the NVO software tools and likely yield further surprises.

The NVO would enable searches for transients which occur on time-scales of months to years where the greatest returns are likely to come from analyzing data taken for other purposes. One program in this category is the search for flares from the tidal disruption of stars by the massive black holes which lie at the center of many galaxies. Ulmer (1999) has shown that such events should have thermal spectra which peak in the UV, absolute magnitudes in excess of $V = -19$, and time-scales of 0.1–1.0 years. The rate of these events depends on the stellar density profile in the host galaxy with cuspy dwarfs having higher rates than smooth L_* galaxies. Magorrian & Tremaine (1999) predict a rate of $0.11h^{2/3}(z/0.3)^3$ per square degree per year which is approximately 1% of the Type-Ia supernova rate. Given the low rate and long time-scale of these events, the NVO may provide the only practical search mechanism.

4. Real-Time Interaction with the NVO

The deployment of wide-field imaging systems on 4-m class telescopes has been accompanied by the development of software pipelines which reduce and often analyze the data on-site in real-time. Some of these efforts have arisen from the science requirements of particular projects such as the asteroid and supernova searches where rapid follow-up of new discoveries is essential. Many observatories have also begun to implement pipelines for real-time reduction and analysis of image data from general observer programs with the aims of significantly reducing the workload for the end-user, ensuring a uniform quality of data-reduction, and preparing the data for archiving. Even for the largest mosaic cameras currently planned, real-time analysis is well within the capabilities of current computing technology.

A large fraction of the information in astronomical images often goes unused as the observer is interested only in a few target objects and nearby calibration stars. For small objects (point sources or distant galaxies) mature software already exists to extract into the catalog domain most of the relevant information such as position, brightness, and shape. There is an increasing trend, at least in the optical, to generate these catalogs in real-time, regardless of the primary science goal.

The CFHT 12-k Mosaic imager consists of 12 2048 × 4096 CCDs and covers a 48′ × 28′ portion of sky in a single 200 Mbyte image. Real-time reduction and analysis is accomplished with a software package called ELIXIR (G. Magnier, private communication) which flat-fields the images, detects all objects more than 3-σ above background, performs astrometry, and matches the detected objects either with cataloged data or previous overlapping images of the same field. The entire process takes approximately 150 seconds on a 500-MHz Pentium III computer, and seven such machines are available in the analysis cluster. The real-time data stream is also used to monitor image quality and instrumental zero-points. Although the science goals of many projects will likely require additional downstream analysis once the data leaves the mountain, the individual observer is spared the difficulty of reducing the large volumes of data from scratch.

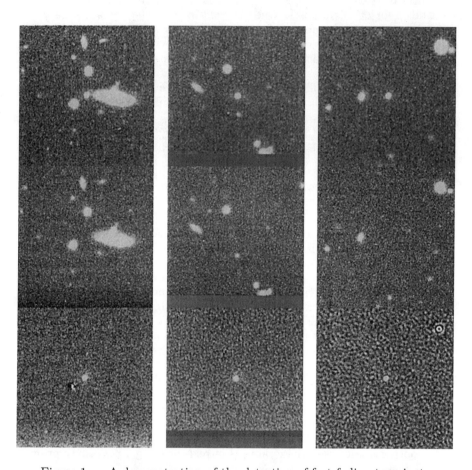

Figure 1. A demonstration of the detection of fast-fading transients from the Deep Lens Survey. These images show three transients with $R \sim 21.5$ which were discovered during the KPNO campaigns. The top and middle images are from the January 2000 and December 1999 runs respectively. The bottom image in each sequence is the PSF-matched "difference image". All of these objects had fallen below the detection limit (faded by approximately 3 magnitudes) by the end of the January run, 3 days later.

This "on the fly" reduction of general observer data represents a tremendous opportunity for the NVO. If a standard was in place, it would require only a modest additional effort to make data from such a system ready for immediate comparison with NVO data. One simple application of such a system to transient studies would be to immediately identify objects in an image whose properties differed from their cataloged values or were not present in either the new data or the archive. With current practices, objects which are not of immediate relevance to the observer are generally ignored.

5. Summary

The NVO offers significant opportunities to advance the study of high-energy transients, primarily by assisting in rapid identification of new events at longer wavelengths and through archival searches for transients in existing data. The highest payoffs for transient science, however, will be achieved if the NVO can accommodate the interaction of new imaging data with the archives in real-time since most of the basic parameters of an observation can now easily be made available immediately after the data is collected.

References

Alcock, C. et al. 2000, AJ, 119, 2194

Djorgovski, S.G., et al. 1998, in Wide Field Surveys in Cosmology, eds. S. Colombi & Y. Mellier (France: Editions Frontieres)

Gherles, N. 1999, American Astronomical Society Meeting 195, #92.08

Heise, J. 1999, oral presentation, 5th Huntsville Gamma-ray Burst Symposium

Koehn, B. W. 1999, DPS meeting #31, #12.02

Magorrian, J. & Tremaine, S. 1999, MNRAS, 309, 447

Perlmutter et al. 1998, ApJ, 516, 2

Lamb, D. & Reichart, D. 2000, ApJ, 536, 1

Renault, C. et al. 1998, A&A, 324, L69

Ricker, G. 1997, in All-Sky X-Ray Observations in the Next Decade, eds. M. Matsuoka & N. Kawai, 366

Schmidt, B. P. et al. 1998, ApJ, 507, 46S

Tyson, J. A. 2000, Physica Scripta, vol. T, 85, 259

Udalski, A., Kubiak, M., & Szymanski, M. 1997, Acta Astron., 47, 319

Ulmer, A. 1999, ApJ, 514, 180

Viggh, H. E. M., et al. 1998, in Proceedings of the Sixth International Conference and Exposition on Engineering, Construction, and Operations in Space, eds. R. G. Galloway & S. Lokaj (Virginia:ASCE), 373

York, D. & the SDSS Collaboration 2000, AJ, September 2000, in press

Searches for Rare and New Types of Objects

S.G. Djorgovski, A.A. Mahabal, R.J. Brunner, R.R. Gal, S. Castro

Palomar Observatory, California Institute of Technology, Pasadena, CA, 91125

R.R. de Carvalho

Observatorio Nacional, CNPq, Rio de Janeiro, Brasil

S.C. Odewahn

Department of Physics & Astronomy, Arizona State University, Tempe, AZ, 85287

Abstract. Systematic exploration of the observable parameter space, covered by large digital sky surveys spanning a range of wavelengths, will be one of the primary modes of research with a Virtual Observatory (VO). This will include searches for rare, unusual, or even previously unknown types of astronomical objects and phenomena, *e.g.*, as outliers in some parameter space of measured properties, both in the catalog and image domains. Examples from current surveys include high-redshift quasars, type-2 quasars, brown dwarfs, and a small number of objects with puzzling spectra. Opening of the time domain will be especially interesting in this regard. Data-mining tools such as unsupervised clustering techniques will be essential in this task, and should become an important part of the VO toolkit.

1. Introduction: Mining the Sky

The great quantitative increases in the amount and complexity of information harvested from large digital sky surveys, with information volumes now measured in multiple Terabytes (and soon Petabytes), with billions of sources detected and tens or hundreds of parameters measured for each of them, pose some fundamental questions: Will this quantitative increase lead to a qualitative change in the way we do astronomy? Will we start asking new kinds of questions about the universe, and use new methods in answering them? How to exploit this great riches of information in a systematic and effective way, and how to extract the scientific essence and knowledge from this mass of bits and pixels? This is indeed what a Virtual Observatory (VO) idea is all about.

There may be two (or, better yet, at least two) main streams of the new, VO-enabled astronomy:

First, there will be statistical astronomy "done right", *i.e.*, studies such as the mapping and quantification of the large scale structure in the universe, of the Galactic structure, construction and studies of complete samples of all kinds of objects (stars or galaxies of particular types or particular ranges of properties, AGN, clusters of galaxies, *etc.*). This is the "bread-and-butter" of astronomy, the way to map and quantify our universe in a systematic, statistically sound fashion, and to feed and constrain our basic theoretical models and understanding. We should never again be limited by the Poissonian errors from small samples of objects; of course, understanding of possible systematic errors and biases in the sky surveys now becomes even more important. Both the numbers of sources and the wide-angle coverage are important for such studies. In some sense, this will be a direct extrapolation of the type of astronomy we have been doing all along, but brought to a higher level of accuracy and detail by the sheer information content of the new, digital sky.

The second stream, where we may expect more novelty and surprises, is a systematic exploration of the poorly known portions of the observable parameter space, and specifically searches for rare types of astronomical objects and phenomena, both already known, and as yet unknown. Here we can use the large numbers of detected sources to look for rare events which would be unlikely to be found in smaller data sets: if some type of an interesting object is, say one in a million or a billion down to some flux limit, then we need a sample of sources numbering in many millions or billions in order to discover a reasonable sample of such rare species. Rare objects may be indistinguishable from the more common varieties in some observable parameters (*e.g.*, quasars look just like normal stars in images), but be separable in other observable axes (*e.g.*, the shape of the broad-band spectral energy distribution). This type of new astronomy with large digital sky surveys (and a VO) is the subject of this review.

2. Exploring the Parameter Space

Some axes of the observable parameter space are obvious and well understood: the flux limit (depth), the solid angle coverage, and the range of wavelengths covered. Others include the limiting surface brightness (over a range of angular scales), angular resolution, wavelength resolution, polarization, and especially variability over a range of time scales; all of them at any wavelength, and again as a function of the limiting flux. In some cases (*e.g.*, the Solar system, Galactic structure) apparent and proper motions of objects are detectable, adding additional information axes. For well-resolved objects (*e.g.*, galaxies), there should be some way to quantify the image morphology as one or more parameters. And then, then there are the non-electromagnetic information channels, *e.g.*, neutrinos, gravity waves, cosmic rays . . . The observable parameter space is enormous.

We can thus, in principle, measure a huge amount of information arriving from the universe, and so far we have sampled well only a relatively limited set of sub-volumes of this large parameter space, much better along some axes than others: We have fairly good sky surveys in the visible, NIR, and radio; more limited all-sky surveys in the x-ray and FIR regimes; *etc.* For example, it would be great to have an all-sky survey at the FIR and sub-mm wavelengths, reaching to the flux levels we are accustomed to in the visible or radio surveys, and with an arcsecond-level angular resolution; this is currently technically difficult and expensive, but it is possible. The whole time domain is another great potential growth area. Some limits are simply technological or practical (*e.g.*, the cost issues); but some are physical, *e.g.*, the quantum noise limits, or the opacity of the Galactic ISM.

Historically, the concept of the systematic exploration of the universe through a systematic study of the observable parameter space was pioneered by Fritz Zwicky, starting in 1930's (see, *e.g.*, Zwicky 1957). While his methodology and approach did not find many followers, the core of the important ideas was clearly there. Zwicky was limited by the technology available to him at the time; probably he would have been a major developer and user of a VO today! Another interesting approach was taken by Harwit (1975; see also Harwit & Hildebrand 1986), who examined the limits and selection effects operating on a number of axes of the observable parameter space, and tried to estimate the number of fundamental new ("class A") astrophysical phenomena remaining to be discovered. While one could argue with the statistics, philosophy, or details of this analysis, it poses some interesting questions and offers a very general view of our quest to understand the physical universe.

So, it is not just the space we want to study; it is the parameter space (in the cyber-space). Much of the total observable parameter space which is in principle (*i.e.*, technologically) available to us is still very poorly sampled. This is our *Terra Incognita*, which we should explore systematically, and where we have our best chance to uncover some previously unknown types of objects or astrophysical phenomena — as well as reach a better understanding of the already known ones.

This is an ambitious, long-term program, but even with a relatively limited coverage of the observable parameter space we already have in hand it is possible to make some significant advances.

3. Looking for the Rare, but Known Types of Objects

Some types of astronomical objects, *e.g.*, particular types of stars or quasars may be relatively rare, or simply be hard to find in the available data sets. But we could use some of their known or expected properties (*e.g.*, typical broadband spectra, or variability) folded through the survey selection functions (*e.g.*, bandpass curves) to design experiments where such objects can be distinguished from the "uninteresting" majority (*e.g.*, normal stars or galaxies). This approach has been used very successfully in the past: most quasars have been found using some such approach, first as "radio-loud stars", then as UV excess objects, *etc.*; ultraluminous IRAS galaxies have anomalously large FIR/visible flux ratios; variable stars and distant supernovæ distinguish themselves with particular types of light curves; and so on.

Sometimes a simple cross-wavelength match can reveal interesting objects or phenomena by indicating those with unusual broad-band energy distributions: recall the discovery of quasars and radio-galaxies, or ULIRGs, or LMXBs, or intra-cluster x-ray gas, or the recent progress on GRBs through the study of their afterglows. This is an obvious area where a VO can be used to construct a detailed, panchromatic view of the universe, and isolate different kinds of objects, with better understanding of the observational biases and selection effects;

Lonsdale's contribution to this volume illustrates such an approach to a general census of AGN.

If objects are spatially unresolved in some sky survey, then the only distinguishing information is in their flux ratios between different bands, *e.g.*, colors. As an example, FIR flux ratios have been used to classify IRAS sources as probable stars or galaxies (*e.g.*, Boller *et al.* 1992).

Even within a given survey with a limited wavelength baseline this approach can be used to separate physically distinct types of objects, or, through some photometric redshift indicator, objects of a given type but in different redshift ranges. This color selection technique is now the principal discovery method for quasars at $z \gtrsim 4$ (Warren *et al.* 1987, Irwin *et al.* 1991, Kennefick *et al.* 1995a, 1995b, Fan *et al.* 1999, 2000a, 2000c, *etc.*), or brown (L and T) dwarfs (Kirkpatrick *et al.* 1999, Strauss *et al.* 1999, Burgasser *et al.* 2000, Fan *et al.* 2000b, Leggett *et al.* 2000, *etc.*).

As an illustration, in Figure 1 we show how the color selection works with the examples of high-z and type-2 quasars discovered in DPOSS (Djorgovski *et al.* 1998; and in preparation). Normal stars form a temperature sequence, seen here as a banana-shaped locus of points in the parameter space of colors. The spectra of these quasars, when folded through the survey filter curves (Figures 2 and 3), produce colors discrepant from those of normal stars.

In the case of high-z quasars, absorption by the intergalactic hydrogen (the Lyα forest) produces a strong drop blueward of the quasar's own Lyα emission line center, and thus a very red $(g - r)$ color, while the observed $(r - i)$ color reflects the intrinsically blue spectrum of the quasars: these objects are red in the blue part of the spectrum, and blue in the red part of the spectrum — unlike any stars. To date, ~ 100 such quasars have been found in DPOSS; we make them publicly available through our webpage[1]. At intermediate Galactic latitudes, there is about one of them per million stars, down to $r \sim 19.5$ magnitude. Thus a good color discrimination and a good star-galaxy separation are essential in order to avoid an excessive contamination of the spectroscopic follow-up samples by mismeasured stars or misclassified galaxies. A variant of this technique (based also on the Lyman-limit drop) is now used to find galaxies at $z \gtrsim 3$ (*e.g.*, Steidel *et al.* 1999, Dickinson *et al.* 2000, and references therein).

A similar "convex spectrum" effect can be caused by the presence of strong emission lines in the middle band (r), as shown here in the example of type-2 quasars discovered in DPOSS (Djorgovski *et al.* 1999, and in prep.). We found a whole population of these long-sought objects (which are now also appearing in considerable numbers in the CXO x-ray data), selected through their peculiar colors. The selection effects are complex, depending both on the [O III] line fluxes and equivalent widths, so we find only a subset of them, those with a mostly unobscured narrow-line region, in the redshift interval given by the width of the DPOSS r band ($z \sim 0.31$–0.38 for the [O III] lines). These objects are sufficiently rare, with surface density $\lesssim 10^{-2}$ per square degree for our selection criteria, that one must have a survey covering a very large area, yet go sufficiently deep to detect the host galaxies (in our survey most of the light in the g and i

---

[1] http://www.astro.caltech.edu/~george/z4.qsos

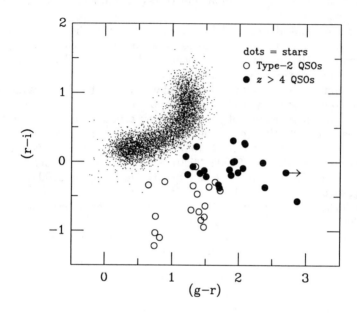

Figure 1. A representative color-color plot for objects classified as PSF-like in DPOSS. The dots are normal stars with $r \sim 19$ mag. Solid circles are some of the $z > 4$ quasars, and open circles are some of the type-2 quasars found in this survey. While quasars are morphologically indistinguishable from ordinary stars, this color parameter space offers a good discrimination among these types of objects. Similar methodology is now also used to discover brown dwarfs in SDSS and 2MASS.

bands is from the hosts). This is why this population was missed in the past, with surveys lacking either the necessary depth or the area coverage.

This simple, but very efficient and demonstrably successful method can be used to isolate other kinds of sources as well, *e.g.*, stars of a particular spectral type, to be used as tracers of Galactic structure, or the samples of mostly unobscured quasars in general (*cf.* Wolf *et al.* 1999 or Warren *et al.* 2000). Multiplicity of bandpasses and the dynamical range of the wavelength baseline help; after all, multicolor photometry can be viewed as an extremely low resolution spectroscopy.

Analogous techniques could be used in other parameter spaces, for example for an objective classification and selection of galaxies of a particular type, when image morphology can be quantified appropriately.

4. Looking for New Kinds of Objects

Perhaps the most intriguing new scientific prospect for a VO is the possibility of discovery of previously unknown types of astronomical objects and phenomena.

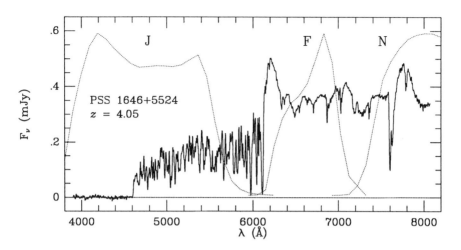

Figure 2. A spectrum of a typical $z > 4$ quasar, with the DPOSS bandpasses shown as dotted lines. The mean flux drop blueward of the Lyα line, caused by the absorption by Lyα forest and sometimes a Lyman-limit system, gives these objects a very red $(g - r)$ color, while their intrinsic blue color is retained in $(r - i)$. This places them in the portion of the color parameter space indicated in Figure 1.

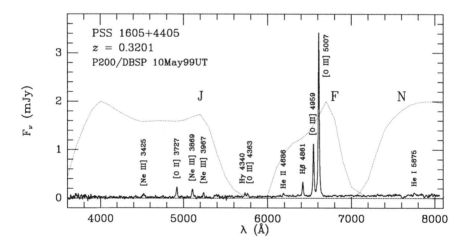

Figure 3. A spectrum of a typical type-2 quasar, with the DPOSS bandpasses shown as dotted lines. The presence of the strong [O III] lines is the r band places such objects in the portion of the color parameter space indicated in Figure 1.

Such things might have been missed so far either because they are rare, or because they would require a novel combination or a way of looking at the data.

A thorough, large-scale, unbiased, multi-wavelength census of the universe will uncover them, if they do exist (and surely we have not yet found all there is out there). Methodology similar to that used to find known rare types of objects, *i.e.*, as outliers in some suitably chosen, discriminative parameter space, can be used to search for the possible new species. This "organized serendipity" can lead to some exciting new discoveries.

Possible examples of new kinds of objects (or at least extremely rare or peculiar sub-species of known types of objects) have been found in the course of high-z quasar searches by both SDSS (Fan & Strauss, private communication) and DPOSS groups. Two examples from DPOSS are shown in Figures 4 and 5. These objects have most unusual, and as yet not fully (or not at all) understood spectra, which cause them to have peculiar broad-band colors. Their colors places them in the designated portions of the color space where high-z quasars are to be found, and clearly other, as yet unexplored portions of this parameter space may contain additional peculiar objects. While some may simply turn out to be little more than curiosities, others may be representative of genuine new astrophysical phenomena.

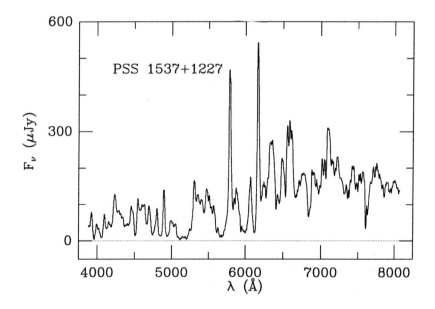

Figure 4. A spectrum of a peculiar object PSS 1537+1227, obtained at Palomar. The object was initially selected as a high-z quasar candidate due to its colors, in a manner illustrated in Figure 1. It turned out to be an extreme case of a rare type of a low-ionization, Fe-rich, BAL QSO, at $z \approx 1.2$. A prototype case (but with a spectrum not quite as extreme as this) is FIRST 0840+3633, discovered by Becker *et al.* (1997).

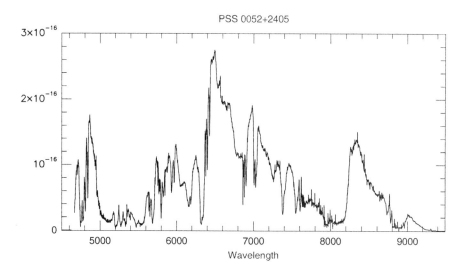

Figure 5. A spectrum of another peculiar object, PSS 0052+2405, obtained at Keck. This object was also initially selected as a high-z quasar candidate. Its nature is still uncertain, but it may be another example of a peculiar BAL QSO — or something completely different.

In order to tackle this problem right, we need proper computational and statistical tools, generally falling in the area of unsupervised clustering or classification, which is a part of the more general and rapidly growing field of Data Mining (DM) and Knowledge Discovery in Databases (KDD). This opens up great opportunities for collaborations with computer scientists and statisticians. For an overview of some of the issues and methods, see the volume edited by Fayyad et al. (1996b), as well as several papers in this volume. Good visualization tools are also essential for this task.

If applied in the catalog domain, the data can be viewed as a set of n points or vectors in an m-dimensional parameter space, where n can be in the range of many millions or even billions, and m in the range of a few tens to hundreds. The data may be clustered in k statistically distinct classes, which could be modeled, e.g., as multivariate Gaussian clouds in the parameter space, and which hopefully correspond to physically distinct classes of objects (e.g., stars, galaxies, quasars, etc.). This is a computationally highly non-trivial problem, approaching Terascale supercomputing, and it calls for some novel and efficient implementations of clustering algorithms. However, not all parameters may be equally interesting or discriminating, and lowering this dimensionality to some more appropriate subset of parameters would be an important task for the scientists actually using such tools to explore the data.

If the number of object classes k is known (or declared) a priori, and training data set of representative objects is available, the problem reduces to supervised classification, where tools such as Artificial Neural Nets (ANN) or Decision Trees

(DT) can be used. This is now commonly done for star-galaxy separation in the optical or NIR sky surveys (*e.g.*, Odewahn *et al.* 1992, or Weir *et al.* 1995), and searches for known types of objects with predictable signatures in the parameter space (*e.g.*, high-z quasars) can be also cast in this way.

However, a more interesting and less biased approach is where the number of classes k is not known, and it has to be derived from the data themselves. The problem of unsupervised classification is to determine this number in some objective and statistically sound manner, and then to associate class membership probabilities for all objects. Majority of objects may fall into a small number of classes, *e.g.*, normal stars or galaxies. What is of special interest are objects which belong to much less populated clusters, or even individual outliers with low membership probabilities for any major class. Some initial experiments with unsupervised clustering algorithms in the astronomical context include, *e.g.*, Goebel *et al.* (1989), Weir *et al.* (1995), de Carvalho *et al.* (1995), and Yoo *et al.* (1996), but a full-scale application to major digital sky surveys yet remains to be done. An array of good unsupervised classification techniques will be an essential part of a VO toolkit.

5. Other Domains of the Parameter Space

Most of the work described so far involved searches in the catalog domain, and specifically in the parameter spaces of colors measured in optical and NIR sky surveys. However, many other domains of the observable parameter space are still wide open and waiting to be fully explored.

The low surface brightness universe (at any wavelength!) is one of the obvious frontiers, and is addressed elsewhere in this volume by Schombert and by Brunner *et al.* ; see also the review by Impey & Bothun (1997), and references therein. Conversely, we may be missing some compact, *high* surface brightness galaxies (a possibility envisioned by Zwicky many decades ago): *cf.* Drinkwater *et al.* (1999); however, a field spectroscopic survey of almost-unresolved DPOSS objects at Palomar (Odewahn *et al.* , in prep.) failed to turn up a substantial number of such objects. In any case, expanding the dynamical range of the limiting surface brightness and angular resolution in digital sky surveys at any wavelength is likely to be one of the key area of research in a VO.

Perhaps the most promising new domain for exploration is the time domain: variability at all time scales, and all wavelengths, be it periodic, eruptive, or chaotic in nature. The subject is addressed by Diercks elsewhere in this volume, and by Paczyński (2000). The importance and the scientific promise of the exploration of the time domain has been recognized through the high recommendation of the NAS Decadal Report, *Astronomy and Astrophysics in the New Millennium*, of the Large Synoptic Survey Telescope (LSST). Other large-scale sky monitoring program are already in progress (*e.g.*, Akerlof *et al.* 2000, Groot *et al.* 2000, Everett *et al.* 2000, and the many searches for the Solar system objects reviewed by Pravdo elsewhere in this volume). Synoptic monitoring of the sky over a range of wavelengths, and mining of the resulting multi-Petabyte data sets may be the most technically demanding and among the most scientifically productive areas for a VO.

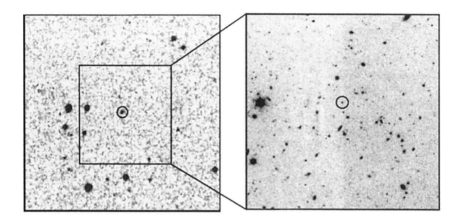

Figure 6. An example of a serendipitously discovered optical transient event from DPOSS. The left panel shows a portion of a DPOSS F plate image with an $r \sim 18.5$ magnitude, starlike object circled. The object was selected due to its apparent peculiar color (bright in r, extremely faint in the other two DPOSS bands); however, this was simply a consequence of the plates taken at different times, with one of them catching it in a bright state. The right panel shows a portion of the corresponding Keck R band image. The DPOSS transient was positionally coincident with an $R \sim 24.5$ magnitude galaxy, with an estimated probable $z \sim 1$. At such a redshift, this object would have been a few hundred times brighter than a supernova at its peak. It may be an example of a GRB "orphan afterglow", or possibly some other, new type of a transient.

Most of the studies described so far involve searches in some parameter or feature space, i.e., catalogs derived from survey images. However, we can also contemplate a direct exploration of sky surveys in the image (pixel) domain. Automated pattern recognition and classification tools can be used to discover sources with a particular image morphology (e.g., galaxies of a certain type). An example from planetary science, an automated discovery of volcanos in Magellan Venus radar images, was described in Fayyad et al. (1996a) and Burl et al. (1998). An even more interesting approach would be to employ AI techniques to search through panoramic images (perhaps matched from multiple wavelengths) for unusual image patterns. For example, it may be possible for a program to discover gravitationally lensed arcs in rich clusters, and possibly some other, as yet unknown phenomena.

Finally, an unsupervised classification search for unusual patterns or signals in astronomical data represents a natural generalization of the SETI problem (Djorgovski 2000).

6. Concluding Comments

A VO, applied on the plethora of large, digital sky surveys would enable a thorough and systematic exploration of the observable parameter space, leading to a more complete understanding of the physical universe. Introduction of novel DM and KDD techniques, developed in collaboration with computer scientists will be essential. In addition to the construction of significant samples of various rare types of astronomical objects which can be used for further studies, we are likely to find some completely new things. In this way, a VO will be a *unique* tool of astronomical discovery.

There is, however, one significant bottleneck which we can already anticipate in this type of studies: the follow-up spectroscopy of interesting sources selected from imaging surveys. While there seems to be a vigorous ongoing and planned activity to map and monitor the sky in many ways and many wavelengths, spectroscopic surveys will be necessary in order to interpret and understand the likely overabundance of interesting objects found. This is something we have to consider in our plans.

Acknowledgments. The processing and initial exploration of DPOSS was supported by a generous gift from the Norris foundation, and by other private donors. Prototyping VO developments at Caltech and JPL have been funded by grants from NASA, the Caltech President's Fund, and several private donors. We are grateful to all people who helped with the creation of DPOSS and our Palomar and Keck observing runs, and especially a number of excellent Caltech undergraduates who worked with us through the years. Work on the applications of machine learning and AI technology for exploration of large digital sky surveys was done in collaboration with U. Fayyad, P. Stolorz, R. Granat, A. Gray, J. Roden, D. Curkendall, J. Jacob, and others at JPL.

References

Akerlof, C., *et al.* 2000, AJ, 119, 1901

Becker, R., Gregg, M., Hook, I., McMahon, R., White, R., & Helfand, D. 1997, ApJ, 479, L93

Boller, T., Meurs, E., & Adorf, H.-M. 1992, A&A, 259, 101

Burgasser, A., *et al.* 2000, AJ, 120, 1100

Burl, M., Asker, L., Smyth, P., Fayyad, U., Perona, P., Crumpler, L., & Aubelle, J. 1998, Mach. Learning, 30, 165

de Carvalho, R., Djorgovski, S., Weir, N., Fayyad, U., Cherkauer, K., Roden, J., & Gray, A. 1995, in *Astronomical Data Analysis Software and Systems IV*, eds. R. Shaw *et al.* , ASP Conference Series, 77, 272

Dickinson, M. *et al.* 2000, ApJ, 531, 624

Djorgovski, S.G., Gal, R.R., Odewahn, S.C., de Carvalho, R.R., Brunner, R.J., Longo, R., & Scaramella, R. 1998, In *Wide Field Surveys in Cosmology*, eds. S. Colombi *et al.*, Gif sur Yvette: Eds. Frontières

Djorgovski, S.G., Brunner, R., Harrison, F., Gal, R., Odewahn, S., de Carvalho, R., Mahabal, A., & Castro, S. 1999, *B.A.A.S.*. 31, 1467

Djorgovski, S.G. 2000, in *Bioastronomy '99*, eds. G. Lemarchand & K. Meech, ASP Conference Series, 213, 519

Drinkwater, M., *et al.* 1999, ApJ, 511, L97

Everett, M., *et al.* 2000, MNRAS, in press [astro-ph/0009479]

Fan, X. *et al.* (the SDSS Collaboration) 1999, AJ, 118, 1

Fan, X. *et al.* (the SDSS Collaboration) 2000a, AJ, 119, 1

Fan, X. *et al.* (the SDSS Collaboration) 2000b, AJ, 119, 928

Fan, X. *et al.* (the SDSS Collaboration) 2000c, AJ, 120, 1167

Fayyad, U., Djorgovski, S.G., & Weir, W.N. 1996a, in *Advances in Knowledge Discovery and Data Mining*, eds. U. Fayyad *et al.*, p.471, Boston: AAAI/MIT Press

Fayyad, U., Piatetsky-Shapiro, G., Smyth, P., & Uthurusamy, R. (eds.) 1996b, *Advances in Knowledge Discovery and Data Mining*, AAAI/MIT Press

Goebel, J., Volk, K., Walker, H., Gerbault, F., Cheeseman, P., Self, M., Stutz, J., & Taylor, W. 1989, A&A, 222, L5

Groot, P., *et al.* 2000, MNRAS, in press [astro-ph/0009478]

Harwit, M. 1975, QJRAS, 16, 378

Harwit, M., & Hildebrand, R. 1986, *Nature*, 320, 724

Impey, C., & Bothun, G. 1997, ARA&A, 35, 267

Irwin, M., McMahon, R., & Hazard, C. 1991, in *The Space Distribution of Quasars*, ed. D. Crampton, ASP Conference Series, 21, 117

Kirkpatrick, D., *et al.* 1999, ApJ, 519, 802

Leggett, S., *et al.* (the SDSS Collaboration) 2000, ApJ, 536, L35

Odewahn, S.C., Stockwell, E., Pennington, R., Humphreys, R., & Zumach, W. 1992, AJ, 103, 318

Paczyński, B. 2000, PASP, 112, 1281

Steidel, C., Adelberger, K., Giavalisco, M., Dickinson, M., & Pettini, M. 1999, ApJ, 519, 1

Strauss, M., *et al.* (the SDSS Collaboration) 1999, ApJ, 522, L61

Warren, S., *et al.* 1987, Nature, 325, 131

Warren, S., Hewitt, P., & Foltz, C. 2000, MNRAS, 312, 827

Weir, N., Fayyad, U., & Djorgovski, S. 1995, AJ, 109, 2401

Wolf, C., *et al.* 1999, A&A, 343, 399

Yoo, J., Gray, A., Roden, J., Fayyad, U., de Carvalho, R., & Djorgovski, S. 1996, in *Astronomical Data Analysis Software and Systems V*, eds. G. Jacoby & J. Barnes, ASP Conference Series, 101, 41

Zwicky, F. 1957, *Morphological Astronomy*, Berlin: Springer Verlag

Exploring the Multi-Wavelength, Low Surface Brightness Universe

R.J. Brunner, S.G. Djorgovski, R.R. Gal, A.A. Mahabal

Department of Astronomy, California Institute of Technology, Pasadena, CA, 91125

S.C. Odewahn

Department of Physics & Astronomy, Arizona State University, Tempe, AZ, 85287

Abstract. Our current understanding of the low surface brightness universe is quite incomplete, not only in the optical, but also in other wavelength regimes. As a demonstration of the type of science which is facilitated by a virtual observatory, we have undertaken a project utilizing both images and catalogs to explore the multi-wavelength, low surface brightness universe. Here, we present some initial results of this project. Our techniques are complimentary to normal data reduction pipeline techniques in that we focus on the diffuse emission that is ignored or removed by more traditional algorithms. This requires a spatial filtering which must account for objects of interest, in addition to observational artifacts (*e.g.*, bright stellar halos). With this work we are exploring the intersection of the catalog and image domains in order to maximize the scientific information we can extract from the federation of large survey data.

1. Introduction

Looking at large scale images (*i.e.*, several degrees or more), one is immediately drawn to the high density of small galaxies, especially at high Galactic latitude, which are often strongly clustered. Interestingly enough, the vast majority of these galaxies have optical surface brightness distributions that are nearly identical to the terrestrial sky (Freeman 1970, Disney 1976). This interesting point, unless it is the manifestation of a cosmic coincidence, is most easily explained by accepting that current surveys suffer from an implicit surface brightness selection effect (see Disney 1998 for a stimulating discussion). As a result, untold numbers of galaxies remain uncataloged with many important consequences.

For example, while the theoretical predictions of models of hierarchical structure formation have been successful in predicting the observed properties of the high redshift universe, they tend to over-predict the number of observed local group galaxies (Klypin *et al.* 1999). This situation can be viewed as either a failure of the models, or a failure of the observations, possibly due to selection effects. In addition, galaxies which have low surface brightness (LSB) distributions have enormous cosmological implications (*e.g.*, Impey & Bothun

1997), yet, they are relatively unexplored, primarily due to the intrinsic selection effects in finding them. For example, LSB galaxies constitute an unknown fraction of mass (both baryonic and dark matter) which must be accounted for when determining Ω or Λ.

In order to address these questions, we have initiated a project to reprocess the Digitized Palomar Observatory Sky Survey (DPOSS, Djorgovski et al. 1998) optical photographic plate data in effort to find previously unknown, low surface brightness objects (Brunner et al. 2001). Our techniques are complimentary to normal data reduction pipeline techniques in that we focus on the diffuse emission that is ignored or removed by more traditional algorithms. This requires a spatial filtering which must account for objects of interest, in addition to observational artifacts (e.g., bright stellar halos). As part of this project, we have developed a novel background enhancement technique to look for new low surface brightness sources (see Figure 1 for a demonstration). Additional aspects affecting low surface brightness galaxy research in the context of a virtual observatory are addressed elsewhere (see, e.g., Schombert, J. in these proceedings).

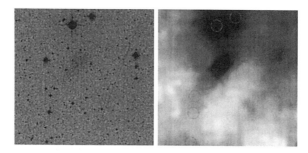

Figure 1. The left image is an approximately 17 arcminute square cutout of the DPOSS F plate image containing the dwarf Spheroidal Andromeda III. The image on the right is the background enhanced image generated by our software pipeline. The elongated object in the background image, which is Andromeda III, is clearly detected via this technique.

2. The Technique

Overall, the software pipeline we developed utilizes publicly available software tools, such as SExtractor (Bertin & Arnouts 1996), to generate our final candidate lists. With this technique, we can easily apply the same process to additional datasets, such as the SDSS, either individually or jointly in a full multi-wavelength exploration of parameter space.

Briefly, our software pipeline performs the following steps.

- Pull raw DPOSS F and J plate scan footprints off on-line storage.
- Mosaic full plate images.
- Apply Vignetting correction to full plate mosaics.

- Process the full plates to produce a background map, an object map, and a bright star catalog.

- Process the background map using optimized convolution kernel, using the object map as a pixel mask, to detect background variations.

- Eliminate bright stellar halos using the bright star catalog.

- Combine candidate lists from J and F plates to remove candidates that arise from individual plate defects.

- Visually classify candidates according to assigned classes (*e.g.*, planetary nebula, dwarf spheroidal galaxy, low surface brightness galaxy, *etc.*)

In the past, non local group LSB galaxies were identified by visually inspecting POSS-I or POSS-II sky survey plates (*e.g.*, Schombert *et al.* 1995). Previously, this process, particularly automated approaches, was hampered by the unknown vignetting corrections, the plate mosaicing process, as well as the sheer amount of data that needs to be explored. As part of the Digital Sky project, a technology demonstrator for a future National Virtual Observatory, these types of problems are being tackled, and algorithmic solutions are close to being implemented. We, therefore, plan to extend the automated low surface brightness survey to target the unexplored population of low surface brightness disk galaxies (see Figure 2 for a demonstration).

The technique we have developed can be easily applied to other large imaging surveys, *e.g.*, the SDSS and 2MASS surveys, in an effort to further explore the low surface brightness universe. In addition, this work can be naturally extended to included additional wavelength information from supplemental surveys in order to improve the source classification (see, *e.g.*, Brunner, R. these proceedings) since astronomical objects have different source characteristics, including surface brightness and morphology, at different wavelengths. Other, complimentary projects are also being applied to the DPOSS data in order to extract the maximal amount of information from this photographic plate data (see, *e.g.*, Sabatini *et al.* 2000).

3. Contaminants

Sometimes the background enhancement procedure picks up objects which, while interesting in their own right, are not the objects of primary interest. Primarily these objects are either planetary nebulae, stellar clusters, interacting galaxies, or galaxy clusters (see Figure 3 for a montage). All of these candidates are flagged due to the presence of low surface brightness features: the nebula itself, the combined stellar halos, the tidal interactions, and the cD envelope, respectively.

Acknowledgments. This work was made possible in part through the NPACI sponsored Digital Sky project and a generous equipment grant from SUN Microsystems. RJB would like to acknowledge the generous support of the Fullam Award for facilitating this project. Access to the DPOSS image data

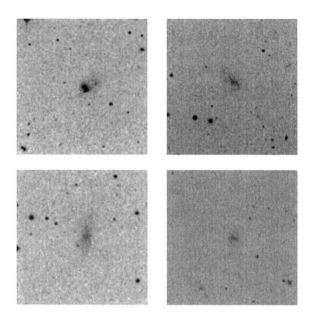

Figure 2. Cutouts of low surface brightness galaxies first detected visually from POSS-II plates by J. Schombert, and later detected in HI at Arecibo.

stored on the HPSS, located at the California Institute of Technology, was provided by the Center for Advanced Computing Research. The processing of the DPOSS data was supported by a generous gift from the Norris foundation, and by other private donors.

References

Bertin, E., & Arnout, S. 1996, A&AS, 117, 393.
Brunner, R.J., Roth, N., Hokada, J., Gal, R., Mahabal, A.A., Odewahn, S.C., & Djorgovski, S.G. 2001, AJ, in preparation.
Disney, M. 1976, Nature, 263, 573.
Disney, M. 1998, In B. Mclean, editors, *The Proceedings of the 179th IAU on New Horizons from Multi-Wavelength Sky Surveys*, IAU Conference Series, Kluwer Academic Press.
Djorgovski, S.G., Gal, R.R., Odewahn, S.C., de Carvalho, R.R., Brunner, R.J., Longo, R., & Scaramella, R. 1998, In *Wide Field Surveys in Cosmology, 14th IAP meeting*, Eds. Frontieres.
Freeman, K.C. 1970, ApJ, 160, 811.
Klypin, A., Kravtrov, A.V., Valenzuela, O., & Prada, F. 1999, ApJ, 522, 82.
Impey, C., & Bothun, G. 1997, ARA&A, 35, 267.

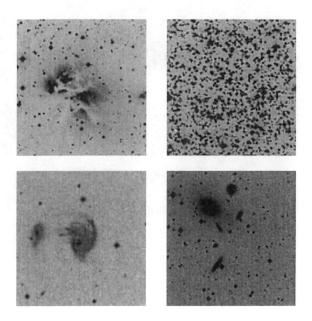

Figure 3. Examples of some diverse types of contaminants generated by the LSB software pipeline. The upper left figure is a Planetary Nebula, the upper right figure is an open cluster. The figure on the lower left is a pair of interacting galaxies, while the figure on the lower right is a galaxy cluster.

Sabatini, S. Scaramella, R., Testa, V., Andreon, S., Longo, G., Djorgovski, S.G., de Carvalho, R.R. 2000, In the Proceeding of the XLIII SAIt National Conference Mem. Soc. Astr. It., in press, astro-ph/9909436.

Schombert, J.M., Pildis, R.A., Eder, J.A., & Oemler, A. 1995, AJ, 110, 2067.

Virtual Observatories of the Future
ASP Conference Series, Vol. 225, 2001
R.J. Brunner, S.G. Djorgovski, and A.S. Szalay, eds.

The ROSAT Optical X-ray Cluster Survey: A Comparison of X-ray and Optical Cluster Detection Methods and Preliminary Results.

Megan Donahue, Caleb Scharf, Jennifer Mack, Marc Postman, Paul Lee, Mark Dickinson

Space Telescope Science Institute, 3700 San Martin Drive, Baltimore, MD 21218

Piero Rosati

European Southern Observatory, 85748 Garching bei München, Germany

John Stocke

University of Colorado, CASA, CB 391, Boulder, CO 30901

Abstract. How a sample of cluster of galaxies is defined may affect the conclusions of any study using that sample. In this work, we investigate how using X-ray cluster detection methods such as wavelets compares, on the exact same region of sky, with the optical method known as the "matched filter" technique (Postman et al. 1996.) In the past, different conclusions about cluster evolution since z=1 were made about optical and X-ray cluster samples. We see that the matched filter method detects more cluster candidates per unit sky area than do X-ray techniques for clusters $z < 1$. With our study we are able to evaluate on a one-to-one basis whether the detection techniques "see" the same clusters, which cluster candidates does each technique miss, and whether the explanation of the difference is simply that the optical method has many more spurious candidates. Our survey, comprised of 23 ROSAT fields with long exposure times and I-band 4-meter imaging of the complete fields, can answer these questions.

1. Introduction

Clusters of galaxies are the most massive gravitationally-bound structures in the universe, $10^{13} - 10^{15} M_\odot$. By virtue of their vast scale and gravitational attraction, the study of clusters of galaxies provides insights into a host of global properties of the universe, including the mean density of matter and the baryonic fraction, the formation of large-scale structure, the evolution of the gaseous fraction of the universe, galaxy formation, and even star formation from the metal and entropy content of the intracluster medium (ICM).

Clusters are the massive repositories not only of hundreds of galaxies but of hot, X-ray emitting, intergalactic plasma. The total mass of the intracluster medium (or ICM) of the optically richest, most massive clusters of galaxies is significantly more than that contained in the galaxies. Approximately $0.17 h_{50}^{-3/2}$ (Evrard, 1997) of the mass of the clusters is in hot gas, where $h_{50} = H_0/50$ km s^{-1} Mpc^{-1} inside a radius where the overdensity is 500 times the critical density. Finding and defining clusters of galaxies have been historically classic astronomical problems: a cluster does not have a well-defined "edge" (as hinted at in the definition of the cluster gas fraction) or shape, and the crud along the line of sight towards the cluster could contaminate the observation of that cluster. Contamination is particularly a problem in the optical, where contributions along the line of sight could cause a poor cluster to look like a rich cluster. X-ray selection has always been thought to be more reliable, since bright, extended, extragalactic X-ray structures are nearly always either clusters of galaxies or relatively nearby galaxies.

However, X-ray cluster detection assumes that the cluster have a well-developed ICM. For the most massive and nearby clusters, this assumption is probably pretty good, based on the similarity of X-ray and optically-selected local catalogs of clusters of galaxies. But we do not know whether or when this assumption might break down. At high redshift, the evolutionary state of the cluster, such as the presence of a shock front or a central cooling flow, may affect the detectability of the ICM. At lower masses, galactic winds driven by supernovae and galaxy-ICM interactions may increase in importance relative to the gravitational energy inputs, and decrease the X-ray luminosity of the ICM with respect to the optical luminosity. The optical luminosities of high redshift clusters may be more dominated by blue galaxies or by disk galaxies, whose colors and magnitudes may differ substantially from the bulge-dominated/elliptical galaxy populations of low-redshift populations.

2. Methodology and Results

We imaged the central fields of 23 ROSAT PSPC observations with the KPNO 4-meter telescope, to $I < 23.5$ magnitudes in the central $30' \times 30'$ of each field. We created galaxy catalogs using FOCAS. We then produced a cluster candidate catalog using the matched filter technique of Postman et al. (1996). This technique identifies clusters, photometric redshifts, the probability of detection in units of sigma, and an optical luminosity (Λ). We detected 125 candidates with $\sigma > 3$ and $z = 0.2 - 1.2$. Based on Postman et al. (1996) and other experience, we expect that $\sim 30\%$ of these candidates are spurious. An independent cluster search in the V-band in 5 of these fields reveals the same cluster candidates with $z_{est} < 1.0$.

The X-ray fields were analyzed in exactly the same way as described in Rosati et al. (1995). A wavelet detection algorithm identified extended sources. We found 52 X-ray cluster candidates, of which 37 were included in our accessible survey (not obscured by bright stars or scattered light), for a factor of at least 3 fewer than identified optically. Twenty seven of these have optical counterparts with centers within a few arcminutes, most within 1.5'; ten have no optical counterparts. Approximately 10 clusters with optical counterparts have been identified spectroscopically, proving the photometric redshifts reliable to better

±0.1, surprisingly. None of the cluster candidates without optical counterparts have been identified spectroscopically.

The Relationship between L_x and L_{opt}. The scatter between Λ, a measure of optical luminosity corresponding to an equivalent N of L^* galaxies, and L_x is large, even for the clusters for which we have estimates for both. The best fit relation fits the relation expected for gas in hydrostatic equilibrium in the cluster and a constant M/L ratio: $L_x \propto L_{opt}^2$, but with large dispersion.

Complete analysis and results are being written up in a Letter and a catalog paper for the ApJ(Donahue et al. in preparation.)

An X-ray Filament. This survey also produced the first discovery of a purely intergalactic, X-ray large scale structure filament (Scharf et al. 2000). The filament was discovered via cross correlation analysis of the X-ray background remaining after all detected point sources and extended sources were excised and the I-band-detected galaxies. The detection has a significance of 3-5 σ, and a surface brightness of $2 - 4 \times 10^{-16}$ erg s^{-1} cm^{-2} arcmin^{-2}, similar to the surface brightnesses of filaments as predicted by e.g., Cen & Ostriker (1999). The estimated diffuse component of this filament is $\sim 75\%$. One possibly non-intuitive conclusion we made based on this discovery is that the detection was possibly not because the field was filament-rich but because it is filament-poor! The idea here is that most ROSAT fields are probably crossed by 3-4 filaments, making the detection of any single filament difficult. If a field is crossed by only one filament, singling it out is less difficult.

X-ray Optical Angular Cross Correlation Function for Galaxies. The unresolved X-ray background was cross-correlated with the I-band galaxy catalogs to probe X-ray/optical correlations significantly below the usual sensitivity limits. By dividing the data into magnitude bins, we measure the evolution of positively correlated X-ray emission using independent models of optical galaxy evolution. We have shown our data probes redshifts up to $z \sim 1$. (Scharf et al. in preparation.)

3. NVO-Relevance and Discussion

We studied cluster properties, sample selection effects, and the X-ray background with archival ROSAT PSPC data and our own ground-based CCD images. Our study is an example of what could be done on a far grander scale with a National Virtual Observatory with X-ray data and a uniform optical catalog of galaxies. We did, in the end, require direct access to the X-ray data to test and quantify cataloging algorithms and to ascertain the detection sensitivity which not only varied from field to field but as a function of position in the detector. Comparisons between the optical and X-ray catalog required detailed knowledge of the optical fields: our fields were completely imaged, but even moderately bright stars would obscure the sky around them or cause scattered light effects that impedes our view. Because of the location of these stars we could not have detected some of the X-ray selected clusters of galaxies no matter how good our matched filter algorithm was. Therefore it was critical for us to have this knowledge. We also required robust calibrations (reliable exposure maps, photometric calibrations) and uniform, productive, and well-studied selection techniques. We were able to leverage the considerable work already done for the individual techniques by Postman et al. (1996) on matched filter selection and by Rosati et al. (1995) on wavelet selections. However, note that both of these individuals are on our team. Their participation in this project was essential to its success. There can be a lag time between the development of the technique and the deployment of software which can be used by individuals other than the inventor.

4. Conclusions

Clusters with a given X-ray luminosity or mass have a wide range of optical luminosities. The dispersion between the optical luminosity and the cluster mass appears to be larger than the dispersion of the X-ray luminosity and cluster mass, but quantifying this difference is yet under investigation. X-ray selection seems to be more efficient at finding clusters, but optical selection appears to find the same clusters. The matched filter algorithm finds real clusters with $z < 1$. The number densities of the cluster candidates are consistent in both the X-ray and the optical if the (1) the dispersion in the L_x-Λ (optical luminosity) relationship is large, and (2) if spurious fraction of the optical clusters is almost zero for $\Lambda > 80$ and $\sim 30\%$ for $40 < \Lambda < 80$.

References

Cen, R. & Ostriker, J. P. 1999, ApJ, 514, 1
Evrard, A. 1997, MNRAS, 292, 289
Postman, M., et al. 1996, AJ, 111, 615
Rosati, P., et al. 1995, ApJ, 445, L11
Scharf, C., et al. 2000, ApJ, 528, L73

The Luminosity Function of 80 Abell Clusters from the CRoNaRio catalogs

S. Piranomonte, G. Longo, S. Andreon, E. Puddu
Osservatorio Astronomico di Capodimonte, Via Moiariello 16, 80131 Napoli

M. Paolillo
D.S.F.A., Universitá di Palermo, Via Archirafi 36, 90123 Palermo

R. Scaramella
Osservatorio Astronomico di Roma, Via Frascati 33, 00040 Roma

R. Gal, S.G. Djorgovski
Department of Astronomy, Caltech, Pasadena, CA 91125

Abstract. We present the composite luminosity function (hereafter LF) of galaxies for 80 Abell clusters studied in our survey of the Northern Hemisphere, using DPOSS data in the framework of the CRoNaRio collaboration. Our determination of the LF has been computed with very high accuracy thanks to the use of homogeneous data both for the clusters and the control fields and to a local estimate of the background, which takes into account the presence of large-scale structures and of foreground clusters and groups. The global composite LF is quite flat down to M^*+5 and it has a slope $\alpha \sim -1.0 \pm 0.2$ with minor variations from blue to red filters, and $M^* \sim -21.9, -22.0, -22.3$ magnitude ($H_0 = 50$ km s^{-1} Mpc^{-1}) in g, r and i filters, respectively (errors are detailed in the text). We find a significant difference between rich and poor clusters supporting the existence of an LF's dependence on the environment.

1. Introduction

The galaxy luminosity function measures the relative frequency of galaxies as a function of luminosity. Cluster LF's can be determined as the statistical excess of galaxies along the line of sight of the clusters with respect to control field directions due to the fact that clusters appear as overdensities with respect to the field. This approach assumes that the contribution of the fore-background galaxies along the line of sight of the clusters is equal to an "average" value. An hypothesis that is rather weak since very often nearby groups, clusters or superclusters happen to lay in same the direction and therefore affect the determination of the LF.

This problem is even more relevant when sampling the cluster outskirts due to: i) the low galaxy density of these regions is strongly affected even by small contaminations; ii) the large observing area which makes more probable the presence of contaminating structures. These outer regions are very relevant due to the fact that they are the putative places for galaxy evolution to occur (van Dokkum et al. 1998).

Non zero correlation function makes therefore very time consuming the accurate determination of LF's using traditional CCD imagers (due to their small field of view) and therefore leads most authors to use comparison average field counts taken from the literature. These average counts are usually obtained from regions of the sky completely unrelated to the cluster's position. Alternative routes are either to observe small comparison fields or to recognize individual cluster membership either on spectroscopic or morphological grounds.

Wide field (hereafter WF) imagers such as Schmidt plates or large format CCD's are the ideal tools to perform accurate determinations of LF's for statistically significant samples of clusters. In this paper we present results from a long term project aimed to derive LF's for a large sample of Abell clusters selected accordingly to the criteria described below.

The work is done in the framework of the CRoNaRio collaboration aimed to produce the first complete catalogue of all object visible on the Digitized Palomar Sky Survey (hereafter DPOSS).

2. The CRoNaRio Project and the cluster sample

The CRoNaRio Project is a joint enterprise among Caltech and the astronomical observatories of Roma, Napoli and Rio de Jainero, aimed to produce the first general catalogue of all objects visible on the DPOSS. The final Palomar-Norris North Sky Catalogue will include astrometric, photometric (in the three Gunn-Thuan bands g, r and i) and rough morphological information for an estimated number of 2×10^9 stars and 5×10^7 galaxies (Djorgovski et al. 1999).

Our final goal is to derive individual cluster LF's and therefore a statistically robust cumulative LF for all Abell (1958) clusters fulfilling the following criteria: the cluster must fall in a plate triplet with available individual photometric zero points, it must not be close to the edges of a plate and it must have at least one reliable spectroscopic redshift estimate. Moreover clusters must not present anomalous structures such as double density peaks, discordant redshift determinations *etc.*

In what follows we discuss the results for a subsample of 80 clusters, an extension of Paolillo et al. (2000).

3. Individual LF determination

In order to compute the cluster LF the first step is to subtract the fore-background contamination. The method, together with the errors involved in the subtraction process, is detailed in Paolillo et al. (1999) and in Paolillo et al. (2000).

Field counts are measured around each cluster, thanks to the wide field coverage of DPOSS plates, after removing density peaks.

A "local field", measured all around the studied clusters, is the adopted estimate of the background counts in the cluster direction. It is a better measure of the contribution of background galaxies to counts in the cluster direction than the usual "average" field, since it allows to correct for the presence of possible underlying large-scale structures both at the cluster distance and in front or behind it.

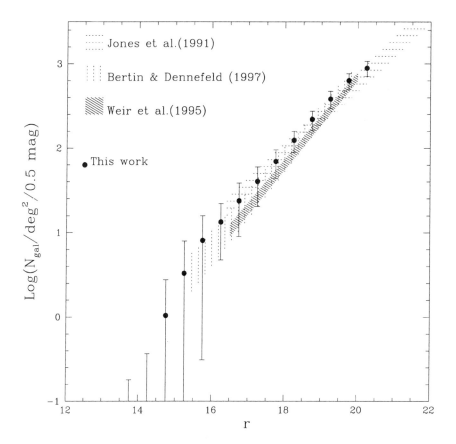

Figure 1. Comparison of our background counts with some previous determinations.

In Figure 1 we show the background counts measured all around the studied clusters (solid dots). We agree fairly well with the measurements by Weir *et al.* (1995) who also used DPOSS data but were marginally higher since we include in the background counts also galaxies belonging to the supercluster in which the studied clusters are embedded in.

Once we have searched for the central 1.5σ density peak in the central region (*cf.* Paolillo *et al.* 2000) we then derive the cluster LF by subtracting from the galaxy counts the local field counts, rescaled to the effective cluster area. This approach allows to take into account the cluster morphology without having to adopt a fixed cluster radius, and thus to apply the local field correction to the region where the signal (cluster) to noise (field and cluster fluctuations) ratio is higher, in order to minimize statistical uncertainties.

4. Results

We combine individual LF's of many clusters to obtain a composite LF for all clusters in our sample. We adopted the method in Garilli et al. 1999. In practice, the composite LF is obtained by weighting each cluster against the relative number of galaxies in an opportunely chosen magnitude range (thus taking into account the different degree of completeness). The final LF for a first sample of 80 clusters is shown in Figure 2.

The fit of the LF's to a Schechter (1976) function gives the values: $\alpha = -1.04^{+0.09}_{-0.07}$; $-1.01^{+0.09}_{-0.07}$; $-0.99^{+0.12}_{-0.11}$ and $M^* = -21.99^{+0.13}_{-0.17}$; -22.02 ± 0.16; -22.30 ± 0.20 mag, respectively in g, r and i where M^* is the characteristic knee magnitude and α is the slope of the LF at faint magnitudes. Figure 2 also shows the three best-fit functions together with the 68% and 90% confidence levels.

In Figure 3, we compare our LF with that found by Garilli (1999) and collaborators and believe that this comparison is particularly significant since i) they derived their LF from a largely overlapping set of clusters and in exactly the same photometric system; ii) they used a completely different criteria for removing interlopers (color changes due to K correction terms).

In Figure 3 we also compare our LF with that obtained by Trentham (1998). At magnitude between -22 and -17 the two LF's are in good agreement. We cannot say much about the sharp rise of the LF found by Trentham at magnitudes fainter than $M_g = -18$, since our data reach the dwarf range only in the last magnitude bins. On the bright end side of the LF we find instead a strong difference which may be explained as a result of the fact that Trentham is underestimating the contribution of bright galaxies to the LF.

The relatively large number of clusters used in the present study allowed us to investigate the dependence of the LF on the cluster richness parameter. We find (see Figure 4) that with a statistical significance of 3σ, rich (R>1) clusters have shallower faint end than poor (R≤ 1) ones. This confirm what suggested by Driver, Couch, & Phillips (1988) that poor clusters host a larger fraction of dwarfs.

References

Abell, G.O. 1958, ApJS, 3, 211

Bertin, E., & Dennefeld, M. 1997, A&A, 317, 43

Djorgovski, G., Gal, R., Odewahn, S.de Carvalho, R.R., Brunner R., Longo G., Scaramella R. 1999 in Wide Field Surveys in Cosmology, Editions Frontieres, 89

Driver, S. P., Couch, W. J., & Phillipps, S. 1998, MNRAS, 301, 369

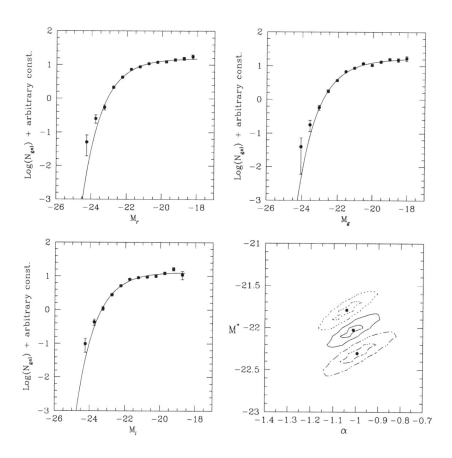

Figure 2. The luminosity function for 80 clusters obtained excluding the brightest member of each cluster (filled dots). The best fit Schechter functions are represented by the continuos line, with the 68% and 99% confidence levels of the best fit parameters in the bottom right panel (g:dotted line; r continuos line; i: dashed-dotted line).

Garilli, B., Maccagni, D., & Andreon, S. 1999, A&A, 342, 408

Jones, L.R., Fong, R., Shanks, T., Ellis, R.S., & Peterson, B.A. 1991, MNRAS, 249, 481

Paolillo, M., Andreon, S., Longo, G., Puddu, E., Piranomonte, S., Scaramella, R., Testa, V., De Carvalho, R., Djorgovski, G., & Gal, R. 1999 in the proceeding of the XLIII SAIt national conference Mem. Soc. Astr. It.

Paolillo, M., Andreon, S., Longo, G., Puddu, E., Gal, R. R., Scaramella, R., & Djorgovski, S. G. 2000, A&A, submitted

Schechter, P. 1976, ApJ, 203, 297

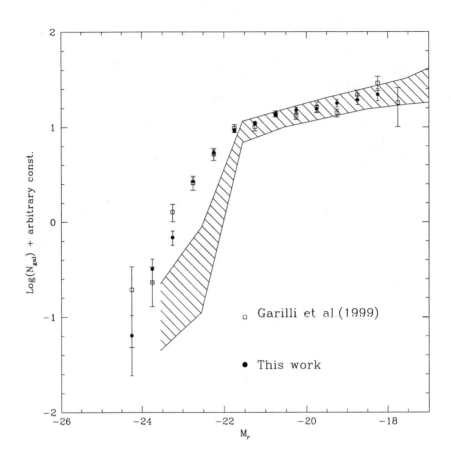

Figure 3. Comparison between our LF, the Garilli et al. (1999) and the Trentham (1998) LF (shaded region) based on CCD data.

Trentham, N. 1998, MNRAS, 294, 193

van Dokkum, P., Franx, M., Kelson, D.D., Illingworth, G.D., Fisher, D., & Fabricant, D. 1998, ApJ, 500, 714

Weir, N., Fayyad, U.M., Djorgovski, S.G., & Roden, J. 1995, PASP, 107, 1243

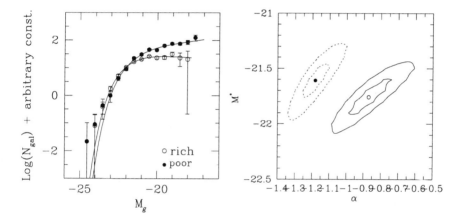

Figure 4. Left panel: The LF and the best fit Schechter functions of the rich (R >1) and poor ($R \leq 1$) subsamples in the g band. Right panel: 68% and 99% confidence levels relative to the fit parameters ($R > 1$: continuous line; $R \leq 1$: dotted line).

Group and Cluster Detection in DPOSS Catalogs

E. Puddu, V. Strazzullo, S. Andreon, G. Longo

Osservatorio di Capodimonte, Via Moiariello 16, Napoli, Italy

E. De Filippis

Astrophysics Research Institute, Twelve Quays House, Egerton Wharf, CH41 1LD Birkenhead, Wirral UK

R. Gal, S.G. Djorgovski

Dept. of Astronomy, Caltech, Pasadena, CA 91125

R. Scaramella

Osservatorio di Monte Porzio, Roma, Italy

Abstract. We discuss the implementation and validation of a procedure aimed to detect groups and clusters of galaxies in DPOSS catalogs.

1. Introduction

The Digitized Palomar Sky Survey (hereafter DPOSS) offers an unique data set to explore the large scale structure of the nearby ($z < 0.2$) universe due to its virtually unlimited sky coverage, to its deepness (21.5 in g) and the availability of three colors (J, F and N).

In this poster we discuss a first application, on a small part of the whole data set, of a newly implemented procedure to search for candidate galaxy clusters and groups. Final goal of this project is a robust measurement of the multiplicity function (hereafter MF) of galaxies.

The MF is a powerful tool to test the various cosmological scenarios. For instance, if the protostructures form from a hierarchical gravitational instability caused by primordial density fluctuations:

$$\frac{d\rho}{\rho_m} \propto A \cdot M^{(-1/2+n/6)}$$

where n is the initial spectrum index at era of baryonic recombination and if one identifies groups of galaxies with mean density $r_g > k$ times the average cosmological density ρ_M, then the function of the mass distribution doesn't depend on the cosmic density parameter Ω but only on the spectrum index n (Gott & Turner 1977). The groups which are seen now at a density enhancement k would therefore correspond to density fluctuations of amplitude A at recombination.

Therefore, a measurement of the MF at the present epoch derived from a catalogue of groups of galaxies identified for a given k, gives a measure of the initial spectrum index n.

2. Detection of Candidate Groups

Even though on rather arbitrary ground, we define as a group all galaxy ensembles having less then 15 members having total magnitude within 3 magnitudes from that of the brightest members. The implemented procedure is described in detail in De Filippis (1999).

In order to compile a catalogue of candidate groups of galaxies in absence of redshift information, we implemented a modified version of the van Albada algorithm (Soares 1989). This algorithm makes use of apparent magnitudes and projected positions on the sky only, and gives for each pair of adjacent galaxies their probability of being physically bound.

Assuming a Poisson statistic, the probability that the distance of a fixed galaxy to its nearest non-physical companion lies between Θ and $\Theta + d\Theta$, is given by:

$$P_I(\Theta)d\Theta = exp[-\pi\Theta^2 n]2\pi\Theta n d\Theta$$

The introduction of a dimensional distance x allows us to combine the angular separation of different pairs to a single distribution removing the effects of clustering in the background.

3. Detection of Candidate Clusters

Starting points are the individual J, F and N catalogs produced by SKICAT for a given DPOSS field after conversion to the g, r and i Gunn-Thuan system. The three catalogs are then matched (assuming a maximum matching distance of 7", for details see Puddu et al. 1999).

After matching, all objects having $r > 19.5$ are filtered out and the resulting matched catalogue is binned into equal area square bins of 1.2' to produce a density map. Then, S-Extractor (Bertin & Arnouts 1996) is run in order to identify and extract all the overdensities having number density 2σ above the mean background and covering a minimum detecting area of 4 pixels (4.8' square). In this way, as in the Schectman (1985) approach, we are not assuming any a priori cluster model. An application to DPOSS field 610 is shown in Figure 1.

All the previously known Abell and Zwicky clusters are recovered and many new candidates detected. For each candidate cluster we then measure the S/N detection ratio and the Abell richness class.

4. Validation of the Algorithms

In absence of a suitably complete samples of galaxies with known redshift we were forced to validate the algorithms shortly summarized above using either simulations or photometric techniques. In what follows we shall discuss first the groups and then the clusters.

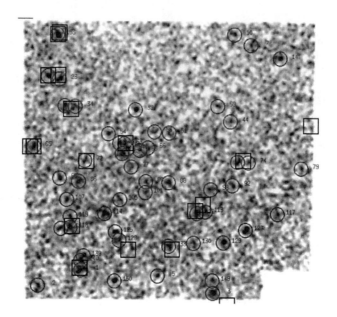

Figure 1. Results for DPOSS field n. 610 (1^h,+15°). Circles mark the position of detected candidate clusters while squares mark already known Abell and Zwicky clusters.

4.1. Group Algorithm Validation

In order to test the accuracy (lost groups) and reliability (spurious groups) of the algorithm we tested it on 70 realistically simulated fields. We first produced the galaxy background by assuming uniform distribution and the field luminosity function given by Metcalfe et al. 1995; then we added groups of galaxies according to the multiplicity function by Turner & Gott (1976), with redshift computed according to:

$$N(z) = \frac{32\pi\rho_0 c^3}{3H_0}[\frac{1}{z+1}[1 - \frac{1}{\sqrt{(1+z)}}]]^3$$

and absolute magnitude of the brightest galaxy in the group taken from the cumulative LF for groups of galaxies:

$$\Phi(M)dM = \Phi^*[10^{0.4(M^*-M)}]^{(\alpha+1)} \cdot exp[-10^{0.4(M^*-M)}]dM$$

where $M^* = -20.85$ and $\alpha = -0.83 - +0.17$. Other parameters varied in the course of the simulations were: i) the diameter of the group inside a Gaussian distribution centered at $D_0 = 0.26 Mpc$; ii) the maximum possible redshift for a group (chosen in the range $0.2 < z < 0.7$). Different simulations have been performed changing the lower limit of the probability for which two galaxies are considered physical companions by the algorithm. The best compromise between accuracy and reliability is found at $p = 0.6$.

In Figure 2 we show the multiplicity function obtained from 12 DPOSS plates (covering a total solid angle of 144 square degrees) compared to six other MFs taken from the literature: MF derived from magnitude limited surveys making use of redshift information, the CfA survey (Geller & Huchra 1983; Garcia et al. 1993; Ramella & Pisani 1997; Ramella et al. 1998), MF derived from diameter limited survey (Maia & Da Costa 1989) and MF derived from limited magnitude survey without redshift informations (Turner & Gott 1976).

As it can be clearly seen, the MF derived by applying our method to DPOSS material is at 98% confidence level indistinguishable from literature MF obtained from redshift surveys.

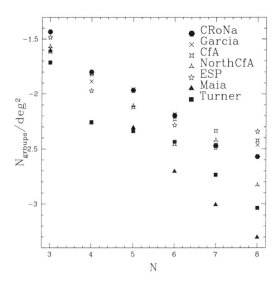

Figure 2. Comparison of our MF against those derived by other authors.

4.2. Cluster Algorithm Validation

We took two different approaches: through simulation and through magnitude - color diagrams. In order to produce realistic simulations we followed a rather lengthy procedure. First of all, we needed a fairly accurate multiplicity function to be used as input parameter for the simulation. Since there is no good MF available in literature, we used a set of 20 DPOSS triplets to identify clusters and derive a preliminary MF (we used only very high S/N clusters for which there is no doubt about their physical nature).

In order to simulate the fields we first distribute randomly the cluster centers and we assume the cluster distribution to be uniform in the volume element and randomly assign to each cluster a redshift in the range $[0.02, 0.2]$. Then we grow the clusters around their centers randomly selecting a radius in the range $[2-3]$ Mpc and weighting the number of members accordingly to the above derived MF.

Absolute magnitudes for galaxies are then extracted from a Schechter function ($M^* = -21.41, \alpha = -1.24$) and scaled to apparent magnitudes taking into account the K-correction term. We finally add field galaxies using the Metcalfe field counts and LF. Even though simulations are still in progress, preliminary results show that our algorithm can recover 95% of the clusters at $z < 0.15$ and 70% of the clusters in the range $[0.15, 0.2]$.

4.3. Photometric Validation

As stressed above, DPOSS data are calibrated via independent CCD frames taken at various telescopes in the $1-2m$ range. This means that we have at our disposal a large number of deep fields, where galaxy photometric properties can be measured with high photometric accuracy. In the last two years we selected calibration fields in order to largely overlap with our candidate clusters sample. These data were used to derive color - magnitude diagrams.

To calibrate the CCD frames taken in the g, r and i Gunn-Thuan system, we use SExtractor to identify galaxies and derive the color-magnitude diagram (r vs. $g-r$) for the candidate cluster region and for a test field (background) well selected. Then we statistically subtract the field contamination and identify the excess of galaxies as cluster members. The existence of an excess of galaxies in the putative cluster line of sight and the fact that they obey to the usual color–magnitude relation is taken as confirmation of the existence of the putative cluster. These galaxies are then used to build the cluster radial profile. The color-magnitude diagram allows also to derive a rough photometric redshift (which, due to the broad band, has however a rather large error).

Preliminary results confirm that for $z < 0.15$ the candidate clusters found by our algorithm are 95% "true" clusters. At higher redshift ($0.15 < z < 0.2$) this fraction drops to 70%.

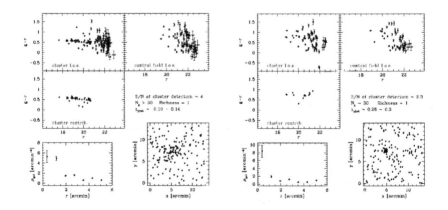

Figure 3. Color-magnitude diagram for the candidate cluster obtained accordingly to the procedure described in the text. Right (high S/N candidate): the cluster turns out to be of richness class I and to be at $z = 0.10 - -0.12$. Left (low S/N candidate): the cluster turns out to be of richness class I and to be at $z = 0.26 - -0.3$.

We plan to improve the simulation by adding non-circular symmetry for the simulated cluster and to study the dependence on the zero-order correlation functions. We also plan to apply the color-magnitude diagram test to all clusters available in our calibration data set.

References

Bertin, E., & Arnouts, S. 1996, A&AS, 117, 393

De Filippis, E., Longo, G., Andreon, S., et al. 1999, MmSAI, in press (astro-ph/9909368)

Garcia, A.M. 1993, A&AS, 100, 47

Geller, M.J., & Huchra, J.P. 1983, ApJS, 52, 61

Gott J.R., & Turner E.L. 1977, ApJ, 216, 357

Maia, M.A.G., & Da Costa, L.N. 1989, ApJS, 69, 809

Maia, M.A.G., & Da Costa, L.N. 1990, ApJ, 352, 457

Metcalfe, N., Shanks, T., Fong, R., & Roche,N. 1995, MNRAS, 273, 257

Puddu, E., et al. 1999, MmSAI, in press (astro-ph/9909367)

Schectman, S.A. 1985, ApJS, 57, 77

Soares, D.S.L. 1989, PhD Thesis

Turner E.L., & Gott J.R. 1976, ApJS, 32, 409

Weir, N., Fayyad, U.M., Djorgovski, S.G., & Roden J. 1995, PASP, 107, 1243

Virtual Observatories of the Future
ASP Conference Series, Vol. 225, 2001
R.J. Brunner, S.G. Djorgovski, and A.S. Szalay, eds.

The WARPS Galaxy Cluster Survey: A Case Study Using Present-Day Multi-Wavelength Archives

Caleb A. Scharf
Space Telescope Science Institute, 3700 San Martin Drive, Baltimore, MD 21218

Abstract. The Wide Angle ROSAT Pointed Survey (WARPS) has used X-ray and optical archives to construct a statistically complete X-ray survey of some 150 galaxy clusters and groups, including more than 20 systems at redshifts higher than 0.6. With this survey, and follow-up XMM and Chandra data, we are pursuing cosmological and gas thermodynamical measurements based on the evolution of the cluster population. We discuss the survey methodology for large datasets and multiple source parameters, and consider the lessons that may point the way to the use and design of future, virtual observatories, for similar science goals.

1. Introduction

The WARPS cluster project uses a novel X-ray detection algorithm (Voronoi Tessellation and Percolation, VTP) to find serendipitously observed objects in archival ROSAT PSPC data, and in particular to seek out extended, galaxy cluster sources (Scharf et al. 1997). Over 400 ROSAT fields were used, with a total of over 80 million photons processed. Approximately 8000 source detections were made, each described by 23 parameters, at a minimum of three detection thresholds. For all sources above a well-defined flux limit, optical data from the Digitized Sky Survey (DSS) and Automated Plate Measuring (APM) machine were retrieved, along with any NASA Extragalactic Database (NED) identifications. Unidentifiable candidates were then followed up using the KPNO, CTIO, MDM, Lick, CFHT, Keck and ROSAT observatories.

Our approach was to automate as much as possible the initial data mining process, while allowing for human intervention, and notification of problems.

2. The Virtual Observation Pipeline

In Figure 1 a simplified overview of the WARPS observation pipeline is shown.

3. WARPS Science

To date the WARPS has yielded many new science results. The evolution seen in the number counts and X-ray luminosity function of clusters to redshifts $z \sim 0.8$ (Jones et al. 1998) is very mild, providing strong evidence for a low density cosmology. We are obtaining Chandra and XMM observations of high-z

 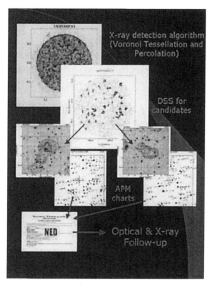

Figure 1. In the above panels an overview of the WARPS data pipeline is presented.

WARPS clusters to measure the cluster temperature (and hence mass) function evolution in order to estimate the universal density parameter to within 10% random errors.

An intriguing degree of substructure is seen in the high-z WARPS cluster population (Ebeling et al. 2000a), suggesting that we may be witnessing the formation epoch of massive systems. However, the X-ray luminosity-temperature relationship (which probes the interplay between cluster mass and intra-cluster gas physics) shows little or no evolution to $z \sim 0.8$ (Fairley et al. 2000). While this latter result is well modeled by current cluster formation theory which includes non-gravitational heating, combined with the former result it presents an interesting theoretical challenge.

Finally, we have identified what appears to be the most luminous ($L_x > 10^{45}$ erg s^{-1}), well relaxed, cluster currently known at z=0.9 (Ebeling et al. 2000b). The existence of just this one massive system is very strongly at odds with the predictions of a high-density cosmology. Together with the existence of other massive systems at high-z this constrains the probability of a critical density universe to be *less than* $\sim 10^{-10}$.

4. Lessons for the Future

What did we learn with WARPS ?

- The ability to automatically retrieve data on a highly targeted, piecewise basis was critical in order to minimize local storage requirements and maximize flexibility in dealing with unusual cases.

- The ability to directly extract imaging and/or visual representations of the data reduced effort and improved uniformity

- The existence of heterogeneous, or moderately calibrated catalogs (NED, WGACAT, APM) greatly expedited the prioritizing of WARPS sources for follow-up and reduced unnecessary expenditure on known objects.

- For WARPS the principle bottleneck was human-hours and the need for targeted ground and space follow-up.

Overall conclusions:

- With science aims requiring statistical completeness it was critical that we were able to readily access, and understand low level data products (raw photon maps, exposure maps *etc.*).

- A project such as WARPS could *only* be completed on a short timescale with the use of "virtual" data.

- Highly interconnected datasets (*e.g.*, threads, parallel searches) would not have offered any real gain in this case, given the need to make individual and often unique decisions for detected sources (*e.g.*, deciding what to do about noise contamination). Furthermore, the ability to retrieve highly specific data *without* descending any hierarchies, or going through pre-existing "mining" tools was a plus.

In the context of a virtual observatory, the requirements of a serendipitous, but statistically controlled, survey such as the WARPS, appear to be straightforward; easy access to raw data, good meta-data to enable correct processing, and easy access (but not necessarily in parallel) to existing object catalogs.

As we carried out the WARPS it was interesting to discover, or re-discover, that the availability of imaging data for easy human inspection was invaluable to the process. It seems unlikely that *any* survey aiming for statistical completeness will not have at least partial (by sampling, for example) human inspection of the process at all stages. For WARPS this was a critical way by which we could monitor and evolve our strategy and techniques. Transparent, multi-wavelength, visualization tools could greatly enhance this process. Furthermore, if a virtual observer were to submit their detection/analysis algorithm(s) to a "centralized" storage system (to avoid large data traffic) then some means of occasionally retrieving particularly problematic data is required. Additionally, a range of sample data could be made available, in order to iron out the most frequently occurring problems.

Acknowledgments. CAS thanks the WARPS collaborators: Laurence Jones, Harald Ebeling, Eric Perlman, Don Horner, Bruce Fairley, Gary Wegner & Matt Malkan. This work has made use of data obtained through the HEASARC Online Service, provided by NASA/Goddard.

References

Ebeling, H., *et al.* 2000a, ApJ, 534, 133

Ebeling, H., *et al.* 2000b, in preparation
Fairley, B., *et al.* 2000, MNRAS, 315, 669
Jones, L., *et al.* 1998, ApJ, 495, 100
Scharf, C., *et al.* 1997, ApJ, 477, 79

Scaling and Fluctuations in the Galaxy Distribution: Two Tests to Probe Large Scale Structure

Francesco Sylos Labini

Dépt. de Physique Théorique, Université de Genève, Quai E. Ansermet 24, CH-1211 Genève, Switzerland, & INFM Sezione Roma1, Roma Italy

Andrea Gabrielli

Ecole Polytechnique, 91128 - Palaiseau Cedex, France

Abstract. We present a concise introduction to the statistical properties of systems with large fluctuations. We point out that for such systems the relevant statistical quantities are scaling exponents and the nature of fluctuations is completely different from the one belonging to system with small fluctuations. We propose then two test to be performed on galaxy counts data as a function of apparent magnitude, which may clarify in a direct way the nature of the large scale galaxy clustering

1. Scaling and Fluctuations in the Galaxy Distribution

In Figure 1 it is shown an homogeneous distribution with Poissonian fluctuations in the two-dimensional Euclidean space. In such a distribution the average density is a well defined property if it measured on scales larger than the mean particle separation Λ. If there are some *small amplitude fluctuations* up to

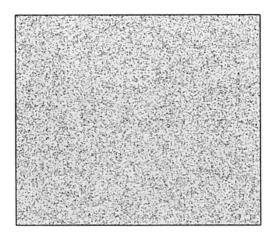

Figure 1. Homogeneous distribution with Poisson fluctuations

a certain scale, or up to the sample's size, the average density is still a well defined property. For example in Figure 2 it is shown a distribution where the fluctuation structures are extended over the whole sample. In such a case the average density is still a well defined property, and it can be measured at large enough scales $(r > \lambda_0)$, when the *amplitude* of the fluctuations become smaller than the average density itself, *i.e.*, when $\delta N(r)/\langle N(r)\rangle|_{\lambda_0} \leq 1$. These two

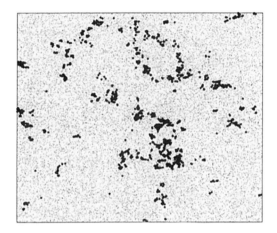

Figure 2. Homogeneous distribution with small amplitude long range correlations

systems belong to the class of distributions with small amplitude fluctuations. The average density is well defined in the sense that the one point properties are well defined. For example in a ball of radius $r > \lambda_0 > \Lambda$ centered in a *randomly* chosen point, in both the distributions, the number of points is constant, a part a fluctuations. When the size of the ball becomes $r \gg \lambda_0 > \Lambda$ the number of points is the same, no matter where the ball in centered. The average density is related to the one-point properties of the distribution. If these properties exist, then it is possible to study the statistical properties of the fluctuations around the average density. For example, given an occupied point, the number of points at distance r is given by

$$N_c(<r) = \langle n\rangle V(r) + \langle n\rangle \int_V \xi(r) d^3r \qquad (6)$$

where

$$\xi(r) \equiv \frac{\langle n(r)n(0)\rangle}{\langle n\rangle^2} - 1 \qquad (7)$$

Note that the volume integral of $\xi(r)$, over the whole space where the average density $\langle n\rangle$ (one point property) has been estimated, must be zero by definition. The volume can be finite or infinite.

A completely diffcrent situation happens in the system shown in Figure 3. In fact, for such a distribution, it is not possible to define the average density, or

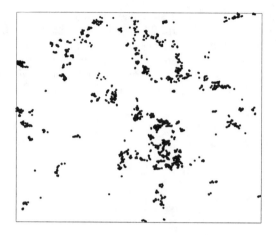

Figure 3. Fractal distribution with $D = 1.47$ in the two dimensional Euclidean space

more generally speaking the one point properties. This is because the system is characterized by having structures and voids of all sizes and at all scales (up to the sample size). Hence if we make the simple exercise we did before, by putting a ball of radius r centered in a randomly chosen point, the number of points inside the ball will strongly fluctuate from a position another. For example, if the center is in a void of size larger than the ball's diameter, we do not find any point, while if the center is in a large scale structure we find a huge number of points. This happens as long as the distribution has fractal properties, *i.e.*, as long as the size of the voids increases in a proportional way to the size of the sample. Hence in such a situation one will never find a scale λ_0 above which the fluctuations become small and it is not possible to measure the one-point average density. Fractals are then distribution with large fluctuations. The proper way to characterize the statistical properties is to focus on the scaling behavior of two-point quantities. For example one may study the two-point conditional average density (Pietronero 1987, Sylos Labini *et al.* 1998) defined as

$$\Gamma(r) = \frac{\langle n(r)n(0) \rangle}{\langle n \rangle} \sim r^{D-3} \tag{8}$$

This function measures the average density at distance r from an occupied point. The term $\langle n \rangle$ is just a normalizing factor equal to N/V which is useful in order to compare the results from different distributions where the number of points N and the volume V can be different. The last equality is Equation 3 gives the point of this discussion: if $D < 3$ the conditional average density has a scaling behavior as a function of distance and hence: (i) the one-point average density is not a defined quantity and (ii) the only charactering quantity is the scaling exponent or fractal dimension D. If, otherwise, $D = 3$ then the system has an average density defined, and hence once can focus the attention to the statistical properties of the fluctuations about the average density itself. In the fractal case the distribution is characterized by *large fluctuations*.

These simple but fundamental properties of systems with large fluctuations change our perspective on the studies of the statistical properties of a given distribution. Instead of assuming that the one-point properties are indeed defined, as it is done in the standard analysis (Peebles 1980),we test whether the distribution has such a property (Pietronero 1987, Sylos Labini *et al.* 1998).

The application of these method to the redshift space distribution of galaxies has been discussed at lengthly and we refer to Sylos Labini *et al.* (1998) for a comprehensive review of the subject. Recently (Sylos Labini & Gabrielli, 2000) we have proposed to study two properties of galaxy counts as a function of apparent magnitude, in order to discriminate between a small scale (~ 5–$20 h^{-1} Mpc$) homogeneous distribution, and a fractal structure on large scales (~ 100–$300 h^{-1} Mpc$): (i) The slope α of the counts can be associated, in the bright end where $z \ll 1$, to a possible fractal dimension in real space, by simple arguments. (ii) Fluctuations of counts about the average behavior as a function of apparent magnitude in the whole magnitude range $B \gtrsim 11^m$. These fluctuations are related to the very statistical properties of the spatial distribution, independently on cosmological corrections. While in an homogeneous distribution they are exponentially damped as a function of apparent magnitude, in a fractal they are persistent at all scales. Large field-to-field fluctuations in the counts have been reported both in the bright and in the faint ends and in different photometric bands: they can be due to some systematic measurement errors or to a genuine effect of large scale structures. For this reason we need better calibrated photometric samples. Therefore, the application of these tests to the forthcoming photometric surveys is then *fundamental* to understand and characterize the large scale galaxy clustering.

References

Peebles, P.J.E. 1980, *Large Scale Structure of the Universe*, Princeton University Press
Pietronero L. 1987, Physica A, 144, 257
Sylos Labini F., Montuori M., & Pietronero L. 1998, Phys.Rep.,293 ,66
Sylos Labini F., & Gabrielli A. 2000, astro-ph/0003215

Part 2
Astronomical Approaches to a Virtual Observatory

NASA Mission Archives, Data Centers, and Information Services: A Foundation for the Virtual Observatory

R. J. Hanisch
Space Telescope Science Institute, 3700 San Martin Drive, Baltimore, Maryland 21218

Abstract. NASA's astrophysics data centers and information services will be cornerstones in the emerging Virtual Observatory, both for the data they hold and for the technologies that have been pioneered to enable and improve interoperability. This paper gives an overview of NASA's astrophysics data management systems.

1. Overview

NASA's Office of Space Science has embraced the systematic archiving of data from space astrophysics missions for nearly two decades. This farsighted vision for comprehensive data management has culminated in an astrophysics data system composed of primary science archive research centers (SARCs): a high-energy SARC, an optical/UV SARC, and an infrared SARC. The High Energy Astrophysics SARC (HEASARC) is located at NASA/Goddard Space Flight Center, the optical/UV SARC, known as the Multimission Archive at Space Telescope (MAST), is located at the Space Telescope Science Institute in Baltimore, and the infrared SARC, known as the Infrared Science Archive (IRSA), is located at the Infrared Processing and Analysis Center (IPAC) at Caltech. The National Space Science Data Center (NSSDC) provides permanent archive services to the three SARCs and provides direct archival services for the COBE and IRAS mission data sets.

Active missions such as SIRTF, HST, and Chandra process and manage their data directly at the associated science operations center (SIRTF Science Center, Space Telescope Science Institute, Chandra X-Ray Center). Active missions coordinate with the relevant SARC to assure access to their data during the mission lifetime and arrange for long-term access after the mission is over. In some cases active missions contract directly with one of the SARCs to provide archive and data distribution services (*e.g.*, the FUSE mission data are processed at Johns Hopkins University and archived at MAST/STScI).

NASA has also supported the development and operations of complementary astronomical information services, such as the Astrophysics Data System (ADS) abstract and bibliographic database at SAO, the Astronomical Data Center catalog collection (a component of the Astrophysics Data Facility at GSFC), and the NASA Extragalactic Database (NED) bibliographic and data service at IPAC/Caltech. NASA has also supported U.S. access to the SIMBAD database at the Centre Données astronomiques de Strasbourg (CDS) in France. These

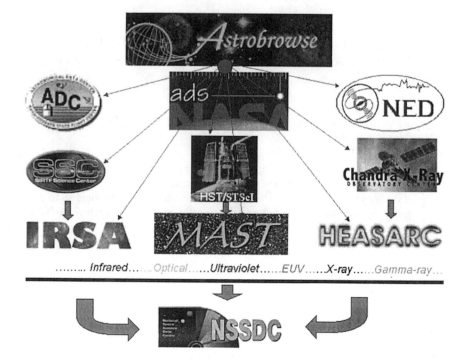

Figure 1. NASA's astrophysics data management structure. NASA's astrophysics data centers, catalog services, and bibliographic and thematic databases are elements in a complementary and comprehensive data and information service. Data from active missions (SIRTF, HST, Chandra) are archived and distributed from the associated science center, then either migrates or are linked to the relevant archive center (IRSA, MAST, HEASARC). The NSSDC provides permanent archive services. ADC, ADS, and NED provide catalog and bibliographic services. Astrobrowse allows users to search all information services for data of interest.

services provide many cross-links to the data in the SARCs; ADS, for example, is including direct links from its bibliographic database to the underlying data.

The relationship between the permanent archive, the active archive centers or SARCs, the active missions, and the ancillary information services in astrophysics are shown in Figure 1.

2. NASA's Science Archive Research Centers

2.1. Infrared Science Archive

The Infrared Science Archive (IRSA) is located at the Infrared Processing and Analysis Center (IPAC) at Caltech. IRSA and IPAC are notable for their experience in managing the data from survey instruments, beginning with IRAS and continuing now with 2MASS. Both the underlying scan or imaging data are available as well as the extracted source catalogs. IRSA staff has been working with Caltech's "Digital Sky" project and are developing experience in indexing and cross-correlating very large databases (*e.g.*, 2MASS *vs.* DPOSS; the 2MASS archive already exceeds 12 TB in size). IRSA is the U.S. portal for data from the ISO mission and will host the SOFIA and SIRTF data archives.

2.2. Multimission Archive at Space Telescope

The Multimission Archive at Space Telescope (MAST) (Imhoff *et al.* 1999) is located at the Space Telescope Science Institute in Baltimore. MAST was established as the IUE mission closed down, recognizing the close scientific relationship between HST and IUE data. MAST is the repository for optical, UV, and near-IR data, including, in addition to HST and IUE: FUSE, ASTRO (HUT, UIT, WUPPE), ORPHEUS (BEPS, IMAPS), and Copernicus. In partnership with the HEASARC, MAST provides access to the EUVE data that are physically stored at HEASARC. MAST also is the primary provider of the STScI Digitized Sky Survey and Guide Star Catalogs and is one of several sites hosting the VLA FIRST (Faint Images of the Radio Sky at Twenty centimeters) survey data. In the future MAST will host the GALEX and CHIPS mission public archives.

2.3. High Energy Astrophysics Science Archive Research Center

The High Energy Astrophysics Science Archive Research Center (HEASARC) (Angelini *et al.* 1999) is located at the Goddard Space Flight Center. HEASARC hosts data from space missions covering the spectral range from the far UV to gamma rays. HEASARC holdings include: Ariel 5, ASCA, Beppo-SAX, CGRO, Einstein, EXOSAT, EUVE, ROSAT, XMM, and RXTE. Data from the Chandra mission will be archived at the Chandra X-ray Center during the active mission, but an access portal will be incorporated into HEASARC. HEASARC staff has led in the development of the Astrobrowse data location service and of Skyview, a first generation data integration service (McGlynn, Scollick, & White 1998).

3. Catalog, Bibliographic, and Thematic Information Services

3.1. Astronomical Data Center

The Astronomical Data Center (ADC) is a part of the Astrophysics Data Facility at Goddard Space Flight Center. The primary role of the ADC is to act as custodian of the many hundreds of standard catalogs that astronomers use in support of their research. ADC has all catalogs on-line and available for search and cross-correlation, either through ADC-provided user interfaces (Impress, CatsEye, ADC Data Viewer) or through the ADC External Query (AEQ) service, which allows HTTP-formatted queries against ADC catalogs to be received from other sites. Staff at the ADC has also taken on a leadership role concerning the development of XML standards for the astronomy community.

3.2. Astrophysics Data System

The Astrophysics Data System Abstract Service, located at the Smithsonian Astrophysical Observatory in Cambridge, has become a fundamental tool in astronomical research (Kurtz et al. 2000). ADS indexes all of the primary astronomical research literature, including both the peer reviewed journals and the Los Alamos astro-ph preprint archive (Grant et al. 2000). Direct links are provided from the abstracts to the full text of articles, either via the on-line journals or via scanned bit-maps of the historical literature. Increasingly the ADS is providing links to the data underlying the literature.

3.3. NASA Extragalactic Database

The NASA Extragalactic Database (NED), hosted by IPAC/Caltech, is a combined bibliographic and database service. It provides a thematic view of extragalactic astronomy, bringing together at one site extragalactic object catalogs, a name resolution service, links to important data resources, complete bibliographic information, and a suite of on-line tools (coordinate conversions, extinction calculations, velocity calculations). NED has recently introduced the "Level 5" service in which collections of the most significant publications related to key topics in extragalactic astronomy have been assembled and enhanced by adding current links to associated data or catalogs.

3.4. International Partners

A notable partner in astrophysics information services is the Centre Données de astronomiques de Strasbourg (CDS) (Genova et al. 2000), whose catalog and bibliographic services are used widely in the NASA astrophysics community. The SIMBAD database of astronomical objects and their bibliographic references (Wenger et al. 2000), Vizier catalog browser (Ochsenbein, Bauer, & Marcout 2000), and Aladin correlation tool (Bonnarel et al. 2000), are important elements of the international astrophysics information system.

4. Interoperability Initiatives

The NASA archive and information services operate in an informal confederation known as the Astrophysics Data Centers Coordinating Council (ADCCC). The goals of the ADCCC are to maximize interoperability among the data centers and services, minimize redundancy of services and development efforts, and provide excellent information services to the entire astronomical community.

An example of improved interoperability is illustrated by access to resources such as data from the EUVE and ROSAT missions. The EUVE data are physically stored at the HEASARC. EUVE data are of interest to astronomers working with X-ray data and UV data, and thus interfaces to the EUVE observation catalog have been implemented at both HEASARC and MAST. These interfaces are designed to be similar to the interfaces for other data sets at each facility, and thus there is a minimal learning curve for users of either facility to gain access to the EUVE data. MAST users can retrieve EUVE data transparently, that is, without ever being aware that the data are physically stored somewhere other than in MAST. Similarly, MAST has recently added a direct link to ROSAT data, also at HEASARC, to enable easy cross-correlation between UV/optical and X-ray data. The HST observation catalog is routinely updated and provided to ADC and CDS to enable local catalog cross-correlation, but with direct links to the HST archive for preview images and making data requests. The Digital Sky project, managed through IRSA, has constructed the first portable services for positional cross-comparison of large catalogs.

Astrobrowse (Heikkila, McGlynn, & White 1999; McGlynn & White 1998) is a front-end search facility that is aware of over 1500 astronomical resources at some 80 sites worldwide. It is a data location service that can format queries for information on a given object in selected bandpasses or in selected data types. Users can located sources of potentially interesting data with one query and need not visit each of hundreds of web sites manually. Development of Astrobrowse has been fostered through ADCCC collaborations.

Members of the ADCCC have worked to improve links between the scientific literature and the underlying data. The ADC includes in its catalog collections many key tables from the literature, and takes on a curatorial role: reviewing tables for errors, making corrections, and then providing access via its catalog browsing and comparison tools. The ADS has pioneered efforts to interlink distributed services (electronic journals, the SIMBAD database, the ADC catalogs, and NED) and is now including links from its abstracts to archival data, enabling researchers to quickly browse—and retrieve for detailed analysis as desired—the data on which a published paper is based. ADS also links to the NED and CDS databases for many abstracts. The NED and SIMBAD name resolvers are now used throughout the community, integrated into virtually all astronomical on-line services.

The ADCCC has been fostering new technological developments aimed at enabling interoperability among astronomy and space science data centers. For example, the ADC at NASA/GSFC has led community efforts in developing XML standards for astronomical catalogs, complementing efforts at IRSA, STScI, and CDS. ADCCC data centers have joined in the ISAIA (Integrated System for Archival Information Access) initiative, whose goal is to define protocols and standards for queries and responses from space science data services (Hanisch 2000, Hanish 2001, these proceedings).

5. Summary

The NASA astrophysics data centers will be cornerstones in the emerging Virtual Observatory. We have already begun to establish the sort of inter-center partnerships and distributed data services that are requisite to the success of the NVO. The work of the ADCCC has demonstrated the feasibility of the NVO, through support for new technology and establishment of standards. We must now take the next steps in realizing the vision of the Virtual Observatory.

Acknowledgments. The author thanks his colleagues at each of the data centers described for their collaboration in the ADCCC and their contributions to this paper.

URLs

ADC	http://adc.gsfc.nasa.gov/
AEQ	http://tarantella.gsfc.nasa.gov/viewer/AEQdoc.html
Aladin	http://aladin.u-strasbg.fr/aladin.gml
Astrobrowse	http://heasarc.gsfc.nasa.gov/ab/
CDS	http://cdsweb.u-strasbg.fr/CDS.html
HEASARC	http://heasarc.gsfc.nasa.gov/
IMPReSS	http://tarantella.gsfc.nasa.gov/impress/
ISAIA	http://heasarc.gsfc.nasa.gov/isaia/
IRSA	http://irsa.ipac.caltech.edu/
MAST	http://archive.stsci.edu/mast.html
NSSDC	http://nssdc.gsfc.nasa.gov/
SkyView	http://skyview.gsfc.nasa.gov/

References

Angelini, L., Breedon, L., Garcia, L., Hilton, G., Stollberg, M., & White, N. 1999, AAS Meeting, 194, #83.01

Bonnarel, F., Fernique, P., Bienaymé, Egret, D., Genova, F., Louys, M., Ochsenbein, F., Wenger, M., & Bartlett, J. 2000, A&AS, 143, 33

Genova, F., Egret, D., Bienaymé, O., Bonnarel, F., Dubois, P., Fernique, P., Jasniewicz, G., Lesteven, S., Monier, R., Ochsenbein, F., & Wenger, M. 2000, A&AS, 143, 1

Grant, C. S., Accomazzi, A., Eichhorn, G., Kurtz, M. J., & Murray, S. S. 2000, 2000, A&AS, 143, 111

Hanisch, R. J. 2000, Computer Physics Communications, 127, 177

Heikkila, C. W., McGlynn, T. A., & White, N. E. 1999, in ASP Conference Series, 172, Astronomical Data Analysis Software and Systems VIII, ed. David M. Mehringer, Raymond L. Plante, & Douglas A. Roberts (San Francisco: ASP), 221

Imhoff, C., Abney, F., Christian, D., Donahue, M., Hanisch, R., Kimball, T., Levay, K., Padovani, P., Postman, M., Smith, M., & Thompson, R. 1999, AAS Meeting, 194, #83.02

Kurtz, M. J., Eichhorn, G., Accomazzi, A., Grant, C. S., Murray, S. S., & Watson, J. M. 2000, A&AS, 143, 41

McGlynn, T., & White, N. 1998, in ASP Conference Series, 145, Astronomical Data Analysis Software and Systems VII, ed. R. Albrecht, R. N. Hook, & H. A. Bushouse (San Francisco: ASP), 481

McGlynn, T., Scollick, K., & White, N. 1998, in IAU Symposium 179, New Horizons from Multi-Wavelength Sky Surveys, ed. B. J. McLean, D. A. Golombek, J. J. E. Hayes, & H. E. Payne (Dordrecht: Kluwer), 465

Ochsenbein, F., Bauer, P., & Marcout, J. 2000, A&AS, 143, 23

Wenger, M., Ochsenbein, F., Egret, D., Dubois, P., Bonnarel, F., Borde, S., Genova, F., Jasniewicz, G., Laloë, S., Lesteven, S., & Monier, R. 2000, A&AS, 143, 9

An Overview of Existing Ground-Based Wide-Area Surveys

B. McLean

Space Telescope Science Institute, Baltimore, MD 21218

Abstract. In recent years there has been a tremendous growth in the amount of data that astronomical survey projects across the spectrum have obtained. Coupled with the technological advances in computing infrastructure and on-line access to these datasets, the techniques of observational astronomical research have been changing. We present a short overview of a number of ground-based surveys which will provide a foundation for the construction of a *Virtual Observatory*. By interconnecting and enabling data mining of these multi-wavelength observations, we will provide a rich framework for discovery.

1. Introduction

Astronomy has a tradition of performing surveys but historically the data was usually distributed in the form of catalogs. The first survey which was generally available in image form was the original Palomar Sky survey, which was both incredibly useful and difficult to use due to its publication on photographic media. The combination of advances in computer technology and Internet connectivity allowed the creation of on-line data centers where catalogs and individual images could be accessed, but a significant turning point was the publication of the Digitized Sky Survey (DSS), both on CDROM and via the web. This dataset provided the community with the ability to retrieve and visualize and manipulate any field on the sky. Since then, a large number of datasets and surveys have become available throughout the entire electromagnetic spectrum.

The paradigm of on-line access to data has fundamentally changed the nature of observational astronomy. Most telescopes routinely use DSS data or the HST Guide Star Catalog as part of their operations, and most astronomers will check if their favorite targets have prior observations, particularly at other wavelengths. In fact, in recent years the whole concept of "Archival Astronomy", where one uses existing data for scientific projects different from that originally proposed, has proven to be increasing popular. This is indeed extremely cost-effective provided the data is well documented and calibrated, and most new or ongoing surveys recognize the value of archiving the data and making it available to the community.

In this short review, it is impossible to mention all the ground-based surveys that are contributing to this data volume explosion, so we shall only provide an overview of the major radio, infrared and optical surveys that are representative of what is happening in our community. Shorter wavelength, higher energy surveys need to carried out from space-based platforms in order to be effective and are addressed elsewhere in this conference.

2. Radio

The Radio sky has been surveyed a number of times over the years at different frequencies and increasing sensitivities. One of the most significant in recent years was performed using the Very Large Array. The NVSS (NRAO VLA Sky Survey) is covering the sky north of $\delta > -40°$ at 1.4 GHz and is now 99.9% complete. Over 1.8×10^6 sources with a limiting sensitivity of 2.5 mJy ahve been identified. The catalog and over 2TB of image data are all available on-line from NRAO and NCSA.

A second VLA survey is also well underway (60% complete) of the 10^4 deg^2 around the North Galactic Cap in order to complement the optical Sloan Digital Sky Survey. This FIRST survey (Faint Images of the Radio Sky at 20-cm) is also being performed at 1.4 GHz but with greater resolution (1.8") and to a deeper limiting sensitivity (1mJy). These data are also being processed and the image data and derived catalogs being made publically available on web sites at STScI, LLNL and NRAO.

The southern hemisphere counterpart to the NVSS is being carried out by the Molonglo Observatory Synthesis Telescope which is observing the sky south of $\delta < -30°$ at 843 Mhz. The SUMSS survey (Sydney University Molonglo Sky Survey) has observed approximately 35% of the area and 10% of the data has been released on-line at the University of Sydney using the same data format adopted by the NVSS.

3. Infrared

The 2MASS survey (2-Micron All Sky Survey) was the first project to perform a ground-based all-sky survey using digital detectors. Two dedicated 1.3m telescopes at Mt.Hopkins and CTIO have been imaging (1" resolution) the entire sky in the infrared J (1.25m), H (1.65m) and K_s (2.17m) bands. Complete coverage was recently achieved and 50% of the data have already been released by IPAC. There will ultimately be over 10TB of image data and a catalog with about 10^8 sources.

In the southern hemisphere the ESO 1m telescope is performing a slightly deeper survey than 2MASS in the I_g (0.8m), J (1.25m), K_s (2.16m) bands. The DENIS survey (Deep Near Infrared Survey of the Southern Sky) has observed 80% of the southern hemisphere. Although only 2% of the data has been publically released at this time, eventually 3TB of images and a source catalog of 5×10^8 will be available on-line from the CDS.

4. Optical

Schmidt telescopes have been performing surveys for several decades. Until recently, these were the only practical means of obtaining all-sky images to a reasonable depth with a reasonable resolution. Palomar observatory in the northern hemisphere, and the UK and ESO Schmidt telescopes in the southern hemisphere have obtained a number of surveys (Table 1).

Survey	Epoch	Emulsion/Filter	Depth	Dec.Zone	Plates	Obs
POSS-I E	1950-58	103aE	20.0	+90:-30	935	100
POSS-I O	1950-58	103aO	21.0	+90:-30	935	100
Pal QV	1983-85	IIaD+W12	19.5	+90:+06	613	100
POSS-II J	1987-98	IIIaJ+GG395	22.5	+90:+00	897	100
POSS-II F	1987-99	IIIaF+RG610	20.8	+90:+00	897	100
POSS-II N	1987-xx	IV-N+RG9	19.5	+90:+00	897	88
SERC J	1975-87	IIIaJ+GG395	23.0	-20:-90	606	100
SERC EJ	1979-88	IIIaJ+GG395	23.0	-00:-15	288	100
SERC ER	1984-00	IIIaF+OG590	22.0	-00:-15	288	96
AAO SES	1990-98	IIIaF+OG590	22.0	-20:-90	606	97
SERC I	1990-xx	IV-N+RG715	19.5	-00:-90	894	94
ESO R	1974-87	IIIaF+OG590	21.5	-20:-90	606	100
AAO Hα	1999-xx	4415+Hα	21.0	s.gal.plane	233	75

Table 1. Major Photographic Surveys

These surveys were distributed to the community on either glass plate or film, and a number of different scanning machines (Table 2) have been used to digitize most of these to date. Each of these scanning machines have different characteristics, project goals and processing algorithms such that one can use their different results as quality control checks to validate each other. Most of these groups have placed the available images and derived object catalogs on-line.

The era of wide-field photographic surveys is coming to an end. The ESO Schmidt was shutdown a few years ago and the Palomar Schmidt is being converted to accommodate a CCD mosaic camera. The UK Schmidt is using a few nights each lunation to complete the remaining IV-N survey fields, but is now primarily being used for wide-field multi-object spectroscopy.

The Sloan Digital Sky Survey (SDSS) is designed to image approximately π steradians centered on the northern galactic cap in 5 bandpasses (u', g', r', i', and z') and perform followup spectroscopy to determine one million redshifts. This is being carried out with a dedicated 2.5m telescope at Apache Point Observatory. Routine operations started in 2000 and will continue for at least 5 years. The scientific potential of this survey has already been demonstrated with an analysis of the test data taken during the commissioning. This has included the discovery of several high redshift quasars and methane dwarfs.

Even though only 10^4 deg^2 will be covered by the SDSS, this is the type of survey that will be the successor to the photographic surveys with deeper imaging and higher resolution pixels (0.4" pixels). When completed, there will be approximately 15TB of calibrated images available and a catalog with 10^8 objects. The data is planned to be incrementally released to the community through MAST (Multi-mission Archive at STScI), following a proprietary period for consortium members.

Institution	Scanner	Major Products
STScI	Γ	DSS-I:[a] POSS-I E, SERC-J/EJ, Pal-QV GSC-I[b] DSS-II:[c] POSS-II J/F/N AAO-SES, SERC-ER, SERC-I GSC-II[d]
USNO	PMM	USNO-A2.0[e] POSS-I O, POSS-I E SERC-J, ESO-R Scanning POSS-II[f]
U.Minnesota	APS[g]	POSS-I O, POSS-I E
ROE	COSMOS[h]	SERC J/EJ, SERC SR
	SuperCOSMOS[j]	SERC J/EJ, AAO SES/SERC-ER SERC I, POSS-I E, ESO-R
IfA Cambridge	APM[j]	POSS-I O, POSS-I E SERC-J, AAO SES/SERC-ER

Table 2. Major Scanning Machines

---

[a] 1 TB on-line, CDROMs
[b] 2.5×10^7 objects, $m_v = 15.5$
[c] 7 TB on-line, CDROMs
[d] 10^9 objects, $m_v = 20$
[e] 526,230,881 Objects
[f] Includes all reject plates
[g] 664 fields $|b| > 20$, Images/Catalogs on-line
[h] 5×10^8 objects on-line
[i] 25% scanned, Images/Catalogs on-line
[j] 75% scanned, 3×10^7 objects on-line

The spectroscopic component of the SDSS is a natural extension of previous and current galaxy redshift surveys which have been important in our understanding of large scale structure.

Survey	Current Coverage (deg^2)	Numnber of Objects
CfA	5000	2×10^4
Las Campanas	700	2.6×10^4
2dF galaxy/QSO	800	10^4
ESO Slice	23	4×10^3
Durham/UKSTU	1500	2.5×10^3
SDSS	10000	10^6
6dF	17000	1.2×10^5

Table 3. Major Redshift Surveys (ongoing/planned)

5. Summary

In this short (and incomplete) review, the most important point I want to emphasise is that on-line archives and catalogs are now integral activities for all major surveys. There is an estimated 50TB of ground-based (radio-ir-optical) survey data available and an atmosphere of cooperation between the different groups so that the "virtual observatory" concept has a solid base to build upon. There is also a large amount of non-survey data available in observatory archives and data centers around the world. Improving connectivity and accessibility of all these data will enrich the entire community who will use this *Global Virtual Observatory*.

Astronomical Data Centers, Information Systems, and Electronic Libraries

Daniel Egret

CDS, Observatoire de Strasbourg, 11 rue de l'Université, 67000 Strasbourg, France

Abstract. We review the current organization of the astronomical information, from archives, to data bases, data centers, information systems, and electronic libraries. We discuss some technical challenges involved in attempting to provide general search and discovery tools, and to integrate them in the Virtual Observatories of the future.

1. Introduction

In this short review of existing data centers and information systems we will briefly describe how the astronomical community is organized in terms of archives, data centers and electronic libraries.

We will not try to be exhaustive, but rather to give a feeling of the current status of the activity in the field, and how this activity can contribute to foster the initial development of the Virtual Observatory. For more complete recent descriptions, see, *e.g.*, Egret & Albrecht (1995), Andernach (1999), and the discussion in Egret *et al.* (2000).

We will also describe some strategies that have to be developed for building cooperative tools which will be essential in the research environment of the decade to come.

2. On-Line Astronomy Resources

We outline in this section the current status of the main categories of on-line astronomy resources, pointing to meta-resources (*i.e.*, organized lists of resources) when they are available.

2.1. Observatory Archives

These are collections of observations from an Observatory, or from a space mission. Data are generally categorized according to their reduction levels, from raw data to calibrated results. Data which are not calibrated or not processed through a standard pipeline are generally not useful for astronomers outside of the community which produced the data. It may be however important to keep them in a safe place for further reprocessing!

For this reason all the space missions, and more and more large ground-based observatories are producing archives of high level reduced and calibrated data, and are providing access to their archives after a short proprietary period.

The key for an even larger usage of these archives is the adoption of standard formats (FITS being the best example of fruitful standard for images and data tables) and of friendly interfaces (currently a WWW interface is imperative).

Adoption of a standard query syntax at the level of the Web server may greatly simplify future attempts of interoperability: this is the case of the ASU protocol[1]), for instance, which allows direct queries from an information service towards an archive.

The observatory archives are generally residing close to the data producers, where the expertise remains. But there are also central repositories, such as NSSDC, especially for past space missions when PI teams are dispersed.

Space mission archives are mostly organized by wavelength ranges (see Hanisch 2001, in this conference, for the NASA organization).

2.2. Data centers

Data centers collect and distribute published catalogs and tables. They constitute an international network, involving CDS, NASA ADC, NAO Japan, INASAN Russia, India, *etc.* Exchange agreements ensure that all published catalogs are available simultaneously at any site. Catalogs can be retrieved in full, by FTP, or can be queried through database management systems, such as VizieR, provided by CDS, with mirror copies available in several places.

Note that collecting and preserving data catalogs is a long-term endeavor, involving astronomical expertise for the selection and documentation of observational data, as well as computer expertise for the data management.

2.3. Databases and Information Systems

Databases are organized collections of data relevant to astronomical objects. The value added when folding data into databases may be very significant: source diagnostic and cross-identification, data calibration and homogenization, data organization and indexes.

Among the most popular databases, one can cite SIMBAD[2] and NED[3].

User statistics are impressive: in both cases, each month, more than 5,000 different users from all over the world submit queries to these databases.

Astronomy data and information centers are becoming increasingly interconnected, with both explicit links to other relevant resources and automatic cross-links that may be invoked transparently to the end-user. More and more information services use these features to provide the user with integrated view of several archives and databases.

2.4. Bibliographic Resources and Electronic Libraries

The NASA Astrophysics Data System (ADS: Kurtz *et al.* 2000), the electronic version of all major astronomy journals, and the preprint services *astro-ph*, are

---

[1] http://cdsweb.u-strasbg.fr/doc/asu.html

[2] http://simbad.u-strasbg.fr/

[3] http://nedwww.ipac.caltech.edu/

powerful tools that altogether constitute a full complete electronic library of our discipline.

Here also a virtual network is being organized, as exemplified by the *Urania*[4] initiative, or by the coordinated efforts to create links between ADS and other services. All contributors to *Urania* share the same *bibcode* (a 19-digit code used for uniquely identifying a published article): this fact has contributed to make the move towards a full electronic library much easier.

Note that a subscription may be required for recent articles published in electronic journals.

3. Yellow-Page Services

How to help the users find their way through the jungle of information services is a question which has been raised since the early development of the WWW (see *e.g.*, Egret 1994), when it became clear that a big centralized system was not the efficient way to go.

Searching for a resource (either already visited, or unknown but expected), or browsing lists of existing services in order to discover new tools of interest implies query strategies that cannot generally be managed at the level of a single data provider.

At a basic level, there is a need for road-guides pointing to the most useful resources, or to compilations or databases where information can be found about these resources. Such guides have been made in the past, and are of very practical help for the novice as well as the trained user, for example: Egret & Albrecht (1995), Heck (1997), Grothkopf (1995), Andernach (1999).

Two on-line yellow-page services have to be cited here:

AstroWeb[5], (Jackson *et al.* 1994) is a collection of pointers to astronomically relevant information resources available on the Internet. The browse mode of AstroWeb opens a window on the efforts currently developed all over the world for making astronomically related information available on-line through the World Wide Web.

The keyword indexing of AstroWeb is very helpful for retrieving Web sites of astronomical projects or facilities.

Star*s Family is the generic name for a collection of directories, dictionaries and databases produced and maintained by Heck (1995). These very exhaustive data sets are carefully updated and validated, thus constituting a gold mine for professional, amateur astronomers, and more generally all those who are curious of space-related activities, and want to locate existing resources.

The StarPages can be queried on-line from the CDS Web site[6].

---

[4]http://www.aas.org/Urania/

[5]http://cdsweb.u-strasbg.fr/astroweb.html

[6]http://cdsweb.u-strasbg.fr/starpages.html

4. Towards an Integrated View of Existing Astronomical Resources

In the following we will focus on Internet resources that actually provide data, of any kind, as opposed to those describing or documenting an institution or a research project, without giving access to any data set or archive.

More generally, with the development of the Virtual Observatory, and with an increasing number of on-line services giving access to data or information, it is clear that tools giving coordinated access to distributed services are needed. This is, for instance, the concern expressed by NASA through the Astrobrowse project (Heikkila *et al.* 1999).

In this section we will first describe a tool for managing a "metadata" dictionary of astronomy information services (GLU); then we will show how the existence of such a metadatabase can be used for providing the end user of the Virtual Observatory with a global view of existing resources.

4.1. The CDS GLU

The CDS (Centre de Données astronomiques de Strasbourg) has developed a tool for managing remote links in a context of distributed heterogeneous services (GLU[7], Fernique *et al.* 1998).

The core of the system is the "GLU dictionary" maintained by the data providers contributing to the system, and distributed to all sites of a given domain. This dictionary contains knowledge about the participating services (URLs, syntax and semantics of input fields, descriptions, *etc.*), so that it is possible to generate automatically a correct query for submission to a remote database.

Let us give a practical example of the current GLU usage: in the CDS bibliographic service, the object names (such as 51 Peg, or M31) are tagged. An automatic program uses this tag, together with the GLU dictionary, in order to generate an anchor towards the SIMBAD database. This provides the reader with the opportunity to learn more about the object. Another computer program can use the same abstract in a different context, to generate a pointer towards the ALADIN cutout image of the object.

The GLU can reversely be queried when the challenge is to provide information about who is providing what, for a given object, region of the sky, or domain of interest. Several projects are working toward providing general solutions for such resource discovery. At this stage, they can be considered as working prototypes of a service needed for the Virtual Observatory.

4.2. Resource Discovery Tools

Astrobrowse is a project that began within the US astrophysics community, primarily within NASA data centers, for developing a user agent which significantly streamlines the process of locating astronomical data on the web. Several prototype implementations are already available[8]. With any of these prototypes, a user can already query thousands of resources without having to deal with

---

[7] http://simbad.u-strasbg.fr/glu/glu.htx

[8] http://heasarc.gsfc.nasa.gov/ab/

out-of-date URLs, or spend time figuring out how to use each resource's unique input formats. Given a user's selection of web-based astronomical databases and an object name or coordinates, Astrobrowse will send queries to all databases identified as containing potentially relevant data. It provides links to these resources and allows the user to browse results from each query. Astrobrowse does not recognize, however, when a query yields a null result, nor does it integrate query results into a common format to enable intercomparison.

A similar concept is used for the prototype tool developed at CDS under the name of *AstroGLU*[9] (Egret *et al.* 1998). The aim of this tool is to help the users find their way among several dozens (for the moment) of possible actions or services, using the GLU dictionary as a meta-database of existing resources.

4.3. Future perspectives

To go further, one needs to be able not only to locate relevant services, but also to integrate results of queries provided by heterogeneous services. This is the goal of the ISAIA (Integrated System for Archival Information Access) project[10] (Hanisch 2000, Hanisch 2001 these proceedings).

The key objective of the project is to develop an interdisciplinary data location and integration service for space sciences. This service will allow users to transparently query a large variety of distributed heterogeneous Web-based resources (catalogs, data, computational resources, bibliographic references, *etc.*) from a single interface. The service will collect responses from various resources and integrate them in a seamless fashion for display and manipulation by the user.

In a not so distant future — if the Virtual Observatory proves to be the efficient framework that we expect it to be ! — we shall expect the remote user to have from her desktop computer a complete integrated view of all existing resources related to a given astronomical object, or a given data type, or an astronomical phenomenon, *etc.*

This will imply to solve many questions, along the road, such as the proper quality control and the proper documentation of the extracted pieces of information. An interested experience, in the field of documentation, is the *Level5* project[11], initiated by Madore in complement to the NED database, which aims at providing reference documentation (review articles, lectures, reference tables, catalogs) for a given field (here: extragalactic astronomy and cosmology).

Moving from current databases and information systems to real wide-scope knowledge bases is one of the ambitious challenges of our discipline in the . The Virtual Observatory initiative will be a powerful support for these developments.

---

[9]http://simbad.u-strasbg.fr/glu/cgi-bin/astroglu.pl

[10]http://heasarc.gsfc.nasa.gov/isaia/

[11]http://nedwww.ipac.caltech.edu/level5/

5. Conclusion

The on-line "Virtual Observatory" is currently under construction with on-line archives and services potentially giving access to a huge quantity of scientific information: its services will allow astronomers to select the information of interest for their research, and to access original data, observatory archives and results published in journals. Search and discovery tools currently in development will be of vital importance to make all the observational data and information available to the widest community of users.

Some key elements for the success of the integration of data and information services into the Virtual Observatory can be summarized as follows:

- international cooperation for coordinating access to all archives and databases;
- sharing common protocols and standards;
- developing simple interoperability tools.

A final warning: the data producers will generally not accept heavy and costly constraints from any central Virtual Observatory on their own developments. In this specific context, light standards and simple tools are the best paths towards a successful Virtual Observatory.

References

Andernach, H., 1999, in *Astrophysics with Large Databases in the Internet Age*, Proc. IXth Canary Islands Winter School on Astrophysics, M. Kidger, I. Pérez-Fournon, & F. Sánchez (Eds.), Cambridge University Press, 1

Egret, D., 1994, in *ADASS III*, ASP Conference Series 61, p. 14

Egret, D., Albrecht, M. (Eds.), 1995, *Information & On-line Data in Astronomy*, Kluwer Academic Publ., Dordrecht

Egret, D., Fernique, P., Genova, F., 1998, in *ADASS VII*, ASP Conference Series 145, 416

Egret, D., Hanisch, R.J., Murtagh, F., 2000, A&AS, 143, 137

Fernique, P., Ochsenbein, F., Wenger, M., 1998, in *Astronomical Data Analysis Software and Systems VII*, ASP Conference Series 145, 466

Genova, F., Egret, D., Bienaymé, O., *et al.* 2000, A&AS, 143, 1 (CDS)

Grothkopf, U., 1995, Vistas Astron. 38, 401

Hanisch, R.J., 2000, *Computer Physics Communications*, 127, 177

Heck, A., 1995, in *Information & On-line Data in Astronomy*, D. Egret and M. A. Albrecht, Eds., 195

Heck, A., 1997, *Electronic Publishing for Physics and Astronomy*, Astrophys. Space Science 247, Kluwer, Dordrecht

Heikkila, C.W., McGlynn, T.A., White, N.E., 1999, in *ADASS VIII*, ASP Conference Series 172, 221

Jackson, R., Wells, D., Adorf, H.M., *et al.* 1994, A&AS, 108, 235

Kurtz, M., Eichhorn, G., Accomazzi, A., *et al.* 2000, A&AS, 143, 41

Virtual Observatories of the Future
ASP Conference Series, Vol. 225, 2001
R.J. Brunner, S.G. Djorgovski, and A.S. Szalay, eds.

Astro-IT Challenges and Big UK Survey Programs—SuperCOSMOS, UKIRT WFCAM, and VISTA

A. Lawrence

Institute for Astronomy, University of Edinburgh, Royal Observatory, Blackford Hill, Edinburgh EH9 3HJ, Scotland UK

Abstract. The growth of database sizes in astronomy, and the new style of large database science, present serious technical and sociological challenges. In the UK, these concerns are centered especially around exciting new survey programs, which I describe here.

1. IT Issues in Astronomy

Some trends in astronomy present serious challenges in information technology. For example, data volumes are growing alarmingly. Every observatory has an archive growing at 1–2 TB a year, several current survey projects have 10–20 TB databases (POSS and UK Schmidt digitizations, SDSS, 2MASS), and facilities now being built will generate PB databases in a few years (UKIRT WFCAM, VISTA). Storage technology (RAID arrays, tape robots, FMD disks) is keeping pace but isn't cheap. Astronomy must make sure it is properly provisioned, and very likely only specialized groups will have copies of these data.

New kinds of science require searching, or operating on, large fractions of these databases rather than just extracting tiny "finding chart" subsets (*e.g.*, rare object searches, large statistical manipulations, population analysis). However I/O speed is not subject to Moore's law and has barely improved over two decades—searching the expected VISTA database naively will take months. Some problems are speeded up significantly by using intelligent database structure, but the real solution is brute force—parallel data machines. Again the technology is coming into place (Beowulfs, SMP machines) but many implementation problems remain and these machines will be used by specialized groups who will need to offer database search and analysis services, rather than just the data access service normal today. This is a daunting task.

Unless something dramatically unexpected happens to network speeds, it will not be feasible for remote users to download a significant fraction of these databases. Once again this pushes us towards expert data centers offering a data-mining service to remote users—the motto is *"shift the results, not the data"*. Possibly the database and the attached services could be not just remote but also distributed, leading us to the "grids" idea so popular in particle physics and elsewhere. Depending on what sort of work will dominate use, it is not at all clear whether a computational grid is needed for astronomy. (For example, the relative demand on data throughput, message passing, and computation; and the ratio of peak load to average load). However what we do need is a quite analogous "service grid". Many of the same technical problems discussed in grid-land will arise—metadata standards, resource allocation algorithms, differentiated quality of service and so on.

The other big issue of course is interoperability of archives, *i.e.*, the ability to seamlessly issue joint queries to multiple distributed databases. Ideally we would like to go beyond a few federations and join hundreds of archives in a kind of club (the Virtual Observatory ideal) but of course this is very challenging, and is as much a sociological problem as a technical one.

The above points all relate to the Virtual Observatory concept, but there are other pressing Astro-IT issues. First of course, simulation theorists continue to want more and more FLOPs on proper supercomputers, which don't come cheap. Some theorists now are talking about keeping the timesteps in simulations to make massive datacubes, which can be analyzed subsequently. Such databases could challenge the observational surveys in size. The other challenging area is in fast real-time operations, in particular interferometry links, and multi-laser adaptive optics.

2. UK Survey Programs

In the UK, much of the thought so far on these issues has been driven by the excitement generated by a few key datasets, and the need to capitalize and/or prepare for them. Some of these are international projects in which UK scientists happen to play a strong role—for example Newton (XMM), SOHO and Cluster. (In particular, the UK hosts the XMM Survey Science Center, with a stated mission to enable the community to extract the serendipitous science from Newton.) However there are three optical-IR survey programs which are UK led, but which expect to produce databases of general international use and significance—SuperCOSMOS, UKIRT WFCAM and VISTA. (WFCAM and VISTA do have Japanese and ESO participations respectively.) I will briefly describe each of these optical-IR programs in turn.

2.1. SuperCOSMOS Sky Surveys (SSS)

The SuperCOSMOS plate digitization machine at ROE is several times faster than other scanners, as well as being exceptionally stable, and using smaller pixels than the DSS. It is scanning UK Schmidt and ESO plates. The first 5000 square degrees in three bands and four epochs is now on-line[1]. The southern sky will be complete in two bands by December 2000, and in three bands and four epochs by mid-2002. The H_α survey should also be on line around 2002. The database by then will be 15 TB, bigger than the SDSS. The SSS has several advantages other previous digitizations. (i) Pixels and objects (to the plate limit) are available in the same FITS file, and are simultaneously browsable using GAIA. (ii) Colors, proper motions, and variability information are available for the first time. (iii) Web users can extract uniform catalogs to their own specification up to 10 degrees across.

---

[1]http://www-wfau.roe.ac.uk/sss/

Figure 1. SuperCOSMOS scans of two regions of sky photographed fifteen years apart. The object near the center of the frame can be seen to move with respect to the other faint stars and galaxies by several arcseconds, and has a magnitude of $B = 20.6$, too faint to be in the Luyten catalogue. It is a candidate halo white dwarf. These objects are rare on the sky as they are so faint, but in fact might constitute most of the mass of the Galaxy

Our next goals are to create a seamless southern sky catalogue which goes as close to the plate limit as possible, and to make available systematic searches and analysis of the whole database. This can already be done within Edinburgh, or by personal arrangement, but of course the aim is to make a user-friendly anonymous web interface. An example of the kind of science that can be done is illustrated in Figure 1, where Nigel Hambly has searched for faint high proper motion objects, which are candidate cool white dwarfs in the Galactic halo. These turn up once every $\sim$10 Schmidt plates. Exactly how often will determine whether such objects can explain the halo dark matter (Hambly et al. in preparation).

2.2. UKIRT Wide Field Camera

WFCAM is a new IR wide-field camera being built for UKIRT, using four 2K Rockwell arrays at 90% spacing, and a novel forward-Cassegrain optical design. After 4 micro-steps and 4 macro-steps, it should produce well-sampled images over 0.75 square degrees, reaching K=18.4 in 200 seconds. It is planned to be used for half of all UKIRT time from the middle of 2002 onwards, and will produce several hundred GB of data every night. How much data is kept, and the levels of compression, are issues which are still being debated, but by 2005 it is likely that the stored database will be $\sim$100 TB, and the science database (objects and compressed pixels) will be $\sim$10 TB. The data pipeline will be run in Cambridge, and the archive established in Edinburgh.

Some WFCAM time will be open, but most of it will be used for a large public survey, operated by the UK IR Deep Sky Survey (UKIDSS) consortium

(PI: Lawrence). This will be at least a hundred times deeper than 2MASS, and thus the true counterpart of the optical sky surveys. The survey design is still under construction, but the preliminary Five Year Plan (2003–2008) has (i) An IR sky atlas over 3000 square degrees to $K = 19$ (ii) A Galactic Plane survey over 1500 square degrees to $K = 19$ (iii) A Deep Survey over 100 square degrees to $K = 21$, and (iv) A Very Deep Survey over 2 square degrees to $K = 23$.

2.3. VISTA

VISTA (Visible and Infrared Survey Telescope for Astronomy) is a new 4m wide field telescope in Chile planned by a consortium of eighteen UK universities (PI: Emerson), and currently under design at the ATC (Edinburgh). The provisional design has a two degree optical field of view with fifty CCDs and a one degree IR field of view with nine 2K Rockwell arrays. In collaboration with ESO, who will have a share of VISTA time, it will be sited on Cerro Paranal, and should start operations in 2005. By 2010 the the database will be several PBs, and even the science database will be several hundred TB, as much of the intended science includes massive monitoring programs.

Some fraction of VISTA time will be open to recurring competition, but the majority is intended to be in a suite of pre-planned legacy programs, using a mixture of areas and depths, and monitoring in selected areas on timescales of minutes, nights, months, and years. The program will not be finalized for some time yet, but a preliminary version known as the "Design Reference Program" is being used to optimize the facility design. This has been constructed by competitive debate within the consortium—members make "Key Program Proposals", these are debated and ranked by the VISTA Science Committee (VSC), and merged into a model observing program which satisfies as many of the science goals as possible simultaneously. The top-ranked science goals include dark matter mapping from weak lensing, the high redshift growth of structure, exo-planet transit searches, degenerate object populations, and the low surface brightness Universe.

3. UK Astronomical Grids Work

An opportunity has arisen to coordinate UK work related to the Virtual Observatory concept, as the recent UK government spending review has announced substantial extra funds for "e-science". Prior to the formal announcement, a consortium based around four expert data centers (Cambridge, Edinburgh, Leicester, and RAL) had been developing plans to tackle the problems spelled out in Section 1, especially as applied to the datasets of Section 2. These plans are not as all-encompassing as the NVO white paper, but more narrow and focussed. The first aims are a set of benchmarking experiments on grid tools, database technologies, and machine architectures. Next, we aim to make mini-grids which federate key databases, by collaboration with international colleagues—XMM and Chandra; Cluster and Image; SDSS and UKIDSS. These will be of immediate scientific value, but will also produce lessons for the wider international agenda. Finally we will undertake R&D programs on data mining algorithms and browsing technology. The funding for this project should clarify shortly.

Virtual Observatories of the Future
ASP Conference Series, Vol. 225, 2001
R.J. Brunner, S.G. Djorgovski, and A.S. Szalay, eds.

Solar System Surveys

Steven H. Pravdo

Jet Propulsion Laboratory, California Institute of Technology, Pasadena, CA 91109

Abstract. Astronomical data is a necessary starting point for a virtual observatory. Solar system surveys provide extensive sky coverage and vast amounts of celestial images. We list and briefly describe the ongoing NEO programs. One of these, the Near-Earth Asteroid Tracking (NEAT) program, archives its data, making it available for the science community. A small investment now can ensure that these data are available in the future, not only for the uses described below, but also for uses not yet envisioned.

1. Introduction

Astronomical data is a necessary starting point for a virtual observatory. Solar system surveys sponsored by NASA's Near-Earth Objects (NEOs) program provide extensive sky coverage and vast amounts of celestial images. Their primary goal is the discovery and characterization of NEOs—asteroids and comets, but the data are applicable to many astrophysical studies. Below, we list and briefly describe the ongoing NEO programs. We also describe how the Near-Earth Asteroid Tracking (NEAT) program archives its data, making it available for the science community through the SkyMorph project, part of NASA's Applied Information Systems Research program (AISRP).

2. The Surveys

NASA intends to discover 90% of the Near-Earth asteroids larger than 1 km by 2010. At least 50% of these objects remain to be discovered (Rabinowitz et al. 2000). To fulfill this goal a number of observing groups use meter-class telescopes with electronic CCD cameras to perform large-scale surveys of the sky at least 18 nights each month. The surveys cover many 1000s of square degrees of sky each month. The limiting magnitudes are in the range $V = 17$–20. Thus millions of objects are observed. The projects are listed in Table 1.

Table 1 also shows that the total image data obtained nightly is about 100 GB, but will rise to about 200 GB when the second NEAT telescope comes on line.

NEAT (Pravdo et al. 1999) is the only project that places its image data in an archive that is publicly-accessible. A separate AISRP project called "SkyMorph" enables the NEAT archive. NEO operational funds are too tight to support this activity. The other projects, except for LINEAR, save their data,

Project	Telescope (s)	Nightly Data (GB)	Archive/Access
Catalina (UA)	0.4, 0.7-m	8	None
LINEAR (MIT/LL)	1.0-m (2)	70	None
LONEOS (Lowell)	0.6-m	5	None
NEAT (NASA/JPL)	1.2-m (2)	25-100	Both
Spacewatch (UA)	0.9, 1.8-m	13	None

Table 1. A listing of the various Near-Earth asteroid surveys currently in progress.

but only on tapes that are currently inaccessible. They are moving toward archiving their data but it appears that, like NEAT, an alternate funding source is required.

3. SkyMorph

SkyMorph is an example of how a large and ongoing astronomical data set can be ingested into an archive in near-real time. This is a desirable, perhaps necessary, attribute for a virtual observatory. The existing SkyMorph archive is an asset for the astronomy community, used in purposes as diverse as discovering supernovae and refining asteroid orbits. These data cover the dimension of time in a way that is unique among large optical databases: most of the sky is re-observed with the same instrument on time scales from hours to months to years.

Up to 25 GB/night of data is transmitted from the AFRL 1.2-m telescope site at Maui to JPL. These data are then processed within days and placed in the archive. Figure 1 shows the system architecture. The archive can be accessed at the worldwide web address: [1]

Like a virtual telescope, it gives users the ability to observe the sky, either by time and/or position, or through its Moving Target Detection utility. The latter "follows" a moving object through the sky and gives views of it from the NEAT database or from other currently implemented databases. Other databases include the Digital Sky Surveys and Hubble Space Telescope, with more to come.

SkyMorph also contains an object database. Users can select objects that have been viewed multiple times and, for example, construct light curves. Further utilities are planned for the object database to allow correlations with interesting sources in other wavelength regimes.

4. SkyMorph Data and Uses

Figure 2 shows the observation centers for data in the SkyMorph archive. Much of the sky is covered multiple times with notable exceptions at low Galactic latitudes and the southern sky.

---

[1] http://skys.gsfc.nasa.gov/skymorph/skymorph.html

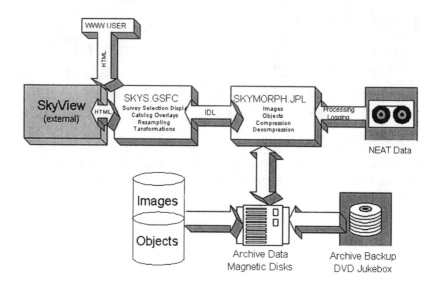

Figure 1. The system architecture for the SkyMorph tool.

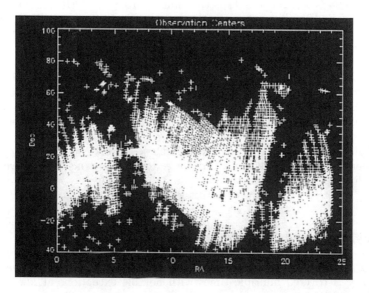

Figure 2. The observation centers for data in the SkyMorph archive.

 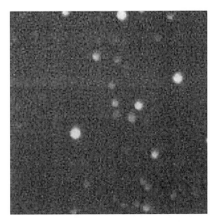

Figure 3. NEAT images of Comet Hale-Bopp (left) and the Near-Earth asteroid Toutatis (right).

The underlying purpose for the NEAT data is to discover NEOs. Thus SkyMorph shows many examples of these objects. Figure 3 shows the comet Hale-Bopp and the Near-Earth asteroid Toutatis. Toutatis is shown as an overlay of three NEAT images taken at different times, and assigned red, green, and blue colors. The three images are aligned so that non-moving star images add to a white result, while the moving asteroid images appear in red, green, and blue.

Figure 4 is an example of how outsiders used SkyMorph to add to our scientific knowledge. These images show the asteroid 1998 MQ moving relative to the fixed stars. A European user successfully searched for pre-discoveries of this asteroid and latter submitted his findings to the Minor Planet Center for inclusion in its databases.

The Near-Earth asteroid 1999AN10 caused a stir soon after its discovery in early 1999. Its poorly known orbit left open the possibility of an encounter with the Earth in 30 years. Fortunately, a pre-discovery of this object on the Palomar plates refined the orbit, eliminating the threat. Figure 5 shows this pre-discovery image obtained using the Moving Target Detection utility of SkyMorph.

Moving further out in distance, Figure 6 shows how astrometric solutions for SkyMorph objects can detect proper motion. GJ 1146 is a nearby star located about 20 parsecs from Earth. The positions extracted from the database describe the proper motion of the star, on order 1 arcsecond per year.

Leaving our galaxy, Figure 7 illustrates why the Supernova Cosmology Project (SCP) is using NEAT data to discover supernovae in nearby galaxies. During two nights of observations in February 1999, the SCP discovered 5 supernovae including the one shown in the figure, SN 1999am. In the next few years the SCP is expecting to discover hundreds more supernovae in a similar fashion.

Finally, the distant quasar 3C 273 was serendipitously observed many times with NEAT. Figure 8 shows the light curve of this object and illustrates the

122   Pravdo

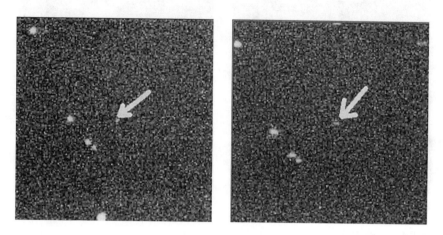

Figure 4. An illustration of the detection process using SkyMorph for the asteroid 1998 MQ.

Figure 5. The pre-discovery image for Near-Earth asteroid 1999AN10.

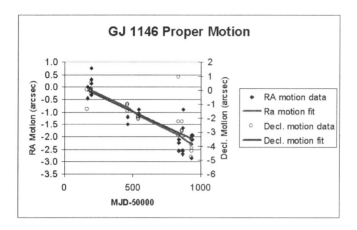

Figure 6. Proper motion determination using NEAT data for GJ 1146.

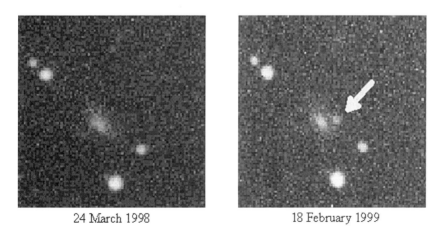

Figure 7. The detection of a distant Supernova using the NEAT data.

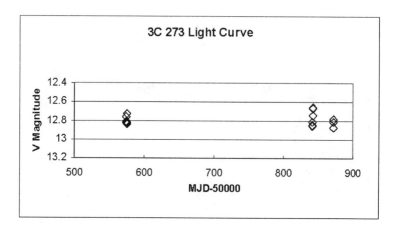

Figure 8. The light curve of the quasar 3C 273 determined using NEAT data.

capability of NEAT to provide long observing baselines for many interesting sources.

5. Conclusions and Plans

Solar system surveys, at least at the current rate, are planned to continue for 10 years. This enormous database is a significant resource for any national or international virtual observatory. Plans should be made for its inclusion. The NEAT data via SkyMorph is expected to increase by a factor of between 3–9 in sky coverage, yielding more than a Terabyte per week. A small investment now can ensure that these data are available in the future, not only for the uses described above, but also for uses not yet envisioned.

Acknowledgments. The research described in this paper was performed in part by the Jet Propulsion Laboratory, California Institute of Technology, under contract with the National Aeronautics and Space Administration.

References

Pravdo, S., Rabinowitz, D., Helin, E., Lawrence, K., Bambery, R., Clark, C., Groom, S., Levin, S., Lorre, J., Shaklan, S., Kervin, P., Africano, J., Sydney, P., & Soohoo, V. 1999, AJ, 117, 1616

Rabinowitz, D., Helin, E., Lawrence, K., & Pravdo, S. 2000, Nature, 403, 165

SkyView: **Experiences Building a Virtual Telescope**

Thomas A. McGlynn

NASA/Goddard Space Flight Center (University Space Research Association), Greenbelt, MD 20771

Laura M. McDonald

NASA/Goddard Space Flight Center (Raytheon), Greenbelt, MD 20771

Nicholas E. White

NASA/Goddard Space Flight Center, Greenbelt, MD 20771

Abstract. During the past decade the *SkyView* virtual telescope has provided a single unified interface with which astronomers — and the general public — can access data from all wavelengths. *SkyView* has addressed many of the issues that face the National Virtual Observatory (NVO) effort: supporting science that spans wavelength regimes, handling very heterogeneous data sets in a consistent fashion, and catering to a very diverse set of users. In this paper we discuss some of the lessons we have learned building and running *SkyView*.

1. What is *SkyView*?

The *SkyView* virtual observatory (McGlynn, Scollick & White, 1997) is a on-line service which allows users to download images quickly and conveniently from survey data at all wavelength regimes. The datasets currently available within *SkyView* include the Digitized Sky Survey (Lasker, *et al.* 1990), VLA FIRST (Becker, *et al.* 1997) and NVSS (Condon, *et al.* 1998) surveys, the 2MASS infrared survey (Skrutskie *et al.* 1995), the ROSAT (Snowden *et al.* 1995) and EGRET (Hartman *et al.* 1999) all-sky X-ray and gamma-ray surveys and dozens of other surveys in all regimes.

Users can retrieve an image from any of these surveys within seconds by selecting the survey and specifying a position using one of *SkyView*'s simple Web forms. While *SkyView* is an archive, it represents a new generation of archives. It does not simply provide data as entered into the archive. Rather *SkyView* generates an image in the user-specified coordinate system, orientation and scaling, automatically mosaics subimages together, and generally addresses all 'geometric' issues so that users can immediately begin to do astronomy.

2. Current Status

SkyView uses the same interface for all wavelength regimes, and for resolutions ranging from a degree or more to an arc-second. If the data has a third, typically energy, dimension, *SkyView* allows the user to average over some range in the third dimension, or to resample data there. The intent of *SkyView* is to provide data that the user can immediately begin to work with. Thus *SkyView* does provide sophisticated image analysis, object detection, photometry, ... tools. Many other packages already exist for this. Some basic smoothing, contouring, color table and image comparison capabilities are included to allow a quick check of the data before the user extracts the data into their own environment.

SkyView can plot overlays of catalog tables from other sites on top of images. About 200 different tables, ranging from the Messier objects through the USNO A-2 catalog are available, most through the HEASARC Web services.

Currently *SkyView* supports over fifty surveys. Users generate over one thousand images each day. *SkyView* uses a local archive for most of the surveys — about 100 GB of information — but also accesses many of the larger surveys over the Web: FIRST from NRAO, 2MASS and NEAT data from IPAC, and APS catalog data from the University of Minnesota. About 2.5 TB of information can be queried through the system.

3. The User Community

Currently *SkyView* generates about 500,000 images for users per year. These users of *SkyView* are diverse. It is difficult to get precise statistics, but using Web logs and information about where users get images we estimate that about 30% of our users are non-astronomers who use the system as interested amateurs, or simply to browse the universe. This was not intended — *SkyView* was planned as a tool for researchers — but was probably inevitable. Since we built *SkyView* to be easy to use for astronomers — with no registration, simple Web interfaces, and immediate results — it was well within the grasp of the interested public. If the NVO is successful in providing clean and clear interfaces to distributed astronomy information it will also become an peerless interface for public outreach. For the first time we shall be able to involve the public directly in our research environment.

Figure 1 shows the central locations for most of the images that users have downloaded from our system. The Galactic plane shows up clearly, but there are also many high latitude sources. Some complex patterns not associated with any source are visible, but we have not attempted to analyze these. The single most popular point is at the position of Sirius which shows up poorly in most surveys. It is either too bright or too faint to be an interesting source in most wavelengths. However it was used as our example for how to enter positions and thus illustrates the need to choose simple exemplars carefully.

The number of requests made by each user is extremely variable. Again we can only estimate this qualitatively, but it appears that there is something like a Poisson distribution of requests, where *SkyView* spends the same total effort on users and projects that request 1 image as 10 images or 100 images.

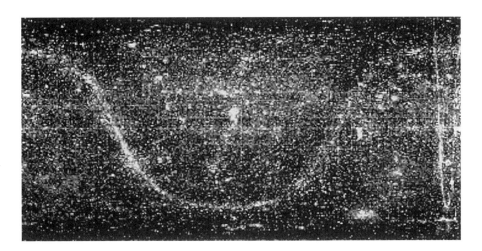

Figure 1.   Central positions of 1.3 million *SkyView* generated images

If the NVO shows a similar behavior then it behooves us to realize that small scale projects will require comparable resources to the few 'grand-challenge' topics that are often discussed with the NVO. Also we need to understand that the NVO will be ubiquitous: It may be that the NVO will be used in the majority of research projects in astronomy — but also that in most of these the NVO will not be the key source of information but a valuable supplier of ancillary and support information. All observatories — from Tycho Brahe's to HST — spend much of their resources in incremental science, and this should be recognized and embraced for the NVO.

4.   Standards

SkyView's ability to export data to other astronomical analysis systems depends upon a fortuitous and fortunate coincidence. At the same time that our system was being developed, the FITS community was developing a clear and unambiguous standard for how geometric information in an astronomical image should be represented. This process has continued throughout the past decade and will soon culminate in the adoption of this World Coordinate System standard (Greisen and Calabretta 1995) by the International Astronomical Union in 2000. Thus FITS provided a concise description of the very aspect of data that *SkyView* wished to address.

The NVO cannot depend upon such fortune. It must be a leader in developing FITS and other data standards for the astronomy community. Only with agreed and used standards will the NVO be able to integrate the archival and software resources of astronomy.

5. Changing Technologies

SkyView has been one of the most successful of tasks initiated in NASA's ADP program. It is now relatively mature in the context of the rapid changes in information technology — there have been three major technology revolutions that have occurred since it first went on line. The coming of the Worldwide-Web greatly enhanced its reach, Java technologies have so far largely passed it by, though we have used Java to provide more interactivity than standard Web protocols admit. Most recently we have embraced the latest wave of changes — the drive towards strongly distributed computing, intelligent servers and customizable interfaces. The next generation *SkyView* system will allow users to add their own resources, be they catalogs, surveys or even analysis tools, into *SkyView* or to use elements of *SkyView* transparently within their own protocol.

One of the reasons that *SkyView* has been able to take advantage of new technologies is that is has had clearly defined—and limited—goals from its inception. As new technologies have become available it is relatively easy to assess their usefulness. If we look at other successful projects which span the discipline domains in astronomy, *e.g.*, NED or the ADS abstract service, it seems that a clear sense of the ambit of the project is a great advantage. Projects which attempted to unify our disciplines too broadly, *e.g.*, the original ADS, have fared less well.

The NVO is and should be a vastly more ambitious project that *SkyView* — and consequently has more loosely defined goals. However our experience implies that we should be careful not to attempt too much with the NVO; that the limits we place on the NVO are as crucial to its success as its goals.

6. *SkyView* in the NVO World

Calling *SkyView* a virtual telescope is apropos the analogy that is suggested by the concept of the National Virtual Observatory. This observatory looks at the universe of data stored in archives throughout the astronomical community. Its telescopes — like *SkyView* — find the datasets appropriate to a given investigation. Its instruments are the sophisticated software systems that correlate and integrate distinct data sets. The invariant is the underlying needs and wisdom of astronomers and astrophysicists who use physical and virtual instrumentalities to further their research goals.

Comprehending the universe of the NVO will be a daunting task. No doubt there will be the equivalent of today's Web search engines, but astronomers will need tools which enable them to view data in ways comparable to the paradigms they are already comfortable with. *SkyView* will be one such tool. It has never been intended as the task that extracts the last iota of science from some some data set. Rather it provides a simple tool with which astronomers can access and visualize very heterogeneous data.

As more and higher quality survey data comes on-line, *SkyView* and services like it will become increasingly important to enable NVO users to browse — to play with — data. While we can hope for ever increasing quality of metadata, it seems unlikely that research projects will be able to use data without some human contact, visualization, quality control. The role of *SkyView* then may be that of the finder scope for the NVO 'main telescopes'. Before users unleash the full power of the NVO's resources on a problem, *SkyView* will help them understand the scope and feasibility of their approach.

Acknowledgments. The *SkyView* project has been funded through NASA's ADP and AISRP programs and is hosted at the High Energy Astrophysics Science Archive Research Center at NASA's Goddard Space Flight Center.

References

Becker, R.H., *et al.* 1997, ApJ, 450, 559

Condon, J.J., *et al.* 1998, AJ, 115, 1693

Greisen, E.W., & Calabretta, M., 1995, in ASP Conf. Proc. 77, (New York: AIP), 233

Hartman, R., *et al.* 1999, ApJS, 123, 79

Lasker, B., *et al.* 1990, AJ, 99, 2019

McGlynn, T.A., Scollick, K.A. & White, N.E., 1998, in IAU Symposium 179, New Horizons from Multi-Wavelength Sky Surveys, ed. B.J. McLean, *et al.* (Kluwer), 465

Skrutskie, M.F., *et al.* 1995, A.A.S., 187, #75.01

Snowden, S.L., *et al.* 1995, ApJ, 454, 643

ISAIA: Interoperable Systems for Archival Information Access

R. J. Hanisch

Space Telescope Science Institute, 3700 San Martin Drive, Baltimore, Maryland 21218

Abstract. We describe initial conceptual design and prototype implementation efforts on the ISAIA project. ISAIA is an interoperability layer (middleware) whose goal is to provide access to distributed space science data resources via common query protocols and metadata standards. The Virtual Observatory will require a mechanism like ISAIA for locating data services relevant to a particular query, for data subset retrieval, and for data integration.

1. Background

The Virtual Observatory will require an interoperability layer to provide transparent access to distributed archives and information services. This layer operates primarily with metadata, translating user queries into a generic transport protocol that can be received by a data service and mapped onto site or service-specific queries. Similarly, query responses are remapped into the transport protocol and returned to an integrator, a program that receives query responses from distributed services and presents them in a coherent, consistent manner to the end-user.

With support from NASA's Applied Information Systems Research Program (AISRP), we have begun design work on such an interoperability service layer in a project called *ISAIA*: Interoperable Systems for Archival Information Access.

A key requirement for a middleware layer such as ISAIA is that it not impose constraints on how information providers manage their data internally. Moreover, if the ISAIA approach is to be successful the threshold for participation must be kept low. It must be simple—a matter of hours—to implement ISAIA-compliant data dictionaries or interfaces. Finally, middleware layer must retain the identity of the participating services so that users understand the provenance of the data they are using and know where to find assistance in using the data.

For additional background on distributed data systems and about ISAIA and its precursor, Astrobrowse, see Hanisch (2000 a, b) and Heikkila (1999).

2. Metadata Standards and Profiles

The key to interoperability—to locating data of interest in distributed archives and information services and then being able to retrieve and integrate that data—is having standard metadata. Metadata is the information that describes the data, the most common example of which in astronomy is the information in a FITS file header. However, the FITS data format standard primarily defines the syntax rules for header information and the associated data structures. There is considerable room for creativity in the definitions of header keywords, such that many similar or identical quantities are described by any number of different terms. Moreover, extracted object catalogs are of critical interest to the virtual observatory, but here again the same measured quantities (a visual magnitude, for example) can have many different labels (Ortiz 2000).

In ISAIA we use the concept of a *profile* to implement metadata standards and to provide a mechanism for the exchange of metadata between data and information services and the integrator. The profile defines standard metadata labels for the data descriptors needed for locating and retrieving information. ISAIA profiles have three components: the resource profile, the query profile, and the returned information profile. The resource profile includes the metadata describing the data and information resources themselves. It characterizes the each service's data holdings and is used to determine which of the hundreds of archives and catalogs should actually be sent a query for certain types of data. For example, if a user is looking for X-ray data sets the resource profile would determine that STScI/MAST would not have data holdings of interest and the query would not be sent to that service. An example showing the elements of a resource profile is shown in Figure 1.

The query and returned information profiles characterize the actual archival data sets and catalogs and form the basis for sending data selection queries to distributed services, and then collecting the results for uniform presentation. To take an obvious example, right ascension and declination are fundamental pieces of metadata describing an astronomical object or observation. However, we know that different astronomical catalogs use different labels, different formats, and different epochs for storing this information. One catalog might use labels "RA" and "DEC", another "right ascen" and "declin", and yet another "CRVAL1" and "CRVAL2" (following the FITS world coordinate system model). The values may be given in hexagesimal format (with variants), decimal degrees, or radians. The query and returned information profiles accommodate this by defining a standard term for right ascension and a format for exchange, say "RA" and decimal degrees.

3. Profile Implementation and Exchange Protocols

The profiles themselves are insufficient for the meaningful exchange of information. There are two approaches to implementing the profiles and the necessary translations from the standard metadata to site and service specific terms and formats. The translations can be implemented for each service, by each service, through a thin interface layer. That is, a data service receives ISAIA-compliant queries, maps the standard metadata into its local terminology, responds to the

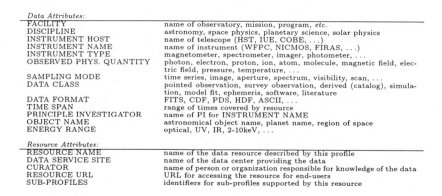

Figure 1. Sample resource profile for space sciences. Terms in the profile at this level are intended to pertain to space science data sets in general. Additional terms are introduced in subprofiles for specific disciplines, allowing users both broad and deep access to space science resources. *Data attributes* are fields which, when a query is forwarded to a data provider, can result in a selection or filtering of data from the provider. *Resource attributes* are fields that describe the resource but which do not result in selection or filtering at the provider's site.

query, and then constructs an ISAIA-compliant response in which the metadata is mapped back into the standard label and format. An alternative approach is to have each site define its mappings of the profile via a data dictionary, and to aggregate the data dictionaries for use by the ISAIA query/response server. The query/response server sends queries to each service preformatted for that service, having translated a generic query into a site-specific one. The server listens for responses, and when they are received translates them back into the ISAIA-compliant terms and formats. The latter approach can be implemented in a straightforward fashion through use of the CDS GLU^1 (Générateur des Liens Uniformes, Fernique et al. 1998). This puts a minimal burden on participating sites: all they need to do is maintain a simple data dictionary. The former approach requires a greater level of effort on behalf of participating sites, but also allows for much higher levels of functionality (*e.g.*, server-side preprocessing of data).

ISAIA also has the concept of hierarchical profiles. A profile can be defined at a very general level, describing, for example, the primary characteristics of astronomical images. Subprofiles could be developed for different classes of images having unique characteristics (CCD images *vs.* radio interferometry data cubes, for example).

In ISAIA the profile definitions will most likely be expressed in XML, and queries and responses can be exchanged via HTTP. This approach has been prototyped by one member of our team for the Planetary Data System and

---

[1] http://simbad.u-strasbg.fr/glu/glu.htx

Palomar Testbed Interferometer (Hughes *et al.*, these proceedings). Data service providers are increasingly using Java for user interfaces given that it has much greater capabilities than HTTP/CGI. Assuming that this trend continues, ISAIA will also need to be able to access distributed services through JDBC or other mechanisms.

4. Lessons Learned

Our original concept for ISAIA was to provide an interoperability layer that would span all of space science—astrophysics, planetary science, and space physics—as having tighter linkages and access mechanisms among these disciplines has long been a goal of NASA's Office of Space Science. In our year of design work, however, we have found it challenging to identify metadata elements that are common to all disciplines *and* adequate discriminators or selectors for specific data sets. For example, almost all astrophysics data searches are based on the position of an object on the sky: the right ascension and declination. Almost all space physics data searches are based on the time of an event and the location of the event relative to the Sun or Earth (magnetotail, corona, mesosphere, *etc.*). Planetary data is most strongly categorized by the name of the object, location in each body's latitude and longitude coordinate system, and by time. Moreover, with a few exceptions there are relatively few scientifically compelling problems that require such a tightly integrated space science data system. Nevertheless, the basic approach to the interoperability layer that has been explored in the ISAIA project remains important for interlinking astrophysics resources and enabling the virtual observatory. The same technology can be used to integrate information services within the other space science disciplines, and could be used to build integrated services as merited by particular interdisciplinary research projects.

We have also recognized that building a distributed system with highly distributed effort is also difficult. The more distributed the effort, the more time must be spent on management oversight and communications. Work must be segmented into independent packages to the extent possible. Staff need to be focused on the task; 10% levels of effort are too easily diluted and distracted. It is critical to constrain the project scope to first meet the primary requirements and to keep the requirements driven by the science goals, not by applying technology for its own sake.

Acknowledgments. The ISAIA project team includes Tom McGlynn and Nick White (NASA/GSFC HEASARC), Joe King (NASA/GSFC NSSDC), Cynthia Cheung (NASA/GSFC ADC), Ray Plante and Bob McGrath (NCSA), Joe Mazzarella (Caltech/IPAC), Arnold Rots (SAO/CXC), Steve Hughes (JPL), Mike A'Hearn (UMd), Reeta Beebe (NMSU), Françoise Genova (CDS), and Paolo Giommi (BSDC). This work has been supported by NASA's Applied Information Research Program under grants to the Space Telescope Science Institute, the National Center for Supercomputing Applications/University of Illinois, and the Goddard Space Flight Center.

References

Fernique, P., Ochsenbein, F., & Wenger, M. 1998, in ASP Conference Series, 145, Astronomical Data Analysis Software and Systems VII, R. Albrecht, R. N. Hook, & H. A. Bushouse, eds., 466–469

Hanisch, R. J. 2000a, Computer Physics Communications, 127, 177

Hanisch, R. J. 2000b, in ASP Conference Series, Astronomical Data Analysis Software and Systems IX, D. Crabtree, N. Manset, and C. Veillet, eds., in press

Heikkila, C. W., McGlynn, T. A., & White, N. E. 1999, in ASP Conference Series, 172, Astronomical Data Analysis Software and Systems VIII, ed. David M. Mehringer, Raymond L. Plante, & Douglas A. Roberts (San Francisco: ASP), 221

Ortiz, P. F. 2000, Computer Physics Communications, 127, 188

The Digital Sky Project: Prototyping Virtual Observatory Technologies

R.J. Brunner

Department of Astronomy, California Institute of Technology, Pasadena, CA, 91125

T. Prince

Department of Physics, California Institute of Technology, Pasadena, CA, 91125

J. Good, T. H. Handley, C. Lonsdale

IPAC, California Institute of Technology, Pasadena, CA, 91125

S.G. Djorgovski

Department of Astronomy, California Institute of Technology, Pasadena, CA, 91125

Abstract. Astronomy is entering a new era as multiple, large area, digital sky surveys are in production. The resulting datasets are truly remarkable in their own right; however, a revolutionary step arises in the aggregation of complimentary multi-wavelength surveys (*i.e.*, the cross-identification of a billion sources). The federation of these large datasets is already underway, and is producing a major paradigm shift as Astronomy has suddenly become an immensely data-rich field. This new paradigm will enable *quantitatively and qualitatively new science*, from statistical studies of our Galaxy and the large-scale structure in the universe, to discoveries of rare, unusual, or even completely new types of astronomical objects and phenomena. Federating and then exploring these large datasets, however, is an extremely challenging task. The Digital Sky project was initiated with this task in mind and is working to develop the techniques and technologies necessary to solve the problems inherent in federating these large databases, as well as the mining of the resultant aggregate data.

1. Introduction

The Digital Sky project is an NPACI (National Partnership for Advanced Computing Infrastructure, an NSF Computer Science initiative) sponsored program to study the role of advanced computational systems (both processor and network oriented) in the distribution of data from multiple large area, digital sky surveys. This project was originally conceived by Tom Prince, the principal investigator for the project, in 1996 as way to leverage high performance computing to tackle some of the incipient problems involved in federating and mining the large amounts of Astronomical information that were beginning to become available. The project was initially funded in 1997, and has since grown to include a large number of participants from Caltech Astronomy, the Caltech Center for Advanced Computing Research, the Infrared Processing and Analysis Center, the Jet Propulsion Laboratory, and the San Diego Center for Supercomputing.

One of the fundamental tenets of the Digital Sky project is the requirement not to develop another analysis tool (which is the last thing needed by the Astronomical community). Instead, we conceived that the project would be able to optimally achieve its goals by serving as a technology demonstrator. Initially, we focused on identifying the set of requirements to create a prototype virtual observatory (*i.e.*, the "Digital Sky"). Afterwards, we researched the core problem of astronomical data federation. Currently, we are exploring the concept of image data mining.

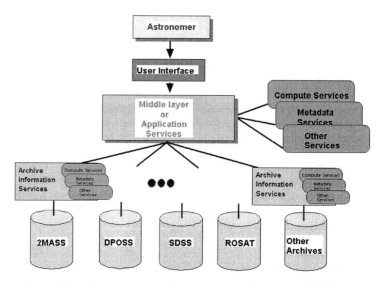

Figure 1. A top-down overview of the architecture for a virtual observatory. The ability to view either a seamless view of the sky or the detailed capabilities of any individual archive is the overriding requirement.

Overall, our approach is designed to employ a minimalist design, that focuses more on integrating existing components in cooperation with community experts (see Figure 1 for an overview). Any knowledge or applications developed in the course of this project have been disseminated to the appropriate experts within the community who are in interested in identifying and attempting to solve the complex problems that arise in federating disparate, large area, digital sky surveys.

2. Towards a Virtual Observatory

The first major task we tackled was the identification of the types of issues which needed to be successfully addressed in order to seamlessly federate highly distributed datasets in order to facilitate knowledge discovery (see also, Szalay & Brunner 1998). This highly desirable end-result is now more commonly known as a virtual observatory. Formally, we split our list of identified issues into two categories: basic and advanced, based on the difficulty of successfully implementing the appropriate service.

The basic services listed below often are provided as part of a commercial database system, which most of the major archive centers already utilize.

Catalog Search Engines. In order to be fully operable within a virtual observatory framework, a dataset must be able to support, at a minimum, basic query functionality, such as spatial range queries.

System Metadata. In order to develop general purpose tools, as well as simplify the learning process, archives need to implement a standardized format for describing both the data contained within an archive and the services which the archive can perform.

Relationship Generators. A virtual observatory must be able to support the cross-identification of billions of sources in both a static and dynamic state over thousands of square degrees in a multi-wavelength domain (Radio to X-Ray).

Image Metadata Search Engines. Image data should be able to be selected based on the actual metadata of the images, for example, spatial location, observational date, *etc.*

Image Archive Access. Archived image data should be accessible to an end user, even if it is merely served from an FTP site.

Query Optimizations. Within a virtual observatory, a query could be distributed to multiple archives. As a result certain optimizations can be performed depending on the status of the underlying network topology (*e.g.*, network weather) in order to balance the resulting server load. Optimistically, a learning mechanism can be applied to analyze queries, and using the accumulated knowledge gained from past observations (*i.e.*, artificial intelligence), queries can be rearranged in order to provide further performance enhancements.

The following set of services are more advanced, requiring extra effort, beyond the more common basic services, to be fully implemented. These services often require post-processing of extracted data in order to be performed, and can, therefore, make use of specialized hardware.

Computational & Data Grid. Numerous, complex queries will swamp traditional archive topologies. Instead, a virtual observatory should be implemented to minimize network traffic by utilizing advanced computational systems to perform complex analysis (*e.g.*, correlation analysis) on the server side as opposed to the client. This solution can efficiently capitalize on parallel I/O and computing resources, as well as replicated and persistent datasets, in order to simplify the implementation of the following advanced services.

Image Processing. Often, a user will want to post-process an image data request in order to obtain a scientifically useful result. In order to implement this feature, basic image processing operations need to be available on the image server, or close to it, to perform, at a minimum, basic image operations such as mosaicing, sub-setting and registration.

Statistical Analysis. In some cases, a user will be more interested in a particular statistical analysis of a query result than the actual resulting dataset itself (*e.g.*, a histogram, statistical measure, or cluster finding code). In such cases, a user of a virtual observatory should be able to filter the query result using either an available toolkit or else custom developed statistical codes as necessary.

Visualization. Undoubtedly, a major component of a virtual observatory is the ability to visualize data, which might stem from simple graphical representations of catalog data (as part of a statistical analysis), seamless serving of image data, or virtual explorations of parameter space. In certain scenarios, such as defining a new aggregate class of objects (*e.g.*, clusters) or to aid in the mining of the aggregate data, this process is integrally linked to the actual analysis which is the desired end-goal.

Machine Learning. When exploring the forthcoming datasets, especially after they have been federated, traditional techniques will quickly be swamped and rendered hopelessly antiquated. The only efficient technique to explore the vast, newly opened portions of parameter space is to capitalize on the inherent capabilities of the very compute resources which are facilitating the construction of the virtual observatory. These capabilities give rise to the adoption of algorithms which let the computer mine for the priceless nuggets in the mountains of data, and include techniques which can be either supervised (*e.g.*, find everything similar to this particular object) or unsupervised (*e.g.*, find interesting things).

The overriding design principle that we have advocated is to encapsulate the archival services (both in design and implementation) which will simplify the effort required to provide interoperability between different archives (*i.e.*, a plug-n-play model). This approach allows for future growth by providing a blueprint for new archives to follow and thereby capitalize on existing infrastructures via the adoption of community standards. This common service approach, reduces the overall cost to the community by providing a standardized code base, and facilitates interoperability for analysis tool providers.

In order to be successful, a virtual observatory must be able to grow through the incorporation of new surveys and datasets in addition to its original tenants. This requirement necessitates the adoption of archival standards for exchanging not only the actual data, but also both metadata and metaservices between constituent archives. This will allow for different analysis tools to be able to seamlessly work with data extracted from different archives. All of this work will culminate in the creation of a National (and eventually Global) Virtual Observatory, which will eventually enable and empower scientists and students anywhere to do important, cutting-edge research.

3. Relationship Generators

Before any advanced data exploration or mining tools can be employed, however, the data of interest must be federated. Indeed, this data federation service is one of the primary requirements for the National Virtual Observatory (NVO). Federating these different datasets, however, is a challenging task. As this project's initial focus, we researched solutions to the problems inherent in the dynamic, multi-wavelength cross-identification of large numbers of Astronomical sources.

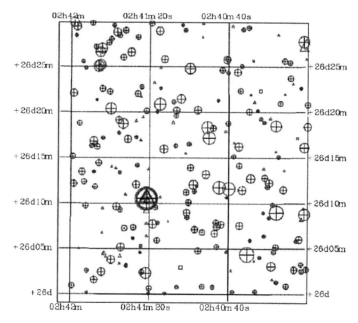

Figure 2. A cross-identification example between 2MASS and DPOSS. In order to provide an uncluttered visual image, both datasets were reduced in size by imposing a magnitude cut. This image demonstrates many of the common problems in multiple-wavelength data federation, namely high source densities, non-detections, and source splitting and deblending.

Specifically, we identified the following topics, which we feel must be addressed in order to develop a robust cross-identification service. First, in order to handle the large amounts of dynamic, multi-wavelength information, including custom user datasets, a data federation framework must utilize the forthcoming national computational grid to optimally perform the necessary calculations. These calculations can be quite complex, particularly in the case of multi-wavelength data which has varying positional accuracies (i.e., beam widths) or involves different physical phenomena. As a result, this framework must be able to incorporate not only user defined probabilistic associations within the federation algorithms, but also multiple associations (i.e., binaries and clusters) and previously published results during the cross-identification process. Finally, the newly generated data must be able to remain persistent so that data mining algorithms can be applied to the (potentially computationally expensive) result.

As an initial technology demonstration, we developed a custom data federation service which utilized an optimal data-chunking algorithm allowing it to easily scale with the size of the resulting datasets (see Figure 2 for a demonstration). The majority of these tests focused on data from the 2MASS (2 Micron All Sky Survey, cf. Skrutskie et al.) and DPOSS (Digitized Palomar Observatory Sky Survey, Djorgovski et al. 1998) projects. As a result of its efficacy, this software was incorporated by the 2MASS survey into its quality assurance pipeline as well as the Infrared Science Archive (IRSA) as the core of its data federation service. We also developed probabilistic association techniques, building on some of the pioneering work by Lonsdale et al. (1998) in order to federate the ROSAT bright source catalog (which has relatively large spatial uncertainties) with available optical datasets (e.g., Rutledge et al. 2000). This work is now being extended to provide an unbiased quantification of quasars and their environments (Brunner et al. 2001, in preparation).

4. Image Mining Operations

With the success of the ongoing cross-identification work, the project has now refocused on the challenges of exploring image (or pixel) parameter space. Ideally, a virtual observatory should be able to generate seamless views of the universe from any available image dataset. This process, however, is complicated by various observational artifacts (see Figure 3 for a demonstration). In order to simplify the development, we have initially separated the overall task into the generation of scientifically calibrated datasets from the generation of visually clean images (which are ideally suited for public outreach).

As a demonstration of the former, we have mosaiced multiple DPOSS plates in an effort to explore diffuse emission over very large angular scales (see, e.g., Mahabal et al. and Jacob et al. this volume). The latter project has developed into an astronomical equivalent of the popular teraserver project, which provides the ability to pan and zoom around space-based images of Earth. This new project, aptly named virtualsky, is accessible online at *http://www.virtualsky.org/* (Williams, R 2000, private communication), allows a user to pan and zoom around various astronomical datasets in an analogous fashion.

Acknowledgments. We wish to thank the members of the Digital Sky project, which is funded through the NPACI program (NSF Cooperative Agreement ACI-96-19020), and the many Pasadena area virtual observatory enthusiasts for stimulating discussions.

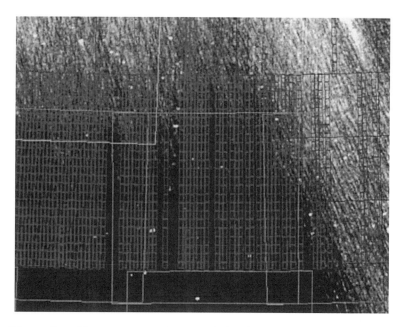

Figure 3. Demonstration of some of the difficulties inherent the image cross-identification project. Superimposed on an IRAS 100 Micron image, approximately 10 degree square are the outlines for available DPOSS plates (large squares) and 2MASS images (small rectangles).

References

Djorgovski, S., de Carvalho, R.R., Gal, R.R., Pahre, M.A., Scaramella, R., and Longo, G. 1998, In B. McLean, editors, *The Proceedings of the 179th IAU on New Horizons from Multi-Wavelength Sky Surveys*, IAU Symposium 179, 424

Lonsdale, C.J., et al. 1998, In B. McLean, editors, *The Proceedings of the 179th IAU on New Horizons from Multi-Wavelength Sky Surveys*, IAU Symposium 179, 450

Rutledge, R., Brunner, R.J., Prince, T., & Lonsdale, C. 2000, ApJS, in press

Skrutskie, M.F., et al. 1997, In F. Garzon et al., editors, *The Impact of Large Scale Near-IR Sky Surveys*, Kluwer, 25

Szalay, A.S., & Brunner, R.J. 1999, Future Generations of Computer Systems, 16, 63

Architecture of the Infrared Science Archive

J. Good, G. B. Berriman, N. Chiu, T. Handley, A. Johnson, M. Kong, W-P. Lee, C. J. Lonsdale, J. Ma, S. Monkewitz, S. W. Norton, & A. Zhang

Infrared Processing and Analysis Center, California Institute of Technology, Pasadena, CA 91125

Abstract. This paper describes the architectural approach that has been used in building the services that make up the Infrared Science Archive (IRSA). This approach is one of loosely coupled archive access and client visualization/integration/interaction where each component is usable (and useful) in its own right but which also form a whole which is much greater than the sum of the parts. We believe that this mix of local autonomy and global interoperability is crucial to the success of any effort such as the NVO.

1. General Philosophy

The IRSA architecture is a collection of permanent data holdings and software toolkits plus specific "applications", which are specialized constructs built to the needs of customers. Berriman *et al.* 2001 (these proceedings) give an overview of IRSA's charter and services while this paper focuses on the underlying IRSA architecture.

Processing modules are constructed from the bottom up to whatever level is appropriate and not from the top down as a fragment of some overarching "system".

This is a more cleanly modularized object-oriented approach than those where each object is only viable in conjunction with a large collection of other objects. This latter approach is only marginally objectified and really a thinly-disguised "Big System".

2. A Partial Inventory

The server-side components of the IRSA system can be classed into three categories: traditional shared-object libraries which handle truly generic functions such as coordinate transformation; standalone executables which handle broader but still common tasks, such as database table subsetting; and user applications, which are focussed on specific science goals.

2.1. Library Modules

These are the lowest-level tools and are used in a wide variety of contexts. Most of them have both C and JAVA implementations. Examples include:

- Table file interface
- FITS image reader
- World Coordinate System (WCS) transforms
- Coordinate system transforms and precession
- Spatial indexing (sky and color)
- CGI support and server workspace management
- Service (child process) control

2.2. "Service" Modules

Middle-level functionality is more appropriately encapsulated in full programs than as a set of subroutines. These can be built and tested independently and used standalone, from scripts, or called from other programs. This last approach is too restrictive, however, unless the interaction with these "services" can be controlled at a finer level than the traditional command-line/argument piping technique.

IRSA service programs all implement a formal command/control grammar (specifically strictures on the syntax of response messages which require them to be parsable text structures). Driver applications can then, if they wish, fork these services off as child processes and interact with them in a command/response loop. We have found this to be an extremely effective way to objectify the system while allowing for the kind of implementation heterogeneity that is inherent in the real world of scientific software.

Examples include:

- Database table ingest
- Catalog spatial proximity cross-comparison
- NED/SIMBAD name resolution
- Catalog subsetting/output table formatting
- Image archive search with exact coverage checks

2.3. Web "Applications"

Realistically, since user needs evolve and grow, one should expect end-user applications to be the most transitory of the component classes. Currently, they are frequently implemented as CGI programs, though we expect this will soon be augmented with true distributed (JAVA) objects.

With this in mind, we have built most of our applications as relatively lightweight constructs making heavy use of the service modules described above.

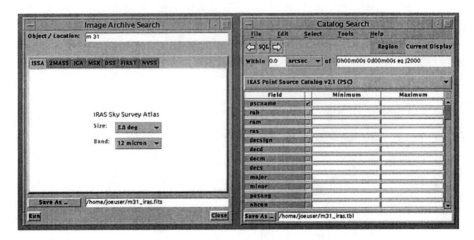

Figure 1. IRSA Archive Access Client Tools

This approach has allowed a high degree of flexibility: The same applications (or minor variations on them) also provide custom functionality to specialized users (the ISO project accesses the ISSA images with a custom map overlay) and supports client-side data integration (see the next section). Examples include:

- Catalog search and user table cross-comparison
- 2MASS survey visualization and image retrieval
- 2MASS extended object images
- 2MASS coverage checks and QA services
- IRAS data (image and catalog) integrated visualization/data retrieval
- IRAS galaxy atlas
- MSX data (image and catalog) integrated visualization/data retrieval

3. Data Integration Client Tools

With the base of services and service construction infrastructure described above, IRSA has the basis for supporting a very wide range of archive analysis tasks. The weak link in this chain is the current (nearly exclusive) reliance on batch-like HTML form/browser result rendering which form the basis for most astronomical archive access.

Over the last two years, IRSA has been constructing a set of client tools (in JAVA) which allow access to archive data and interactive visualization of result sets. These components have been carefully constructed so as to be usable individually, in concert, or under the auspices of any future data reduction/analysis system. A first release of this functionality is planned for November of this year.

The following outlines some of the current (and near future) attributes of this software:

3.1. Data Sources

(see Berriman *et al.* 2001, these proceedings)

- IPAC DBMS-based astronomical catalogs
- IPAC DBMS-based image archive metadata
- IPAC image sets
- IPAC vector datasets (*e.g.*, ISM database)
- Remote catalog archives (Vizier, MAST, *etc.*)
- Remote image archives (DSS, NVSS, VLA FIRST)
- NED/SIMBAD
- Remote citation/documentation services

3.2. Data Types

- FITS images
- Tables (catalogs, image metadata, *etc.*)
- Vector files (contours, scan tracks, *etc.*)
- HTML (Help, third-party Web services, *etc.*)
- XML data structures (*e.g.*, some catalog tables)

3.3. Interaction Environment

- Catalog Query Builder
- Image Metadata Query Builder
- Sky Viewer
  - Layer control
  - Base image
  - Map overlays
    * symbols generated from catalog tables
    * vectors from contour maps *etc.*
    * image outlines from metadata
  - Coordinate grid overlays
  - Coordinate feedback (fully functional transforms)
  - Color table selection and stretch/zoom/pan/range
  - Area examine/pixel functions
  - Focal plane overlays
- Table Viewer

3.4. Framework

- Bean-based. Those components (usually GUI) which are generic enough that they are reusable in a wider context can be constructed in adherence to a set of rules which govern their method naming, configuration setting, and event passing. These "beans", as they are called, can then be manipulated by GUI builder programs. Some, but by no means all, of our components fall in this category and will be made available in this form.
- Intercomponent event mechanism (*e.g.*, "new table", "location change", "region definition")
- Example Uses:
  - Map/Table/Graph interaction
  - Query status monitoring
  - "Pick"-driven background estimation
  - Area-selection driven user model activation
- User-defined events
- Add-on components (a few examples):
  - Background estimators
  - Minor planet ephemerides
  - User-written modules (*e.g.*, interactive galaxy morphological classifier)

The archive data access components (examples are shown in Figure 1) can be used individually or directly coupled to the display functionality (illustrated in Figure 2.). At this time, remote service must run to completion during a session (though the actual access is in separate threads and does not interfere with interactive visualization). In the near future, an archive request environment will be put in place so that users can disconnect and messages regarding processing request completion queued up until they reconnect.

While several of these objects have "main" methods, they are really meant to be used primarily as building blocks for applications.

4. Conclusion

The IRSA architecture has been designed from the beginning with an NVO-style framework in mind. First, each of the individual components has been constructed so that it is easily integrable with other cooperating information systems' components. Second, some specific services (in particular the large scale dynamic catalog ingest and rapid cross-comparison) were specifically constructed as prototypes for a NVO. Finally, our new client tools are designed to be components in an NVO integration environment.

Taken together, these pieces, fragmentary though they may be in the context of a true NVO, define an architectural approach and provide a solid example of what an NVO could and should become.

IRSA Architecture 147

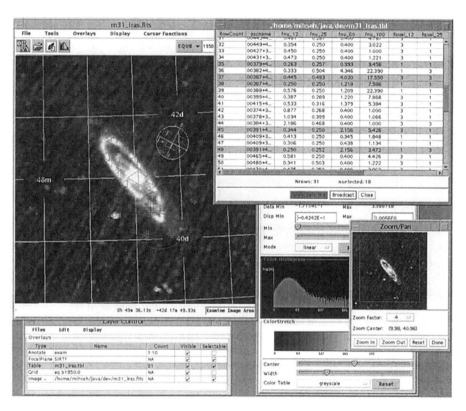

Figure 2. IRSA Archive Data Interaction Client Tools

Education & Public Outreach: A View from Research Institutions and Observatories

C. Christian

Development, Technology & Innovation, Space Telescope Science Institute, 3700 San Martin Drive, Baltimore, MD 21218

Abstract. Imagine a planetarium show in a small school facility or a small science museum in which multi-wavelength data of the sky is projected, with real-time access to the latest observations. Or suppose a web surfer could fly through the Local Group chasing a notorious digital archrival. Consider the science fair projects students could create with access to astronomical data. These possibilities can be realized as astronomers make data and expertise available to the public. A data bank linked to ancillary resources could provide many such opportunities for education and public understanding of science programs, in addition to the benefit provided to the research community for which it is principally intended.

1. Context

Observatories and research institutions share a common basis: both are intended to facilitate excellence in scientific investigations. Education and Public Outreach (EPO) programs in these institutions are designed to take advantage of proximity to the research community in order to bring the benefit of the scientific endeavor directly to the public.

An EPO group can broker information to a variety of audiences such as the news media, journalists, educators, students, science museums, planetaria, libraries and the public. I broadly categorize these varied audiences into 4 main categories: News, Formal Education, Informal Science Education and Public Information. EPO usually refers to the latter 3 enterprises. Access to scientific data, information, and human expertise must be tuned to the needs of each arena.

2. Education Resource Development Model

It is my experience that educators, personnel in science museums, planetaria, journalists and the public expect that scientific and technical information will be produced in a suitably professional way, especially from national/international facilities. Development of products and programs for EPO demands robust processes as well as authorship by individuals with authentic expertise.

2.1. Development Team

The basic model I adopt for the EPO product team includes talented individuals with expertise in the specific content area (scientists), those with expertise in science education (formal or informal), those with practical field expertise (*e.g.*, teachers, science museum personnel), technical expertise (scientific programmers, multimedia developers) and those with presentation skills (communications experts, artists, animators, audio/visual engineers). The production process also requires production expertise and discipline.

As for any good development team, each individual is chosen for a specific area of expertise. I expect all team members to participate in the development and production. Development is driven by user needs based on data drawn not only from user feedback, but also by observing users *in situ*. The model avoids the pitfalls encountered by other techniques in which development is driven directly from the content specialist (*e.g.*, scientist) with insufficient knowledge of presentation techniques, educational standards and other factors. In this model, scientists have the responsibility to insure that content is accurate and up-to-date and to educate fellow team members about research results, controversies, and subtleties. Scientists also are responsible for identifying other researchers with specific expertise required for development and also suitable scientific reviewers. Scientists are not charged with guaranteeing education content, grade level, preconceptions and educational outcomes—educators, curriculum developers and classroom practitioners are. Educators are responsible for pedagogy, standards and learning and guiding the group in structuring materials to achieve educational goals. Other members guide the group in understanding how material can be most effectively presented, visualization principles, *etc.*

2.2. Development Process

The full development process description is similar to the process used for software development, and is beyond the scope of this short communication. Important facets of the development are that user needs assessment is accomplished at the outset, and a review, pilot testing and evaluation plan are formulated early, along with the product conceptual design. This insures that tests and feedback will occur at critical junctures, avoiding multiple iterations on the content and presentation. The feedback includes scientific review by external experts at several stages. Review in other areas is planned as well.

For multimedia development, a conceptual design phase, or "storyboarding" is especially critical. This step is crucial for limiting the scope of the production, insuring the project adheres to the original goals and needs and that development does not proceed prematurely. Too often the development of "web pages" proceeds without any needs assessment or goal setting. The detailed design phase plans out the details, logistics and dependencies of various parts of the product before code is written or graphics are finalized.

Detailed design is the next major step, followed by production, with suitable checks, reviews and tests. "Alpha" testing with representative users and reviewers and "beta" field testing are important steps for creating a robust product, allowing ample time for feedback. Field trials of the publicly released products include evaluation in a variety of settings with varied users. The scientists' role throughout is as co-author in the revision stage and identifying suitable scientific reviewers and testers.

3. EPO in the Observatory

EPO is integrated into the Observatory by managing the effort as a normal observatory function. At the Space Telescope Science Institute for example, the unit responsible for outreach is the functional equivalent of the operational units (scheduling, pipeline processing, calibration, and archive, to name a few). Scientific staff maintain functional duties, along with their protected research time. Scientists change functional duties periodically, moving from one unit to another (including outreach) based on expertise and interest.

The EPO effort benefits from the observatory infrastructure (computer systems, administration, *etc.*), as well as ready access to various scientific and technical expertise. Interaction with observatory science visitors is key as these individuals bring in the latest information on research in progress and breaking news on results. The observatory in turn, benefits from outreach team expertise in science communications, graphic arts, multimedia design, evaluation and other areas, and further benefits from public visibility provided by EPO activities.

4. EPO in the Research Institution

Scientists in a research institution can interact positively with their institution's EPO effort especially as they provide invaluable local scientific expertise. Their connections to the science community at large provide access to a network that other organizations do not have. The discipline of science naturally infuses a culture of experimentation into the organization, encouraging the EPO effort to try new approaches and study effective means of providing public access to science. The research institution, by its existence and activity, maintains a knowledge base of current science questions and science breakthroughs, and supplements data and nuggets of information on scientific investigation with models, simulations, theories and concepts. In the context of a databank, this ancillary information, in addition to scientific expertise itself, should be integrated into the available resources.

5. NASA Educator Resource Network

The Office of Space Science at NASA has created a network of institutions working cooperatively to bring educational and other materials to the public. The location and function of the centers is described on the NASA space science site[1].

---

[1] http://spacescience.nasa.gov/education/index.htm

5.1. Funding Profile

During the creation of the network, NASA articulated in its strategic plan that 1–2% of the total funding in large programs should be devoted to education and outreach. This level of funding would be suitable for a major effort such as the Virtual Observatory. The types of activities being conducted across the space science discipline are educational programs in collaboration with school districts and curriculum developers, major and modest projects with science museums such as the permanent exhibit on Hubble Space Telescope and the Far-Ultraviolet Spectroscopic Explorer in Maryland, other exhibitions on space weather and X-ray astronomy, traveling exhibits on several NASA programs, and infrared camera exhibits at public venues such as national parks, and an in-flight video used on a major airline. This network and expertise exists and would be a natural framework for EPO activities associated with the VO.

5.2. Education Resource Catalog

NASA's Office of Space Science also has developed an education resource catalog for public access to online, near-line and hardcopy materials produced by the NASA missions. My goal in writing specifications for the architecture of the catalog is to create an open system built on the Dublin Core[2] specifications for educational materials. Suitable meta-tags have been constructed in collaboration with a large federally funded consortium. The information in the catalog could be of general use for any EPO project through its open architecture. The public interface into the catalog will be like successful commercial product interfaces[3].

6. NVO—Developer's View

As a developer, if I want to create an immersive environment for astronomical discovery I would need: a) a data and information server; b) data and ancillary information interlinked (publications, press, educational resources); c) access to simulations/models; d) tools tailored to user experience base/user needs; and e) hooks for innovative uses (exploratory and collaborative programs, multimedia games, live-data projection, *etc.*). I will want to build my own interface and applications. I further would benefit from pointers to other infrastructures and to human expertise.

7. Summary

Education and Public Outreach in observatories and research institutions has an enormous potential, when the technical and scientific expertise of the researcher is suitably integrated into programs and product development. A virtual observatory has the opportunity to offer valuable data as well as ancillary resources and access to expertise both to the scientific community and to the public.

---

[2] http://purl.oclc.org/dc/

[3] http://amazon.com

Acknowledgments. This work is supported by NASA Contract NAS5-26555 to the Association of Universities for Research in Astronomy, Inc.

References

Christian, C. A., Eisenhamer, B., Eisenhamer, J., & Teays, T. 2000, Journal of Science and Technology Education, in press
   keywords Education Outreach Resources *

Museums, Planetaria and the Virtual Observatory

Charles T. Liu

Department of Astrophysics, American Museum of Natural History, 79th Street at Central Park West, New York, NY 10024

Abstract. Today, museums and planetaria do much more than merely educate and entertain the public. Increasingly, they house departments of astronomy and astrophysics — scientists with active research programs who also devote a portion of their time to informal education, exhibit creation, and public outreach. This paradigm embodies two basic ideas — the museum as a research center, and the planetarium as a research tool — and meshes perfectly with the goals and implementation of a future Virtual Observatory. Just as the Virtual Observatory will transform astronomy research, it will also transform astronomy education and outreach; museums and planetaria therefore can (and should) be fully integrated into its planning, design and development. The key points to this process will be [1] to make the Virtual Observatory an active, participatory experience; [2] to layer its content and accessibility for children, novices, educators, and experts respectively; [3] to incorporate visualization for analysis as well as illustration; and [4] to integrate outreach and research as fully as possible, to bring Virtual Observatory science to the public.

1. The Big Picture

One day, the Virtual Observatory (VO) will change the paradigm of astronomy research worldwide. On that day, the paradigm of astronomy education and public outreach (E/PO) will also change. As the Virtual Observatory White Paper emphasizes, VO presents a tremendous opportunity to bring astronomy to the worldwide public. To realize this potential, we cannot merely attach education and outreach efforts onto the VO as an afterthought; rather, we must integrate E/PO capabilities into the VO at every stage of design, development and implementation.

Within this overarching context, museums and planetaria can and should be a central component of that integrated effort. Consider, for example, that the Hayden Planetarium at the American Museum of Natural History in New York (Figure 1) is on a pace to draw four million visitors this year; conservatively estimating each visit to be about two hours long, the Hayden alone will focus $\sim 10^7$ person-hours of attention on astronomy — the same order of magnitude as all the person-hours of college astronomy courses taken in North America. When we further add the person-hours of visitors to the Adler Planetarium in Chicago, the Griffith Observatory in Los Angeles, the Chabot in Oakland, and

Figure 1. The front entrance of the new Hayden Planetarium and the Rose Center for Earth and Space at the American Museum of Natural History in New York City. The building is roughly cubical, 120 feet (36 m) on a side. The planetarium theater resides in the top half of the 87-foot (27 m) sphere inside the glass walls and seats 429 people. Below and surrounding the sphere are five major permanent exhibit elements, including the Gottesman Hall of Planet Earth and the Cullman Hall of the Universe. See Liu et al. (2001, these proceedings) for the technical capabilities of this facility. Photo by D. Finnin.

so many other similar facilities, each with millions of visitors as well, the power of museums and planetaria is clear and unequivocal.

Many of us in the astronomical community still find it hard to integrate education and research into our professional activities. Often, the obstacles are reasonable and understandable. This synthesis, however, is truly important and worthwhile at every level of our work, regardless of the difficulties we may face. (Remember that **all** of us are who we are today because of the education and public outreach we have experienced.) Even from a purely pragmatic perspective, keep in mind that government officials and wealthy donors visit museums and planetaria far more often than they visit universities and research laboratories; and even when they do not visit, public policymakers listen very carefully to any group of constituents that numbers in the tens of millions. Over the years, literally billions of dollars have been channeled into astronomy and space science as a direct result of E/PO efforts and projects. As we move forward into the next century, the astronomical community will surely need to optimize even further our ability to energize individuals, governments and the general public to appreciate and support our work.

2. The Extraordinary Experience

Museums and planetaria impact public awareness and support of astronomy by providing the "Extraordinary Experience" — an unusual and memorable event that stays with you long after the event itself is over. This service perfectly complements classroom-based and World Wide Web-based education and outreach. These latter resources make astronomy an accessible and understandable part of everyday life, which is also crucially important in bringing astronomy to the public. That process of making science "ordinary," combined with the "extraordinary" stimulus a museum provides — a school field trip, a family vacation visit, a special lecture, or even a single virtual-reality ride through the universe — helps to create a complete set of experiences that brings home to each person the wonders of the cosmos and the value of studying them.

Perhaps ironically, museums and planetaria also provide for most of the general public the model of an academic institution. Certainly, this perception makes sense. A museum, just like a college or a laboratory, is a place that values knowledge and learning for its own sake; and far more people, especially children, visit museums than university astronomy departments, mountaintop observatories or government research centers. We in the more traditional research community have gained tremendous benefits from this circumstance, and can do even better by embracing museums and planetaria as part of a greater "astronomical academia" and integrating our programs as fully as possible — not just with the largest sites such as the Hayden and the Adler, but with local and regional institutions as well.

One superb example of such synergy resides in Tucson, Arizona, where the Flandrau Planetarium and Science Center has a strong relationship with the University of Arizona's Steward Observatory and Lunar and Planetary Labs, and also with Kitt Peak National Observatory (NOAO). These institutions often provide the planetarium with source material for their exhibitry, which is viewed by thousands of schoolchildren and other public visitors each year. Reciprocally, students taking introductory astronomy classes at the University of Arizona go regularly in groups to the Flandrau, led by professors and teaching assistants, and learn the basics of the night sky and its contents. The instructors gain direct experience with the equipment and techniques of public outreach — skills that have benefitted many a graduate student or faculty member in the years to follow. And as might be expected, student evaluations of these classes have consistently shown that the planetarium trip is very often the most memorable part of the course.

3. The Modern Model of Museums and Planetaria

The connection between traditionally research-oriented centers and traditionally outreach-oriented ones no longer flows only in one direction. Museums and planetaria have advanced rapidly in both technology and academic philosophy, spurred on in part by an increasingly techno-savvy and scientifically sophisticated public audience. Two fundamental ideas have arisen, which together vastly increase the already tremendous value of these E/PO sites, and further bridge what may in the past have often been two parallel but disjoint tracks. These concepts, elaborated below, make it even easier and more important to integrate what goes on in museums, planetaria, and university and government research departments.

3.1. The Museum As Research Center

As museums update their astronomy presentations to reflect modern astrophysical knowledge, many of them have realized the value of having research scientists to guide and inform the exhibit and show productions, in order to create the most interesting and current material possible. If these scientists leave their research careers behind, however, then that important value is lost; so museum administrations have begun to build true departments of astronomy and astrophysics — permanent staff with research time built into their job descriptions, just as at national observatories or government research centers.

The current leading example of this phenomenon is the American Museum of Natural History (AMNH) in New York City, the parent institution of the Hayden Planetarium. As part of the reconstruction of the Hayden and the creation of the Rose Center for Earth and Space, AMNH established a Department of Astrophysics to complement its other nine curatorial departments. There are currently a dozen Ph.D. scientists in the department, half of whom are full-time, permanent staff with 50% to 80% research time; the other half are postdocs with no institutional service commitments. Data analysts, graduate students and administrative personnel complete the staff list — an astronomy department worthy of most medium-sized universities.

It should be noted that there has always been a strong academic tradition at AMNH; the familiar positions of a provost, dean of science, and a grants administration office already existed long before the Department of Astrophysics was established. The new department thus fit nicely into AMNH almost as soon as it was born. But even without such an infrastructure, a number of other institutions with new or renovated planetaria have been following this same model; two examples are the Adler Planetarium and the Denver Museum of Natural History.

Research specialties in the AMNH Department of Astrophysics span planetary, galactic and extragalactic astronomy, computational astrophysics and cosmology (see, *e.g.*, Shara *et al.* 1999; Liu *et al.* 1999; Mac Low & Ossenkopf 2000; Sugerman, Summers & Kamionkowski 2000; Rich *et al.* 2000). Thus, any exhibit element in the entire Rose Center is fully informed by a scientist at the cutting edge. The museum benefits further from the existence of the department; its $200 million investment in the new Hayden Planetarium and Rose Center immediately commands international credibility as an academic center rather than merely a flashy display, and it does not risk the embarrassment of becoming obsolete with the next new discovery. Finally, every member of the department understands the value of public outreach efforts, and works hard to convey the subject matter effectively to visitors of all ages and academic backgrounds. This leads to an interesting and perhaps unexpected result: with the publicity surrounding the new planetarium and center, some of the astrophysicists at AMNH have actually achieved mild celebrity status — unusual for members of our community, but certainly welcome for the exposure and support it engenders for our field and our work.

3.2. The Planetarium As Research Tool

The technology that has revolutionized computer graphics, movies and multimedia has moved into planetarium domes as well. Many new planetarium theaters include not only fiber-optic equipped sky projectors with stars just a few arcseconds across, but also three-dimensional, interactive real-time visualization systems driven by supercomputers. In such cases, the dome can be used as an analysis tool for massive multi-dimensional datasets whenever the theater is not in use.

Clearly, any planetarium with an associated department of astronomers can be the gateway to a wide range of research and collaborations to visualize and analyze very large datasets — *exactly* the kind of system that needs to be associated with the Virtual Observatory. Two planetaria have already established infrastructures to become scientific visualization centers, with tested experience in presenting astrophysical datasets: the Hayden in New York, and the Adler in Chicago. (Both systems are thoroughly described in papers elsewhere in these conference proceedings; I refer the reader to those contributions for the relevant details.) The challenge falls to these institutions to provide convenient and useful user interfaces that will maximize these systems' scientific return, and to the community at large to find innovative and productive ways to use these powerful and revolutionary visualization tools.

4. Education and Outreach With The Virtual Observatory: The View From Museums And Planetaria

As we design, develop and ultimately implement a global Virtual Observatory, I hope that we all keep in mind the vital importance of education and public outreach in the fabric of the VO concept. The many lessons we have learned from museums and planetaria lead us to a set of important points to remember during this complex but exciting process.

• *Make Virtual Observatory an active learning experience.* Why is a planetarium visit so memorable? One of the most important reasons is that a visitor must participate interactively in the event — first making the effort to get there, and then browsing, reading and manipulating the many displays, captions and exhibits. A visit to the Virtual Observatory must also be an active, engaging exercise to achieve a maximum impression.

• *Make Virtual Observatory easy to use for students and educators.* When designing and constructing museum exhibits or planetarium shows, scientific content is best presented in layers: big ideas with big text for younger children, several sentences of elaboration with medium-sized text for most adults, and detailed explanations on computer screens for serious visitors and experts. An E/PO interface for VO should be similarly layered — for children, students, teachers, informed amateurs and professional researchers. This way, VO will be able to reach the full spectrum of the interested public, without being too intimidating or too simplistic.

- *Develop visualization tools for both analysis and illustration.* In E/PO work, there is no substitute for effective visual materials. At this conference, we have heard that simple visualization tools for VO data can be quickly and easily implemented; even if they will not be immediately useful for scientific research, their value for public outreach will be instantly evident and powerfully effective.

- *Integrate Virtual Observatory's research and education efforts.* The public is almost always more interested in the cutting edge — the latest distant quasar or exoplanet, the most recent planetary probe or Hubble Space Telescope images — than in the well-worn, tried-and-true exercises and demonstrations most of us have used before. The VO concept naturally lends itself to public participation in interesting and educational research programs. The Virtual Observatory White Paper cites the tremendous success of the E/PO program "SETI@home"; consider as just one example thousands of schoolchildren, flying through the multidimensional phase space of a massive dataset using their desktop computers. Whether or not they discover anything, these children will remember their participation and their effort for a long time. This one illustration is in fact relatively mundane - the possibilites for E/PO and research integration are limited only by our imagination.

All of these recommendations stem from reasonable logic and common sense. In fact, radical strategies or fashionable pedagogical theories may well prove unnecessary to develop an optimally successful E/PO component to VO. This is the key: we must commit to an integrated E/PO vision, where public outreach has genuine priority in the institutional infrastructure. Museums and planetaria, as both mature public outreach institutions and full-fledged astrophysical research centers, serve as some of the finest examples of this fundamental philosophy. If we can learn from them and adapt their best ideas to the Virtual Observatory, then the rest will follow.

Acknowledgments. I thank the organizers of this conference for inviting me to participate in this important and seminal event.

References

Liu, C. T., Petry, C. E., Impey, C. D., & Foltz, C. B. 1999, AJ, 118, 1912
Mac Low, M.-M., & Ossenkopf, V. 2000, A&A, 353, 339
Rich, R. M., Shara, M. M., Fall, M. S., & Zurek, D. 2000, AJ, 119, 197
Shara, M. M., Moffat, A., Smith, L. F., Niemcla, V. S., Potter, M., & Lamontagne, R. 1999, AJ, 118, 390
Sugerman, B., Summers, F. J., & Kamionkowski, M. 2000, MNRAS, 311, 762

The Virtual Observatory and Education: A View From the Classroom

James C. White II

The Astronomical Society of the Pacific, 390 Ashton Avenue, San Francisco, California 94112

Abstract. The design, construction, and implementation of a national Virtual Observatory will fundamentally change the way a large portion of archival astronomical research is conducted. Such a facility will also offer teachers in grades K-12, and certainly at the collegiate level, a new tool to introduce science to their students. The task of using a virtual observatory in a classroom is not merely one of understanding how to use the facility, however, but of understanding how to use the facility *to teach*. As important to the teacher as the Virtual Observatory itself will be the materials and instruction available on how to turn the facility into a virtual laboratory for student learning and experiment.

1. Introduction

The current formulation and eventual construction of a national, or better yet, international virtual observatory (VO) is exciting to consider, especially since such an inclusive facility will likely enable researchers at smaller institutions to participate more fully in astronomical research. There is another group that can participate, too, with a national VO—students of all ages in educational settings either formal or informal.

The utility of the VO in the classroom can be great when this new portal is opened and used correctly. Children, young adults, and nascent scientists can be brought into a world of real data and compelled to confront the excitement and even the tedium of scientific research. But as we consider an ideal classroom, be it at the elementary-, middle-, high-school, or even college levels, keep in mind the fact that most students do not learn astronomy from astronomers and that the vast majority of science teachers in grades K-12 have little or no experience with astronomy. Indeed, many will confess quite openly their fear of teaching astronomy because of their ignorance of it. Organizations like the Astronomical Society of the Pacific[1] attempt to work directly with teachers to lessen this anxiety and to lead them to see the wonderful opportunity astronomy as science in the classroom affords them and their students.

There are three reasons astronomy is the best of the sciences for general educational development. First, it is a field of human endeavor that is based upon our innate curiosity about the heavens; this alone makes astronomy a ter-

---

[1] http://www.aspsky.org

rific vehicle for learning. As we are all aware, we are more interested in learning something in which we are already interested. Further, "teaching astronomy" means far more than merely teaching Moon phases or mnemonics for planetary ordering—it serves as a stage for students to see how the scientific method works, from *stellar* successes to humbling blunders, and the continuum of the evolution in thought in between.

The second reason is that astronomy as an academic discipline is grandly encompassing science. The physical and, increasingly, natural sciences and mathematics are included, and the manner in which astronomy can touch us opens possibility for cross-disciplinary investigations into the humanities and social sciences, as well.

Finally, astronomy is good for all ages and is ripe as a subject for perpetual, life-long learning and contemplation. Most people (except for scientists or mathematicians!) could not care less about the latest advances in algebra, but bring up the subject of astronomy, and almost everyone is enthralled.

This is what we astronomers have to work with, and the national VO will only make what we do more embracing of the rest of the population. To effectively bring the VO into the classroom so that learning may be enhanced, care on our part must be taken or else we will find ourselves in the oft-sited situation of "plenty of powdered milk, but no clean water." Let me briefly outline the challenges to the VO's use in classrooms at different educational levels; I will also provide some recommendations about how we should meet these challenges.

2. The VO, the Spectrum, and Educational Development

I have no doubt that a fully implemented virtual observatory will be of use to teachers, yet I do doubt the usefulness or even appropriateness of it for all levels of educational development. The VO will provide full accessibility to astrophysical data spanning the entire electromagnetic spectrum. Let us consider as an example, then, how students at different grade levels come to "see" the spectrum.

Primary School. Young children are generally aware of only the visible part of the EM spectrum. Their exposure in the classroom is to visible light as they come to understand the interaction between light and objects and the creation of shadows. Visits to planetaria and science museums can re-enforce the more sophisticated notion that "white" light is made up of "all the colors of the [visible] spectrum." Except for classroom exercises in which a teacher uses data from the VO to exercise his or her students in pattern recognition in nature—and I stretch on the usefulness of such an exercise when using the VO—I see no use of the VO in primary school classrooms.

Middle School. By grades 4 or 5 children have been intellectually exposed to other regions of the EM spectrum. They have been told the warmth they feel from the Sun, say, is due to "invisible" light. Further, many have been introduced to the notion that too much sunlight is not good for you; in fact, teachers can use discussions of the danger of sunburn to introduce the concept of ultraviolet radiation. And if a child or one of his or her family members has had a broken bone, he or she will quite likely have heard of X-ray radiation.

At this educational level, I see the VO as a neat tool to present children with the reality of a Universe that looks different at different wavelengths. Visible, infrared, and ultraviolet views of the same region of space can be used along side similar, multiwavelength views of Earth or living creatures.

High School. In high school, students will have enough background in physical science to actually put the VO to work for them. The EM spectrum, in keeping with the present thread, will now appear as a continuum of energy, and teachers will be able to design or use ready-made, in-class and take-home exercises in which students essentially interrogate nature by using the VO.

College. The VO will be a full resource to students at the post-secondary level, its utility limited only by the students' intellectual ability to use it and that of their instructors to help the students formulate interesting exercises and/or experiments.

3. Using the VO in the Classroom

Teachers currently have three possible ways to bring real, or at least realistic, astronomy into their classes: simulating the sky with commercially available software packages, constructing or using exercises that employ real astronomical data, or using real telescopes. A virtual observatory as classroom learning tool can be a blend of the first two options.

Software packages like TheSky by Software Bisque[2] or Starry Night Pro by Space.com[3] do a superb job of presenting a simulated sky from different times and from different vantages. More importantly, however, these packages are powerful enough to serve a teacher as a virtual astronomy laboratory. Exercises are already available to teachers who want their students to investigate broad topics such as Earth's motion through space and the life cycles of stars, and specific topics like stellar parallax and proper motion.

In the future, the level of interrogation possible with such sky simulators will only increase. Even today some college-level instructors have available to them virtual reality Caves to offer students a more intimate experience with extraterrestrial objects (*e.g.*, Indiana University's Department of Astronomy).

Rather than being merely an observer, a student can be given a taste of astronomical research by a teacher who uses exercises based on real astronomical data. As an example, consider the increasingly robust suite of computer activities known as CLEA[4] (Contemporary Laboratory Experiences in Astronomy), by Larry Marschall and his group at Gettysburg College. The activities are constructed to utilize subsets of digital sky surveys, and students are compelled by the exercises to do science. Further, the overall activity environment is one that simulates an actual observatory experience, from data of varying levels of signal-to-nose to clouds during the simulated observing session.

---

[2] http://www.bisque.com

[3] http://www.siennasoft.com

[4] http://www.gettysburg.edu/academics/physics/clea/CLEAhome.html

The third way for a teacher to get some real-seeming astronomy into his or her classroom is to actually incorporate real telescopes into the lesson plan or syllabus. A discussion of this is beyond the scope of this meeting on virtual observatories, but let me mention one program in which teachers are even freed from the need of having real telescopes on-site. Mount Wilson Observatory's Telescopes in Education program [5] makes available to students and their teachers a 24-inch, research-grade telescope for use in remote observations. A simple though sophisticated user interface makes the remote observations possible with only a classroom computer and telephone line.

Each of these entries into real or simulated astronomy has elements appropriate to discussions of the VO in a classroom. Sky simulators offer teachers, students, anyone examples of easy-to-use, front-end programs like the one that must ultimately be designed for the VO. As a colleague mentioned to me recently, a program like Starry Night Pro could serve as a good, commercial interface to the VO. NASA-GSFC's SkyView offers an example of a non-commercial interface [6] and one that might, with proper guidance from the instructor, be useful to even beginning high-school students.

4. VO Materials for Teachers

Once the national Virtual Observatory is online and ready, there will remain much to do before teachers will be able or even willing to use it. Recall that most K-12 teachers had no or only limited exposure to astronomy when they were students, and it has been estimated that between 80% and 85% of post-secondary astronomy is taught by non-astronomers. Clearly, most people using the VO will need aid to and, quite likely, instruction in its use. There are a few incarnations the aid and instruction can take.

Teachers want any resource that will facilitate learning by their students, but such a resource must be designed for a given teacher's grade level. More directly, however, a teacher must be confident that any resource/activity he or she brings into the classroom satisfies national science education standards. This means that a teacher in grade 8 who wants to use the VO for a class science exercise will (quite likely) require materials designed such they satisfy national standards.

This requirement for standards-satisfying materials is not to be taken lightly. Such materials are critical to the VO's use in a classroom. Grade K-12 teachers have precious little time for new material in their often crowded curricula, especially if it means considerable time for them to learn the material and more time to determine how to integrate it into their lesson plans. Hence, what will be needed for a successful implementation of the national VO as a classroom tool is carefully designed exercises and teachers' materials *at the least*.

To make use of the VO in the K-12 classroom more likely, teachers would ideally also have training in the Observatory's use. Training could take the form of regional teachers' workshops—like those the ASP has been conducting

---

[5] http://tie.jpl.nasa.gov

[6] http://skyview.gsfc.nasa.gov

around North America for the past 20 years [7]—or even regional centers of teacher support, like the ASP's Project ASTRO sites spread around the country [8]. Web-based instruction, be it live or not, and tutorials are other ways to reach teachers and provide requisite training in the use of the VO for education.

I cannot stress strongly enough the importance of materials and training for teachers if the national Virtual Observatory is to work in the classroom. And even with these resources available, we still have a big marketing challenge—making teachers aware of them and the VO and convincing the teachers the VO can work in the classroom! Consider, for example, all the educational materials that have been produced by NASA missions of the past. A good portion of that effort was wasted for a couple of reasons. First, only within the last decade has good effort gone into answering the questions, *"How is the information from [a given NASA mission] useful to teachers?"* and *"How can a teacher use this information to teach [and satisfy national science standards]?"* A second reason for the limited success in turning the science of NASA missions (in keeping with my example) into classroom learning experiences is that the shelf-life of the materials produced was too short. Some excellent materials were produced, but they simply faded from view. They were produced but were never effectively marketed to teachers. Counter examples from the past decade include the successful education and public outreach (E/PO) activities and initiatives of the Space Telescope Science Institute [9] and the SIRTF [10] and SOFIA [11] missions.

It is imperative, therefore, that an E/PO capacity be built into the national Virtual Observatory or that extant organizations be included in the formulation and planning of the VO to ensure that the facility and its educational materials are actually usable by non-professionals. Given that a VO will be brickless, having a formal E/PO staff may be impossible, but it is critical that organizations be chartered with serving as sources of E/PO information and expertise. Much as the ASP's Project ASTRO sites do, regional centers could provide localized workshops for teachers, standardized materials—lesson plans, suggested exercises, etc., all created to satisfy national science standards—for teachers' use, and serve as a repository of expertise in the use of the VO in classrooms.

5. First Light for the Virtual Observatory

Coming within the next several years is, at the least, a modification of the way astronomy can be done. And combined with a greater emphasis on queue observing at national observing facilities, we are quite possibly on the verge of a paradigm shift in the way astronomy has been practiced for quite a long time. As the professional astronomical community confronts these changes, it is important

---

[7] http://www.aspsky.org/education.html

[8] http://www.aspsky.org/project_astro.html

[9] http://oposite.stsci.edu/pubinfo/edugroup/educational-activities.html

[10] http://sirtf.jpl.nasa.gov/SSC_EPO.html

[11] http://sofia.arc.nasa.gov/education/in-dex.html

that we not neglect the on-going advantage our science has in the classroom—be it a class of grade 4 children or a seminar for advanced undergraduates.

The creation of a national Virtual Observatory suggests much promise for the facility as a tool for enriching our children's educations by opening up the Universe to them from their classrooms. Yet this is more easily written than done. Considerable effort will need to be expended to create interfaces between the VO and teachers and students, and materials designed explicitly for classroom teachers' easy use, created under the guidelines of national science standards, are clearly required before the VO becomes a classroom fixture. When such materials are available and when teachers have easy access to instruction and support in their and the VO's use, we will see a powerful new tool for teaching science and for kindling a life-long excitement for astronomy in our children.

Virtual Observatories of the Future
ASP Conference Series, Vol. 225, 2001
R.J. Brunner, S.G. Djorgovski, and A.S. Szalay, eds.

The ISO Data Archive and Links to Other Archives

Christophe Arviset, José Hernandez, & Timo Prusti

ISO Data Centre, European Space Agency, Space Science Department, Villafranca del Castillo, PO Box 50727, 28080 Madrid, Spain

Abstract. The Infrared Space Observatory (ISO) Data Archive[1] now offers a wide variety of links to other archives, like name resolution with NED and SIMBAD, access to electronic articles from ADS and access to IRAS data. Moreover, one can now directly query the IDA via a Java Server Page query and get the postcards, a quick look ISO product, and relevant information back, embedded into an HTML page.

1. Introduction

The ISO Data Archive (IDA), developed by the ISO Data Centre of the European Space Agency (ESA) has been online since December 1998. Through its innovative and powerful JAVA user interface, the general astronomer or the instrument expert can make complex queries on the ISO product observations catalog and ancillary data, then have a quick textual and visual look at the data before retrieving the observations with a wide variety of options.

In the last year, development has been focused to improve the IDA interoperability with other astronomical archives, via accessing other relevant archives or via providing direct access to the ISO data for external services.

2. Name Resolution with NED or SIMBAD

On the IDA Query Panel, one can query against a target name. By entering a target name and choosing the name resolver (NED or SIMBAD), the IDA will make contact with the IPAC or CDS server to resolve the target name into coordinates and then search the ISO observations catalog against these coordinates. This is done completely transparently to the user.

2.1. SIMBAD

This service has been available since the opening of the IDA, in December 1998. It resolves target name into coordinates, calling the CDS server at Strasbourg, France via a specific TCP/IP socket (see Figure 1).

---

[1] http://www.iso.vilspa.esa.es/

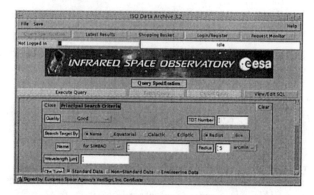

Figure 1. Querying the IDA with Name Resolution with SIMBAD

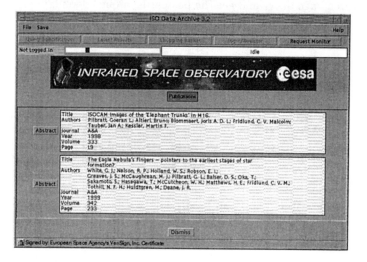

Figure 2. Access to Electronic Journals

2.2. NED

This service has been available since the opening of the IDA, in December 1998. It resolves target name into coordinates, calling the NED server at IPAC, USA via a URL/cgi-bin script. The returned HTML page is then parsed by the IDA to extract only the coordinates. Soon this will be updated to make use of a more efficient server type mechanism provided by NED.

3. Access to Electronic Articles

This service has been available from IDA 3.0, in December 1999. In cooperation with the NED project, all major astronomical journals are scanned for finding reference to ISO observations. Later on, the ISO Data Centre finds the link between the article and the used ISO observations. From the IDA Latest Results

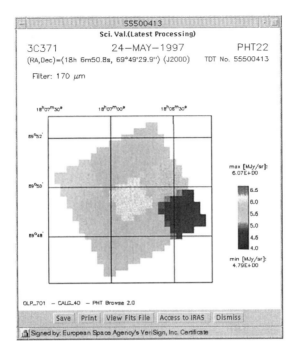

Figure 3. Access to IRAS Data via the ISO Postcard

Panel, the button "Articles" indicates if there are known publications linked to the observation. By clicking on the button, one gets an extra window showing the Title, Authors, Journal, *etc.* By clicking the "Abstract" button (see Figure 2), the applet will launch a browser window with the ADS WWW mirror at Strasbourg, France with the abstract of the article associated with the selected ISO observation. The call is made through a standard URL/cgi-bin script as defined by the ADS interface.

4. Access to IRAS data

This service has been available from IDA 3.0, in December 1999. From the IDA Latest Results Panel, one can see the small icons giving a quick overview of what the observations is about. By clicking on one of them, a bigger window is launched with the postcard giving more information on the ISO observation. By clicking the button "Access to IRAS" (see Figure 3), a browser window will open, from the InfraRed Science Archive (IRSA) WWW page located at IPAC, USA. The window will contain the data covering the region of the sky of the selected ISO observation. The call is made through a standard URL/cgi-bin script as defined by the IRSA interface.

5. Direct Access to the ISO Postcards

This service has been available from IDA 3.0, in December 1999. Through calling a URL/Java Server Page (JSP) containing the ISO observation identifier (so called TDT number). It returns the ISO postcard (GIF image) of this observation an ancillary quality information embedded into an HTML page. The ISO Data Centre can provide the ISO observations catalogue and the syntax for any astronomical application to be able to easily link with the ISO products without going through the standard IDA interface. This will soon be available through SIMBAD at CDS, Strasbourg, France.

6. Plans for the Future

The ISO Data Archive is still being actively and continuously improved and several versions are released every year. In particular, in the area of links with other archives, it is planned to bring the following new features in future releases:

- Once all ISO products are reprocessed and put on hard disk, provide direct access to the FITS products using similar mechanism as to the direct access to the ISO postcards.

- Provide direct access to the ISO Survey Product Display Tool, Java applet allowing some manipulations (panning, zooming, ...) of the ISO FITS products.

- Possibility for the IDA users to enter/modify themselves the links between ISO observations and their publications.

- Possibility for the IDA users to enter "User Reduced Data" into the IDA and to retrieve user reduced data deposited by other observers.

In general, the ISO Data Archive 3-tier architecture is very open and allows easy provision of new services in the future.

References

Arviset, C., et al. 2000, in ASP Conference Series, Astronomical Data Analysis Software and Systems IX, eds. D. Crabtree, N. Manset, & C. Veillet, in press

An Overview of the Infrared Science Archive

G. B. Berriman, N. Chiu, J. Good, T. Handley, A. Johnson, M. Kong, W-P. Lee, C. J. Lonsdale, J. Ma, S. Monkewitz, S. W. Norton, & A. Zhang

Infrared Processing and Analysis Center, California Institute of Technology, Pasadena, CA 91125

Abstract. This paper describes the data mining, catalog-cross comparison and visualization services available at the Infrared Science Archive (IRSA), NASA's archive node for infrared astronomy data. IRSA is a living archive, which maintains contemporary datasets and continuously develops services to exploit these datasets. Over the past three years, IRSA has devoted most of its resources to support the requirements of the massive 2MASS survey datasets. Given the high volumes of 2MASS data, the services and infrastructure supporting them provide insight in understanding how a future NVO may operate.

1. IRSA's Charter

The Infrared Processing and Analysis Center (IPAC) at Caltech was charged with archiving data sets produced by the Infrared Astronomical Satellite (IRAS). The success of this mission and the demand for its data products made IPAC a leading center for archival research and data distribution. This expertise naturally led to the development of the Infrared Science Archive (IRSA) as the archive node for NASA's infrared astronomy missions. IRSA now provides public access to the catalogs and images from the 2MASS and MSX missions, as well as from the IRAS mission. IRSA's requirements are derived on one hand from the specialized needs of projects, and on the other hand from the needs of users analyzing the data. IRSA also holds ancillary catalogs required to allow exploitation of the infrared datasets, including USNO-A 2, NRAO VLA Sky Survey (NVSS) and Faint Images of the Radio Sky at Twenty-centimeters (FIRST). A full list of IRSAs holdings as of August 2000 can be found at the IRSA web site[1]. In the next two years, IRSA anticipates that it will ingest catalogs from the Infrared Telescope in Space (IRTS) and the Cosmic Background Explorer (COBE), and it will provide integrated access to the Infrared Space Observatory (ISO) data archives, held in Vilspa, Spain. Eventually, IRSA will archive catalogs from the SIRTF and SOFIA missions.

---

[1] http://irsa.ipac.caltech.edu

2. Access To IRSA Services

While all IRSA services can be invoked via a program interface or via remote HTTP or Java client interfaces, users generally access them in server mode through a web client. Users can launch services by visiting the IRSA web site. Processing is performed server-side in the IRSA environment, and the results made available to the user. Broadly speaking, the following services can be applied to the data held by IRSA:

1. completely general catalog queries

2. Image queries and visualization, with customization of visualization tools applicable to individual missions.

3. Cross-comparison between catalogs

4. Statistical representation of large datasets

IRSA receives on average over 220 requests for data each day, and over 99% of these requests are successfully processed. The architecture of the IRSA services is described by Good et al. (2001, these proceedings).

3. IRSA As A Living Archive

A key feature of IRSA is that it is a *living* archive. That is, by providing robust, contemporary archives and by continuously developing services that have the power to exploit them, IRSA permits the development of new scientific products and opens up new avenues of research. The support provided for IRSA's largest customer, the 2MASS project, has demonstrated of the power of such an archive. 2MASS is uniformly surveying the entire sky in three near-infrared bands to detect point sources brighter than about 1 mJy in each band, achieving an 80,000-fold improvement in sensitivity over earlier surveys. This deep survey has generated datasets that are by far the largest obtained in any astronomical survey, with roughly 12 TB of images, and an internal catalog of sources now containing over 1 billion entries in an Informix database.

Efficient mining of these huge 2MASS datasets places extraordinarily large loads on IRSA services and required research into special techniques and the development of optimized software. As an example of the new science that can be performed with the help of these services, astronomers culled from the 2MASS catalog the very red candidate objects from which the first brown dwarfs were identified (Kirkpatrick et al. 1999, and references therein).

Future surveys will produce data volumes even larger than those generated by 2MASS, and will certainly produce ever more spectacular scientific advances. The services developed by IRSA and the infrastructure to support them therefore provide a window into how a future NVO is likely to function. The remainder of this paper is therefore given over to discussions of special features of IRSA services that permit efficient datamining, and current research at IRSA that will provide the next generation of datamining, visualization and analysis tools.

4. Special Features of IRSA Services For Datamining Large Volume Datasets

4.1. Catalog Queries and Indexing of Database Tables

Driving the 2MASS datamining requirements is the need to provide efficient and completely general querying methods. Querying has been made efficient in two ways. First, queries are run in parallel fashion across as many processors and I/O channels as possible. IRSA chose a Sun Microsystems E6500 server and an Informix database because they can be highly optimized for parallel querying. Second, IRSA spatially indexes catalogs using nested hierarchies of increasingly smaller bins. Search mechanisms traverse those branches of the tree to isolate database entries that meet the constraints imposed by the query. IRSA employs three spatial indexes:

1. The Hierarchical Triangular Mesh (HTM), developed at Johns Hopkins to support the Sloan Digital Sky survey, divides the sky into a nested series of equilateral triangles.

2. In magnitude/color space, three dimensional box partitioning with logarithmic steps away from the diagonal.

3. R-tree indexing of image metadata where images cover a large area; R-trees take into account the spatial extent of the elements contained within them.

4.2. Catalog Cross-Comparison and Distributed Queries

Here perhaps lies the greatest challenge to the NVO, and here perhaps lies the greatest potential for ground-breaking new science. IRSA has developed tools that allow for efficient distributed queries and which handle the complex DBMS functionality involved in cross-comparison of catalogs. The heart of the methodology is the use of three-way joins between tables, and sets of candidate source associations (known as relationship objects). The paper of Ma *et al.* (2001, these proceedings) describes the power of this method in more detail and describes future research in this area.

4.3. Image Metadata

IRSA separates images from their associated metadata. The metadata reside in a database catalog, with one record per image. Indexed searches can be made on any parameter, as well as spatially indexed searches based on position. IRSA can therefore efficiently locate images in catalogs it does not hold itself.

4.4. Statistical Representation of Large Datasets

Generally speaking, statistical representations of catalogs or query results are a powerful way of initially studying a large volume of data, allowing a user to quickly refine a search to locate objects of interest, such as those with extreme colors. IRSA provides a service to derive a histogram of a pre-binned representation of the IRSA database.

5. Current Research at IRSA and Development of Future Services

The previous section described research into catalog-cross comparisons. A related issue of importance to the NVO concerns the establishment of standards for data exchange. IRSA has cooperated with STScI and CDS to establish an XML output format for catalog search results, and is pursuing similar standards for the transfer of image metadata.

The heavy use of IRSA services has demonstrated the need for astronomical data interaction systems modeled after Geographical Information System (GIS) interfaces. IRSA is developing such an architecture, which will allow efficient and flexible access to astronomy datasets. This work is described separately in the paper by Good et al. (2001, these proceedings).

IRSA has also prototyped a service to access and display the 2MASS Atlas images via the High Performance Storage Systems (HPSS) at the San Diego Supercomputer Center and Caltechs Center for Advanced Computing Research (CACR). The high volume of these images, 12 TB in total, is typical of the data volumes that must be served through an NVO, and the architecture of this service can therefore be considered a prototype of an architecture that can support an NVO.

References

Kirkpatrick, J. D., et al. 1999, ApJ, 522, L65

The SOHO Data and Information System

George Dimitoglou

Emergent Information Technologies. Space Sciences Division. GSFC Code 682.3, Greenbelt, MD 20771.

Luis Sánchez Duarte

European Space Agency. Space Science Department. GSFC Code 682.3, Greenbelt, MD 20771.

Abstract. Since the early stages of the mission, SOHO teams employed WWW-based technologies to provide tools for mission planning, data analysis and public outreach support. Today, this collection of tools can be viewed as an integrated data and information system that plays a significant role in science operations and data distribution. This paper is an overview of some of the components, functions, products and services provided by the SOHO Data and Information System.

1. Introduction: The SOHO Mission

The Solar and Heliospheric Observatory is a mission of international cooperation between ESA and NASA. From the vantage point of its orbit around Sun-Earth's L1 point, a payload of twelve instruments study the Sun without interruption, from its deep interior to the corona, as well as the solar wind and the heliosphere.

The World Wide Web has been used to support mission planning, operations and data analysis since 1994. All the services and data products described below are accessible through the SOHO Internet pages[1].

2. Data Analysis Support

2.1. Data Products

The most significant data sets of the mission are:

Summary Data. Since the procedures to create the science processed data can take some time, the Principal Investigator teams make available, on a daily basis, quickly processed representative data from their instruments. These data provide a synopsis of solar conditions as seen by the SOHO payload, and are used mainly for coordination of science operations.

---

[1] http://sohowww.nascom.nasa.gov/

Synoptic Data. To complement the summary data, we retrieve also a set of synoptic data from other spacecraft and ground-based observatories around the world. There data are also used mainly for planning of science operations.

Ancillary Data. They include everything related to the status of the spacecraft, including attitude and orbit, and of the ground segment.

Science Processed Data. This data set is made by the calibrated data susceptible of scientific analysis and is provided by the Principal Investigator teams once the final version of the telemetry of the spacecraft is processed and analyzed. Science Processed data are typically updated as the calibration of the instruments improve.

2.2. Software library

While some software to display and to do basic manipulation of the different types of data acquired by the SOHO payload has been developed by the Project Scientist Team, most of the software available for scientific data analysis has been contributed by the Principal Investigator teams, and it has been designed to be compatible with the software to analyze data from other solar observing spacecraft, such as Yohkoh.

2.3. Archive architecture

The Archive is designed to accommodate datasets from all SOHO instruments. Access to the archived data sets is provided via the SOHO Catalog which is comprised by a relational database back-end and a WWW-based interface that provides all necessary facilities for data search, access and retrieval.

2.4. Available Data

Available scientific processed data in the SOHO Archive as of June 2000:

INSTRUMENT	Records	Size (GB)
CDS	24,120	121
CELIAS	35,953	57
COSTEP	10,582	24
EIT	151,708	217
GOLF	905	1
LASCO	184,579	160
MDI	25,621	54
SUMER	57,756	99
SWAN	2,603	1
UVCS	25,882	28
VIRGO	3,807	1

3. Operations Support

Support for both real time and scheduled operations, including targeting and real time planning, review of near real time data acquired by the instruments, design of the observing timelines, definition and storage of joint observing programs with other observatories and archive of solar and spacecraft events.

4. Outreach

The SOHO Project Scientist Team maintains two services to facilitate communication between the general public and the participants in the mission:

SOHO Explore!, a collection of resources for educators, including lesson plans, posters, frequently asked questions about Solar Physics and SOHO and a Glossary of terms.

Ask Dr. SOHO service, a line of communication with the general public, who can ask questions about the Sun, our solar system and our mission via e-mail to SOHO scientists and engineers. Questions should be addressed to letters@sohops.gsfc.nasa.gov

References

St. Cyr, O. C., Sánchez Duarte, L., Martens, P. C. H., Gurman, J. B., Larduinat, E. 1995, Solar Physics, 162

The CDS Role in the Virtual Observatory

Francoise Genova, Francois Bonnarel, Pascal Dubois, Daniel Egret, Pierre Fernique, Soizick Lesteven, Francois Ochsenbein, Marc Wenger

CDS, UMR 7550, Observatoire de Strasbourg, 11 rue de l'Université, 67000 Strasbourg, France

Gérard Jasniewicz

GRAAL, UMR 5024, Universite de Montpellier II, CC 72, 34095 Montpellier Cedex 05, France

Abstract. The CDS role in the present astronomy information network, and its future role in the emerging Virtual Observatory, are presented: provision of value-added reference services, and of tools for data integration and interoperability; active participation to the definition of metadata, in particular in the context of XML development.

1. The CDS Role in the Astronomy Information Network

The *Centre de Données astronomiques de Strasbourg* (CDS) has been developing reference astronomical services for more than 25 years (Genova et al. 2000). Traditionally, the CDS main activities have been SIMBAD, the reference database for nomenclature and bibliography of astronomical objects (Wenger et al. 2000), and the catalogue service, maintained and distributed in close collaboration with the international Data Center network. The catalogue service, first devoted to the distribution of catalogs as a whole, was expanded with the development of the VizieR database, which allows users to browse catalogs by constraining any of their fields (Ochsenbein, Bauer, & Marcout 2000). More recently, another trail was added with the development of the ALADIN sky atlas, which adds images to the reference information gathered, homogenized and distributed by CDS (Bonnarel et al. 2000).

With the irruption of the World Wide Web, astronomers' research environment has been revolutionized by the rapid development of on-line access to information, from observational data in observatory or disciplinary archives, to results published in electronic journals. Taking the best advantage of the new possibilities offered by the WWW has been a major concern of CDS activity in the last years : services have progressively been installed on-line, beginning with the catalogue FTP service in 1993, the last major release being ALADIN, with the Previewer mode in late 1998, the Java Applet in 1999, and the Java standalone in early 2000.

Networking of distributed information is another major evolution opened by the WWW. Definition and maintenance of links is however not an easy task, due to the variety of query syntaxes and to the frequent changes of WWW addresses. The GLU (Générateur de Liens Uniformes) system has been developed by CDS to tackle these problems, first as a tool to build efficient links between the CDS services, and between the CDS services and other reference on-line services (Fernique, Ochsenbein, & Wenger 1998). In November 2000, SIMBAD offers links to HEASARC and to IUE INES database, VizieR links to IUE, the VLA/FIRST survey, CFHT and HST observation logs and data archives, and ALADIN to HST and VLA/FIRST images. These links to data stored in observatory archives or disciplinary centers will continue to be developed in the near future in close collaboration with the major astronomy data holders.

GLU is a generic tool, which has also been used to link a large set of highly heterogeneous, distributed resources in the frame of the AstroBrowse project. This led to the development of resource discovery services at HEASARC, STScI, and CDS (AstroGLU) (Heikkila, McGlynn, & White 1999; Egret, Fernique, & Genova 1998). More generally, the definition of common metadata is essential for interoperability (*e.g.*, the ISAIA project, Hanisch 2001, these proceedings), and CDS develops a new tool to browse graphically the contents of a metadata dictionary. Management of XML descriptions is also being implemented in the GLU.

2. Bibliographic Network

Although seldom cited at the present stage of Virtual Observatory discussions, which concentrate on infrastructure, data management and statistical tools, bibliography is indeed a key element of the system, since it gives access to research results, essential for data interpretation. The bibliographic network of astronomy is already operational, owing to the close collaboration between the journals, NASA ADS bibliographic database, and the data centers. A small set of de facto exchange standards is shared by all partners (bibcode, table description). CDS has played its part in this networking, in particular for the definition of standards (bibcode with NED, and then ADS — Schmitz *et al.* 1995; tables, first defined for the electronic publication of the *Astronomy & Astrophysics* tables — Ochsenbein & Lequeux 1995), and as provider of SIMBAD and VizieR, two of the widely used bibliographic reference services (SIMBAD for the bibliography of objects, VizieR for published tables). CDS also has an active R&T policy in this domain, such as the development of tools for visual exploration of astronomical documents, based on the Self Organizing Map algorithm, operational for *Astronomy & Astrophysics* and *The Astrophysical Journal* (Poinçot, Lesteven, & Murtagh 1998), and for (semi–)automatic recognition of object names in texts (Lesteven *et al.* 1998).

3. The CDS Role in the Virtual Observatory

CDS participation to the future Virtual Observatory can be described as follows:

- Provision of high value-added reference services, essential for data interpretation. Each service is a hub towards other distributed reference services: SIMBAD gathers fundamental information about objects, with links to observations and bibliography; VizieR integrates catalogs, published tables, survey result catalogs, including the very large ones (in November 2000, the largest survey in VizieR is USNO A2.0 and the second largest is the second release of 2MASS, with respectively 526 and 162 million objects), and observation logs, with links to original data in the archives. Both services are major sources for data mining purposes, to gather known information about particular objects or given regions of the sky, or to select objects having given properties.

- Provision of tools for data integration. ALADIN is a prototype of a generic data integration tool: it integrates SIMBAD and NED with tabular data (again catalogs, published tables, observation logs, survey results catalogs), found in VizieR or provided by users with a simple formatting rule (tab separated values or XML), and with images, provided by the ALADIN server (Schmidt survey scans of various sampling and in the future NIR and CCD surveys) or by users (with FITS WCS format). Very recent developments of ALADIN, in prototype phase in November 2000, go further in that direction by allowing usage of any on–line image database in FITS WCS, through a GLU description.

- Research and development of innovative services well adapted to the Virtual Observatory purposes, in particular for data mining (prototyped in the frame of the ESO-CDS Data Mining project, Ortiz & Ochsenbein 2001, these proceedings).

- Provision of generic tools for interoperability, such as the GLU or *SIMBAD name resolver*, presently used by most archive services and the ADS.

- Very active participation to the definition of metadata and exchange standards. Metadata definition must be a collective endeavor. CDS participated to the ISAIA consortium assessment of metadata for astronomy, space plasma physics, and planetology. It also defined the Uniform Content Descriptors (UCDs) for tabular data in the frame of the ESO-CDS data mining project. Technically, XML is certainly one of the key tools for metadata management in the near future. CDS has lead the *Astrores* consortium, which defined an XML standard for tabular data in 1999 (Ochsenbein et al. 2000). ALADIN is also fully XML compatible, and VizieR result can also be retrieved in XML format.

4. Conclusion

CDS will be one of the major hubs of the Virtual Observatory, by providing essential services and tools for data management and interpretation, and by playing an active role in metadata definition. The building of the astronomy network has been extremely successful in the last years, and the Virtual Observatory constructors can take advantage of lessons learned about key elements for success : very active collaboration between international partners with different cultures and sometimes different interests, and very active search for consensus on a small number of essential exchange standards.

References

Bonnarel, F., Fernique, P., Bienaymé, O., Egret, D., Genova, F., Louys, M., Ochsenbein, F., Wenger, M., & Bartlett, J. G. 2000, A&A, 143, 33

Egret, D., Fernique, P., & Genova, F. 1998, in ASP Conference Series Vol. 145, Astronomical Data Analysis Software and Systems VII, ed. R. Albrecht, R. N. Hook & H. A. Bushouse (San Francisco: ASP), 416

Fernique, P., Ochsenbein, F., & Wenger, M. 1998, in ASP Conference Series Vol. 145, Astronomical Data Analysis Software and Systems VII, ed. R. Albrecht, R. N. Hook & H. A. Bushouse (San Francisco: ASP), 461

Genova, F., Egret, D., Bienaymé, O., Bonnarel, F., Dubois, P., Fernique, P., Jasniewicz, G., Lesteven, S., Monier, R., Ochsenbein, F., & Wenger, M. 2000, A&A, 143, 1

Heikkila, C. W., McGlynn, T. A., & White, N. E. 1999, in ASP Conference Series Vol. 172, Astronomical Data Analysis Software and Systems VIII, ed. D. M. Mehringer, R. L. Plante, & D. A. Roberts (San Francisco: ASP), 221

Lesteven, S., Bonnarel, F., Dubois, P., Egret, D., Fernique, P., Genova, F., Murtagh, F., Ochsenbein, F., & Wenger, M. 1998, in ASP Conference Series Vol. 153, Library and Information Services in Astronomy III, ed. U. Grothkopf, H. Andernach, S. Steven-Rayburn & M. Gomez (San Francisco: ASP), 61

Ochsenbein, F., & Lequeux, J. 1995, Vistas in Astron., 39, 227

Ochsenbein, F., Bauer, P., & Marcout, J. 2000, A&A, 143, 23

Ochsenbein, F., Albrecht, M., Brighton, A., Fernique, P., Guillaume, D., Hanisch, R. A., Shaya, E., & Wicenec, A. 2000, in ASP Conference Series, Astronomical Data Analysis Software and Systems IX, ed. D. Crabtree, C. Manset, & C. Veillet, in press

Poinçot, P., Lesteven, S., & Murtagh, F. 1998, A&AS, 130, 183

Schmitz, M., Helou, G., Dubois, P., LaGue, C., Madore, B., Corwin, H. G. Jr., & Lesteven, S. 1995, in Information & On-line Data in Astronomy, ed. D. Egret & M. A. Albrecht (Dordrecht: Kluwer Acad. Publ.), 259

Wenger, M., Ochsenbein, F., Egret, D., Dubois, P., Bonnarel, F., Borde, S., Genova, F., Jasniewicz, G., Laloë, S., Lesteven, S., & Monier, R. 2000, A&A, 143, 9

Lessons Learned for the Virtual Observatory from the Scientist's Expert Assistant Project

Sandy Grosvenor

Booz-Allen Hamilton, 7404 Executive Place, Seabrook, MD 20706

Jeremy Jones, LaMont Ruley

NASA Goddard Space Flight Center, Greenbelt, MD 20771

Mark Fishman, Karl Wolf

AppNet Inc., 1100 West Street, Laurel, MD 20707

Anuradha Koratkar

Space Telescope Science Institute, 3700 San Martin Drive, Baltimore, MD 21218

Abstract. During the past two years, the Scientist's Expert Assistant (SEA) team has been prototyping proposal development tools for the Hubble Space Telescope in an effort to demonstrate the role of software in reducing support costs for the Next Generation Space Telescope (NGST). This effort has been a success. The Space Telescope Science Institute is currently building a new set of observing tools based on SEA technology, and new NASA observatories such as SOFIA are also planning to use SEA.

The lessons learned from SEA are valuable for the planning and development of the Virtual Observatory (VO). These include: the need for ongoing close collaboration of ideas, concepts, and designs between scientists and software developers throughout the development lifecycle; the value of early and frequent demonstrations of prototypes to increase the usability of the final product; and the feasibility and need of developing truly cross-platform, cross-observatory visual software with the Java language.

A major lesson that SEA passes on to VO is the concept that the VO should bring the astronomical data pipeline to a full circle. The pipeline currently starts with planning new observations and ends with the archiving of the final images. It should be a full circle: those archive images should feed right back into the planning process. SEA is continuing to develop new visual tools that will enhance these links between archive analysis tools and observation planning.

1. Introduction

The Scientist's Expert Assistant (SEA) has been a prototype effort of a small team of astronomers and computer scientists for the Space Telescope Science Institute (STScI) and Goddard's Advanced Architectures and Automation (Code 588) group. We developed a functional prototype that uses an interactive visual and expert system approach to developing and planning observing programs. The system was initially developed with Hubble Space Telescope's Advanced Camera for Surveys (HST/ACS) as a test bed and currently supports several of its instruments. In SEA, users can retrieve and display previously observed images of astronomical targets by accessing the data archives, overlay any of HST's instrument apertures on the image. They can then visually change their observing parameters. For example, the probability of scheduling can be significantly affected by specifying particular aperture orientations. In SEA users can interactively rotate the aperture and see the impact of their orientation constraints. SEA also supports a visual visit planner, and exposure time calculator.

The NGST-funded part of the SEA project is now complete. SEA continues to develop via support from Goddard's Code 588 and a grant from NASA. In this new phase we are implementing new capabilities that have direct applicability to a future Virtual Observatory. First, we've generalized SEA to better support multi-observatory program development. Second, we are building into SEA the ability to model the light path of an exposure. This will allow us to begin to predict how a hypothetical target might look when observed. This predicted image can then be compared to existing archival image. If we have a model of an observatory's light path, we can take an existing image from one observatory, run it through the SEA to break out the observatory specific contributions, then re-model the image for a different set of observing parameters. This has exciting potential for the Virtual Observatory tool set, as it brings the data pipeline full circle: from observatory planning to archive analysis back to planning.

2. Lessons Learned from SEA for Virtual Observatory Development

2.1. Rapid Prototyping Works, But Don't Cheat on Infrastructure

The technical approach to developing SEA has been an iterative rapid prototyping approach. Informally known as "design-a-little-build-a-little-test-a-little," this approach involves iterating through the design/build/test cycle allowing new ideas to be quickly explored, and if promising to be developed further. If a particular feature does not show effectiveness, it can be abandoned before excessive resources have been invested. We found this to be an extremely effective development paradigm.

There is a potential downfall to this approach, however. Developers cannot cheat, even during early prototyping, on having well designed, well-implemented internal infrastructure for supporting subsequent development. This is the downfall of many prototyping efforts because there is no framework under the visual facade. A good base framework is critical for ongoing development.

2.2. Java Is Ready for "Prime-Time"

At the inception of SEA, Java was just beginning to emerge as a truly viable development language. Its claims of cross platform independence, its extensive libraries, and its ability to run from the web all looked like a good fit for meeting SEA's goals. Therefore, we chose to design and implement SEA as a pure Java product. The decision as proven a smart one. And with each release, Java has gotten faster and more reliable. Its built-in standard libraries are vast and well-designed, saving us considerable development time. This is especially true of the visual libraries such as Java2D, Java3D and the Advanced Imaging.

2.3. Demo Early, Demo Often

SEA's approach to incorporating user input has been multi-layered and iterative. As one of the first steps of the project, we conducted interviews with several astronomers, and STScI staff about their perceptions of the strengths and weaknesses of their current tools. This included some initial brainstorming about what types of new tools might have the most significant impact. Throughout the project life, we have tried to demonstrate SEA at major astronomical conferences. The feedback we've obtained through these conferences has had a significant impact on the project. We always learned a tremendous amount from every demo. Demos of tools still incomplete and "very young" were also valuable. They sparked discussion between astronomers and developers that refined (or sometimes redirected) subsequent development.

2.4. Close Ongoing Interaction between Astronomers and Developers is Vital

Perhaps the most critical component of incorporating user involvement was having a practicing PhD astronomer as a full member of the team. While most of the team is composed of software developers with little formal astronomical training, our alpha-user has several years of experience at STScI in user-support with little formal computer training. An alpha-user acts not only as a primary direct source of end-user input, but also coordinates input from many other users. They act largely as an advocate of end-users, linking the developers with the astronomical expertise of the user community. It is also a critical requirement of an alpha-user that they be interested enough in the technical aspects of the project to participate in brainstorming and design sessions with developers.

2.5. Collaboration: "Talk is Cheap"—Ongoing Efforts Are Harder to Sustain

During the SEA project, we have been participants in several attempts to promote collaboration. The benefits are obvious in shared development time and expertise. Unfortunately, ongoing collaboration is extremely difficult to sustain. There are several very valid reasons: Lack of organizational commitment, and the simple reality of daily pressures exerting priority over other "optional" activities to name a couple of them. Collaborations seem to work best when carefully focused so that the long-term cost of not participating is clearly better than the initial cost of participation. Distance is also a challenge, we've found that leveraging conferences (staying an extra day with a collaborator) an effective way of solving this challenge.

The JSky project is one example of a successful long-distance collaborative effort. JSky is a coordinated library of reusable Java components for use in astronomy. Originally sponsored by the European Southern Observatory, it is now sponsored by the Gemini 8m project. The Visual Target Tuner (VTT) module of the SEA uses JSky components for much of its underlying image rendering engine. In return, the SEA team has contributed code to JSky. While JSky is still a modest library, it has been a successful collaboration for the SEA in reducing development time and promoting cross-observatory relationships. We hope that the community will embrace it and contribute more components.

3. Lessons For The Virtual Observatory

The experiences of developing SEA has relevance for the Virtual Observatory in addition to the clearly applicable lessons described above:

- SEA's Visual Target Tuner already provides a template for easily accessing multiple archives into an interactive visual front-end. A prototype of combining images from multiple wavelengths already exists.

- The new modeling/simulation functionality under development in SEA is designed at the outset to be distributed, observatory and archive independent, and scalable. Astronomers will need to be able to quickly and transparently reconfigure their analytical environment. For example, a scientist must able to perform a visual analysis at their workstation and then *easily* transfer it to a more powerful environment such as the 3D virtual reality walls at the National Center for Supercomputing Applications (NCSA).

A recurring theme of the talks and posters at the June 2000 Virtual Observatory Workshop focused on the need to tie together astronomical data archives. Very little was said about bringing the entire astronomical data pipeline into a true full circle. Traditionally this pipeline has been a linear one: images are proposed, planned, taken, downloaded, processed, and finally archived. The potential of the Virtual Observatory to provide scientific gains through large scale studies of these now vast and growing data archives is clear. *Another major benefit is the power of tying these new analyses of existing archives back to the beginning of the pipeline: the planning and proposing of new observations.*

Acknowledgments. This work was supported by the Next Generation Space Telescope (NGST) technology funding. The SEA is a joint effort between the Space Telescope Science Institute (STScI) and the Advanced Architectures and Automation Branch of Goddard Space Flight Center's (GSFC) Information Systems Center. All of the documents published concerning SEA can be found at the group's website[1].

---

[1] http://aaaprod.gsfc.nasa.gov/SEA

… Virtual Observatories of the Future
ASP Conference Series, Vol. 225, 2001
R.J. Brunner, S.G. Djorgovski, and A.S. Szalay, eds.

The Virtual Solar Observatory

Frank Hill

National Solar Observatory, PO Box 26732, Tucson, AZ 85726-6732

Abstract. The Virtual Solar Observatory is a system of federated databases, search tools, and analysis packages in the context of solar physics. It would provide a unified search tool, an adaptive keyword thesaurus, and context-based search capabilities for easy centralized access to 12–15 existing solar archives.

1. Introduction

Solar phenomena produce effects over the entire electromagnetic spectrum, in energetic particles, and in several physical variables such as magnetic fields, velocity, and temperature. Consequently, observational solar and space physics research requires correlations of disparate data held in many locations and formats. Since Information Technology advances are now enabling distributed data mining, it is feasible to design a Virtual Solar Observatory (VSO) — a software system linking together several existing archives with a single unified search tool and user interface. The VSO is the solar analog of the National Virtual Observatory (NVO).

A VSO would fill a long-standing need in the solar and space physics research communities for a unified data retrieval and analysis system and would provide a superb educational outreach resource A VSO has different technical requirements from an NVO. Some examples include the catalog, which houses many different types of objects in the NVO compared to the VSO; the choice of useful spatial coordinate and transformation systems; the pattern recognition requirements; and the prevalence of time series data in solar physics.

There are many solar research problems that would benefit from a VSO. These include coronal mass ejection (CME) and geomagnetic storm prediction, coronal heating, the origin of solar activity, sunspot structure, solar irradiance variations, active region evolution, local helioseismology and subsurface dynamics, long-term granulation behavior, and the origin of the solar wind.

The concept of linking solar databases is not new. Previous concepts include the Whole Sun Catalog (Sanchéz-Duarte *et al.* 1997, Dimitoglou *et al.* 1998). Several solar databases already exist, such as the NSO Digital Library (Hill *et al.* 2000), BASS2000 (Roudier and Malherbe, 1997), and ARTHEMIS (Reardon, *et al.* 1997a).

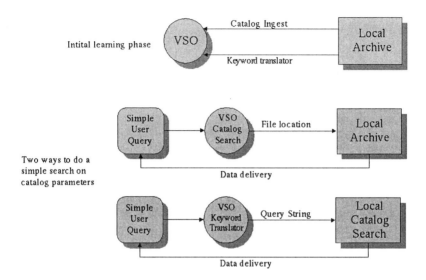

Figure 1. Top: The VSO will have a learning phase during which the catalogs of the component archives will be ingested, and a bi-directional keyword thesaurus will be generated. Centralizing these tasks at the VSO instead of at the component archives will lower the "potential barrier" tor VSO participation. Bottom: The VSO will perform simple catalog parameter searches in 2 ways: performing the search at the VSO server and requesting the local archives to deliver the data to the user; or translating the relevant keywords and performing the search at the local archives.

2. VSO Component Archives

The VSO will initially comprise 104 TB of data from a number of existing archives from a number of organizations. These include: the High Altitude Observatory (HAO), the NASA/Goddard Solar Data Analysis Center (SDAC), the Lockheed Martin Solar and Astrophysics Laboratory, Montana State University, ARTHEMIS/SOLAR (Naples, Trieste, Turin, Florence), the University of Southern California/Mt. Wilson 60-Ft Tower, National Solar Observatory (including GONG and SOLIS), Stanford University/MDI, Big Bear Solar Observatory/New Jersey Institute Of Technology, and HESSI.

These archives contain a wide variety of solar data including full-disk magnetograms, Dopplergrams, and spectroheliograms; high-resolution images, coronal images, vector magnetic field maps, spectra, flare light curves, particle fluxes, and sunspot statistics. The observations span virtually the entire electromagnetic spectrum from gamma rays to radio.

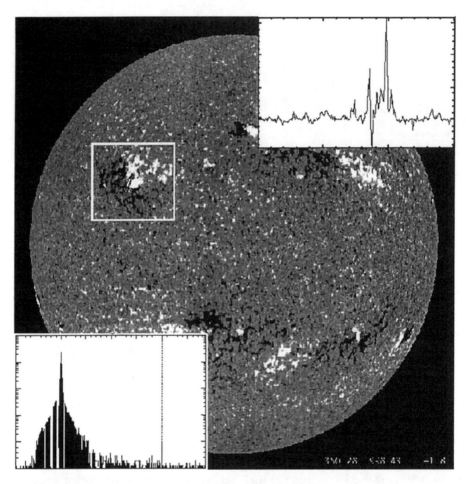

Figure 2. An illustration of a content-based search The image, a magnetogram (magnetic map) of the full solar disk, was selected not by external characteristics (such as date) but by a quantitative internal property, a threshold magnetic flux (left-hand graph) in the selected active region located within the box. Further analysis applied to a group of images thus selected reveals the time evolution of magnetic flux (right-hand graph).

3. Technical Concept

The VSO software will have several features. The core concept is a set of federated distributed data archives connected to a common Internet gateway — a single user interface to search many databases. To facilitate this, the VSO will have bi-directional translation via an adaptive data keyword thesaurus between the local archives and the VSO server. This will provide a low "potential barrier" for local archive integration. Simple catalog searches will be done either by the VSO server, or on any number of local archive platforms. These concepts are schematically outlined in Figure 1. Flexible data delivery formats and media as well as a suite of analysis, graphics and visual computing tools will be available.

The VSO will also perform searches on quantities computed directly from the data content as well as on catalogs of metadata. This will require computational load balancing dynamically distributed across the Internet. Finally, the VSO will provide dynamic usage metrics to measure usage.

4. Management

The VSO management will have two main components. A small core management team (Project Director, Manager/Software Engineer, and Scientist) provides accountability, visibility, and an agency contact point. A larger distributed oversight board contributes scientific, technical, and managerial expertise, and represents member archives, users, and IT professionals. Each member archive will receive support for local VSO integration and operation and contribute labor for overall VSO development.

References

Dimitoglou, G., Mendiboure, C., Reardon, K., & Sanchèz-Duarte, L. 1998, in ASP Conference Series Vol. 155, 2nd Advances in Solar Physics Euroconference: Three-Dimensional Structure of Solar Active Regions, ed. C. E. Alissandrakis & B. Schmieder (San Francisco: ASP), 297

Hill, F., Erdwurm, W., Branston, D., & McGraw, R. 2000, J. Atmospheric and Solar-Terrestrial Physics, in press

Reardon, K., Severino, G., Cauzzi, G., Gomez, M. T., Straus, T., Russo, G., Smaldone, L. A., & Marmolino, C. 1997, in ASP Conference Series Vol. 118, 1st Advances in Solar Physics Euroconference: Advances in Physics of Sunspots, ed. B. Schmieder, J. C. del Toro Iniesta, & M. Vazquez (San Francisco: ASP), 398

Roudier, Th., & Malherbe, J. M. 1997, in Forum THEMIS: Science with THEMIS, ed. N. Mein & S.Sahal-Brèchot (Paris: Obs. de Paris), 45

Sanchèz-Duarte, L., Fleck, B., Bentley, R. 1997, in ASP Conference Series Vol. 118, 1st Advances in Solar Physics Euroconference: Advances in Physics of Sunspots, ed. B. Schmieder, J. C. del Toro Iniesta, & M. Vazquez (San Francisco: ASP), 382

3-D Visualizations of Massive Astronomy Datasets with a Digital Dome

Charles T. Liu, Brian Abbott, Carter Emmart, Mordecai-Mark Mac Low, Michael Shara, Francis J. Summers, & Neil D. Tyson

Department of Astrophysics, American Museum of Natural History[1], 79th Street at Central Park West, New York, NY 10024

Abstract. Three-dimensional, real-time visualizations of very large astronomical datasets are now being achieved using the new Hayden Planetarium's Digital Dome system at the Rose Center for Earth and Space, part of the American Museum of Natural History in New York. Through collaborations with universities, supercomputing centers and other institutions and facilities, this Digital Dome allows users to "virtually observe" computer simulations and observational data with full interactivity and full range of observer motion. This facility (and its descendants) are research tools which can be fully integrated into any realization of a future Virtual Observatory.

1. Introduction

The analysis of terabyte and petabyte datasets — which we all envision will form the backbone of the future Virtual Observatory — will require much more than enhanced versions of currently available hardware and software. Completely new methods must be conceived, developed and implemented. Among these innovations, tools to visualize massive datasets may be the most challenging to integrate into our current data analysis paradigms.

All of us have experience using visualization techniques in two dimensions to analyze data: blinking images to sense variability or motion, plotting data points on various axes to spot outliers, or even just staring at a spectrum to find an emission or absorption feature. At our desks, on our computer screens or on a piece of paper, many among us can effectively pick out minute details in a sea of information. Now imagine shifting our perspective, to add a third spatial dimension — and the ability to view it from *any position or viewpoint*, interactively in real time. The leap in interpretive power may well be as significant as going from photometry to spectroscopy — or to cite a more mundane example, from black-and-white to color.

Such a three-dimensional visualization system is currently operating at the new Hayden Planetarium in New York City. As part of the Rose Center for Earth and Space at the American Museum of Natural History, the Hayden Planetarium theater contains a fully tested, fully operational and robust Digital Dome system that has already visualized massive, multi-terabyte astronomical datasets and is poised to conduct scientific programs — to conduct "virtual observations" for research, discovery and analysis as well as for accurate presentation of specific results. We present here a brief description of the Digital Dome and some of the datasets that have been and will be visualized with the system, and discuss some of the visualization issues relevant to the Virtual Observatory.

2. The Digital Dome System

The view from within the digital dome system, with a test image of the Orion Nebula projected in one direction, is presented in Figure 1. The hardware, software and display specifications of the Hayden Planetarium Digital Dome system can be summarized as follows.

Hardware:	2 Silicon Graphics Onyx supercomputers
(testbed)	3 graphics pipes, 12 MIPS R12000 processors
(full dome)	7 pipes, 28 processors
Software:	C-GALAXY (custom real-time renderer)
	Virtual Director (flight path editor)
	Full compatibility with commercial renderers
	(*e.g.*, Maya, Renderman, Everest)
Display:	9 Megapixel output over 2π steradians
	25-meter diameter viewing screen

Simply put, this Digital Dome system is a functioning model of a "digital observatory," capable of producing a near-virtual reality, immersive environment for any massive three-dimensional dataset. Indeed, it has already been used to visualize and display a number of large astronomical datasets. A few examples include:

- Hipparcos, Yale, Gliese and Henry Draper star catalogs
- The 3-Dimensional Volume of the Orion Nebula (O'Dell & Wong 1996)
- The Milky Way, incorporating numerous stellar and nebular catalogs
- The Nearby Galaxies Catalog (Tully 1988)
- Dynamics of globular clusters (Takahashi & Portegies Zwart 2000)
- Simulations of interacting galaxies (Mihos, Dubinski & Hernquist 1998)
- Cold dark matter models of large scale structure (Cen & Ostriker 1999)

The limits of this system have not yet been approached. Collaborations with the supercomputing centers in San Diego, CA, Pittsburgh, PA and Champaign-Urbana, IL have been established to help provide the necessary computational resources to produce visualizations of almost arbitrary complexity. We are currently constrained only by the number of datasets that the astronomical community can provide.

One small example of the Digital Dome's capabilities is presented in Figure 2. Using impressive Hubble Space Telescope data of the Orion Nebula, O'Dell & Wong (1996) have modeled the structure of the volume centered on the the Trapezium. With their help, and that of the personnel and resources of the San

Figure 1. The Digital Dome system of the new Hayden Planetarium, in the Rose Center for Earth and Space at the American Museum of Natural History in New York. Also shown, in the center of the theater, is the Zeiss Mark IX star projector. (Photo by D. Finnin.)

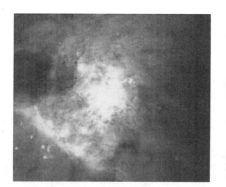 

Figure 2. *Left:* Mosaic of the center of the Orion Nebula, as imaged by O'Dell & Wong (1996) using the Hubble Space Telescope. *Right:* Close-up view of the Trapezium, as visualized using the Digital Dome with the same dataset. The view shown is roughly parallel to the surface of the page in the HST mosaic, from the lower left toward the center.

Diego Supercomputing Center, we have used the Digital Dome to visualize this 3-D dataset (including the stars, gas and dust within the volume). We can now fly into and through the nebula, from any direction and along arbitrary flight paths — not just for education and entertainment, but for data analysis as well.

3. Discussion

At this conference, distributed visualization capabilities (for example, software packages available for single computer screens or one-person "virtual reality helmets") has been acknowledged as an important priority for the Virtual Observatory project. Certainly such systems will have an important place in any VO master plan; and within a decade, they may well be useful research tools. With this in mind, the role of the Hayden Digital Dome and systems like it appears to be twofold:

[1] The Digital Dome provides this 3-D, interactive "virtual observing" environment *today*, to analyze data *now* rather than years from now, and serves as a testbed for future visualization strategies.

[2] The Digital Dome gives a larger experience than can be achieved in a distributed system, in the same way that a musical performance in a concert hall far exceeds the experience of listening to the same music on the radio. Properly utilized, the potential scientific yield should be similarly magnified.

The difficulties of the effective use of these large systems also appear to be twofold. First, they must be easy to use; second, using them must produce worthwhile scientific results. The astronomical community is beginning to rise to these challenges, which are both technical and visionary in nature: in May 2000, a collaboration was formalized between the University of Tokyo, the Institute of Advanced Study and the American Museum of Natural History to do just this. Combining GRAPE-6 computer hardware, newly developed software, and the Hayden Digital Dome, the goal of this project is to study the dynamical evolution of globular clusters with unprecedented detail and perspective. Hopefully, the number of such synergies will increase rapidly with time.

Acknowledgments. Development of the Digital Dome at the new Hayden Planetarium and Rose Center for Earth and Space was supported in part by NASA and the National Science Foundation.

References

Cen, R., & Ostriker, J. P. 1999, ApJ, 514, 1
Mihos, J. C., Dubinski, J., & Hernquist, L. 1998, ApJ, 494, 183
O'Dell, C. R., & Wong, K. 1996, AJ, 111, 846
Takahashi, K., & Portegies Zwart, S. F. 2000, ApJ, 535, 759
Tully, R. B. 1988, Nearby Galaxies Catalog (Cambridge: Cambridge U. Press)

Serving the Sky

A. Mahabal, R.J. Brunner, S.G. Djorgovski, R.R. Gal

Department of Astronomy, California Institute of Technology, Pasadena, CA 91125

Joseph Jacob

Jet Propulsion Laboratory, California Institute of Technology, M/S 126-234, 4800 Oak Grove Drive, Pasadena, CA 91109-8099

Steve Odewahn

Department of Physics & Astronomy, Arizona State University, Tempe, AZ, 85287

Abstract. The Digitized Palomar Observatory Sky Survey (DPOSS) covers the entire Northern sky in three bands. The overlap in the 897 fields that cover the sky allows us to accurately align the different fields, calibrate them with respect to each other, and most importantly, do absolute photometry with the survey.

Our goal is to build a sky server in order to make available the entire Northern sky as a seamless mosaic at different wavelengths and at different resolutions, stored as smaller non-overlapping images which can be juxtaposed on the fly. The smaller images will retain photometric information as well as WCS information. In the following pages we outline some of the technical steps involved.

1. Why Mosaic?

The Digitized Palomar observatory Sky Survey (DPOSS) covers the entire Northern sky by means of 897 plates, each of which is $6°.5 \times 6°.5$. While most objects seen from Earth, including almost all galaxies, fit on a single such plate, there exist objects like the High Velocity Clouds (HVCs) which can cover several plates. In order to get an accurate picture of such objects, multiple plates have to be registered using their overlap region and then combined seamlessly. Additionally, this has to be done for all the different bands (J, F and N in case of DPOSS) in a consistent manner. This involves various steps like (1) Vignetting correction to correct for missing light in the corners, (2) introducing WCS so that the different bands are registered and (3) Photometric Calibration so that the magnitudes are universally understood.

In this paper we outline how these steps are being carried out and the steps being taken in order to serve the sky.

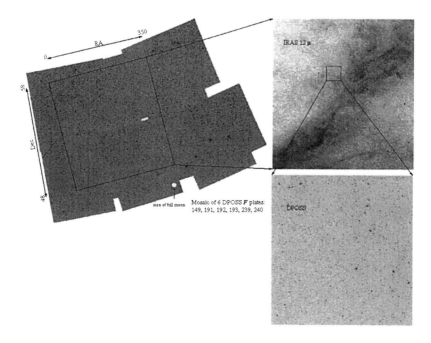

Figure 1. A demonstration of the multiple plate mosaic process.

2. The Steps

2.1. Vignetting Correction

Placing a square plate in a circular telescope aperture results in a nonuniform illumination across the plate, known as vignetting. The pattern must be removed before anything else can be done.

While the pattern is radially symmetric in general, it is also plate and filter dependent. As a result it needs to be evaluated from a large number of plates. For DPOSS plates — each of which is 23040 x 23040 pixels — we perform the procedure described below.

- Take 100 F plates binned 8x8 so that each is now 2880x2880.
- Divide each image by the median of its central 720x720 region.
- Stack the resulting images.
- For each 2880x2880 vertical stack, throw away the 10 brightest and faintest pixels.
- Obtain the median of the remaining 80 pixels.
- Normalize the resulting image such that the maximum in the central 720x720 region is 1.0 (The final photometric normalization will come from CCD calibration anyway).

- That gives the "master vignetting correction" image in F.
- Repeat the procedure for the J and N bands.

Contours of the master vignetting correction images in F, J and N bands are shown in Figure 2. An example of a plate before and after applying vignetting correction is shown in Figure 3.

Figure 2. $F\ J\ N$ contours of the master vignetting correction images

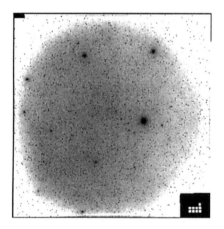 

Figure 3.   Vignetting uncorrected (left) and corrected (right) DPOSS plate F780

2.2. WCS

WCS (World Coordinate System) information is used to identify the same objects in overlapping regions. The digitized plates have WCS accurate to almost one second of arc over the whole plate, degrading towards the edges of the plate. To get accurate astrometry at the plate edges we use WCSTOOLS to compare DPOSS catalogs with the USNO-A2 catalog, which has an astrometric accuracy of better than one arcsecond.

This is done in the following way:

- Divide a plate into chunks of 1024x1024 pixels.
- For each chunk obtain the object catalog from SKICAT.

- For the same region obtain objects from USNO-A2.
- Cross-match.
- Use the cross-match to setup the relation between X-Y and RA-Dec systems in the FITS header.

A small part of plate F193 is shown in Figure 4. USNO-A2 stars are seen overlaid as circles.

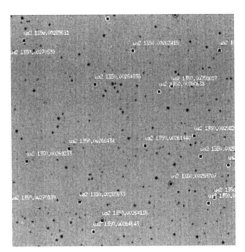

Figure 4. WCS matched for DPOSS plate F193. Note the overlaid USNO-A2 stars which are represented as circles.

2.3. Photometric Calibration

As the plates are taken over a period of years, the observing conditions vary. CCD images obtained at a large number of locations spread over the survey region allow us to transform all the plates to a consistent standard magnitude system.

3. The Path

Once the plates have been vignetting corrected and calibrated, they are cut into smaller 1024x1024 pieces. WCS information is inserted into the FITS headers, overlapping regions merged so that exactly one image corresponds to every part of the Northern sky. The tiles can be conveniently retrieved to form a mosaic of any large piece of sky quickly with a small number of system calls. In addition, the mosaics are stored at multiple resolutions from full resolution down to the entire mosaic fitting within a single tile. In each successive tier of this resolution hierarchy, the pixel width and height are doubled. This permits the images to be quickly served at any arbitrary resolution by resampling from the resolution layer closest to the desired resolution (see also, Jacob *et al.*, these proceedings).

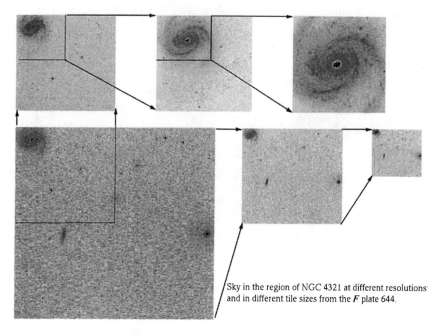

Figure 5.   Serving different resolutions and sizes

Figure 5 shows an example of serving different sizes of the sky around NGC 4321 at different resolutions.

Rapid Cross Identification for the National Virtual Observatory: The Digital Sky Project

J. Ma, T. Handley, J. Good, A. Johnson

Infrared Processing and Analysis Center, California Institute of Technology, Pasadena, CA 91125

R.J. Brunner

Department of Astronomy, California Institute of Technology, Pasadena, CA 91125

T. Prince, R. Rutledge

Space Radiation Laboratory, California Institute of Technology, Pasadena, CA 91125

R. Williams

Center for Advanced Computing and Research, California Institute of Technology, Pasadena, CA 91125

Abstract. One of the necessities for astronomical catalog generations and scientific research is the cross-identification between catalogs in terms of proximity. However, the conventional cross-identification algorithms are prohibitively expensive in terms of CPU requirements for large volume of data. In this paper, we present a fast cross-id service for national virtual observatory, which is based on an $O(N)$ fast algorithm for cross-identifications of large astronomical catalogs, where N is the number of sources involved. The fast cross-id service presented has been successfully deployed within JPL/Caltech's Infrared Science Archive (IRSA) for NASA's 2MASS (Two Micro All Sky Survey) as an initial project. The fast cross-id service has the performance and scalability to handle the anticipated over one billion sources in the final release of 2MASS.

1. Introduction

The Digital Sky Project is funded by National Science Foundation via National Partnership for Advanced Computational Infrastructure. The project is designed to be a technology demonstration to meet the challenges of the explosive growth of astronomical data (Brunner, 2001, these proceedings).

One of the most important needs of a National Virtual Observatory is the cross-identification between catalogs in terms of proximity (Decadal Report[1];

---

[1] http://www.nap.edu/books/0309070317/html/

NVO White Paper, these proceedings). However, the conventional cross-identification algorithms are prohibitively CPU expensive when applied to a large volume of data.

In this paper, we will describe the performance of the cross identification service for rapid cross identifications of large astronomical catalogs under the heterogeneous and distributed data environment in the astronomical community.

The fast cross-id service has been successfully deployed within JPL/Caltech's Infrared Science Archive (IRSA) for NASA's 2MASS (Two Micro All Sky Survey) as an initial project (Berriman, et al. 2001, these proceedings). It has the performance and scalability to handle the anticipated over one billion sources in the final release of 2MASS.

The fast cross-id algorithm has applications in catalog generation, such as identification of duplicates and artifacts. For example, it is used via IRSA services, where a user's table can be cross correlated with on-line catalogs. The 2MASS team use it for quality assurance and source selection. Recently, the process of 2MASS artifact removal has been speeded-up by two orders of magnitudes via the application of the fast cross-id algorithm.

It has also been applied to many catalogs at IRSA, such as 2MASS, DPOSS, IRAS, ROSAT, USNO-A2, NVSS, FIRST, etc. (see Table 2, Figure 1).

A version of the fast algorithm has also been implemented on parallel computer HP exemplar.

2. Data Management and Fast Cross Comparison

The data management scheme to facilitate rapid cross identifications exploits the spherical geometry of astronomical catalogs. To be specific, the sources are organized into non-overlap strips of the same size in declination. Within each declination strip, the data are organized by right ascension in ascending order. The process of organizing the data is called index generation.

The cross comparison is performed on a strip to strip basis. That is, the comparison of a strip of sources in catalog X with catalog Y can be done by comparing the strip of data in X with the neighboring strips of sources in catalog Y only.

3. Performance and Results

Computer programs have been implemented utilizing the algorithms of this paper, and capable of rapid cross comparisons of large astronomical catalogs. The speed of the algorithm is demonstrated by timing results in Tables 1 and 2 below.

3.1. Index Generation

Table 1 shows that the performance of index algorithm is almost linearly proportional to number of sources (tests conducted on Sun Sparc5 with a 4 x 75 Mhz database server).

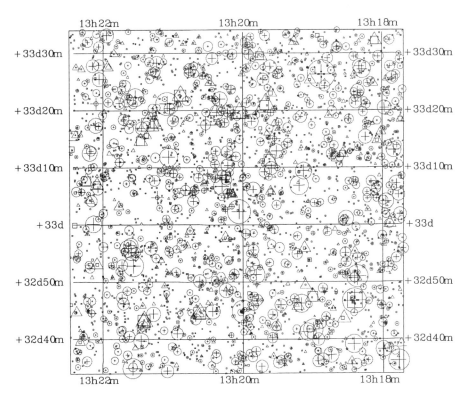

Figure 1. USNO-A2 vs. 2MASS Cross-ID Maps with j_m as 2MASS magnitude and *blue* as USNO-A2 magnitude (Circles: 2MASS matched; Crosses: USNO-A2 matched; Triangles: 2MASS unmatched; Squares: USNO-A2 unmatched)

Number of Sources	Time in Second
2,805	2
196,842	142
308,152	214
12,012,751	7002

Table 1. Timing Tests for Index Generation

3.2. Cross Referencing

Table 2 shows that the performance of fast cross-id algorithm is proportional to number of sources (tests conducted on Sun Sparc 5 with 2 arcsecond matching radius).

Catalog (#src)	Catalog (#src)	#Matches	Time
scan7-2mass 8034	scan8-2mass 8760	642	0.05 sec
DPOSS 308,152	2MASS 246,3223	78883	3.8 sec
FIRST 549,707	2MASS 162,213,354	25850	5m53s
NVSS 1,814,748	2MASS 162,213,354	27622	11m55s
USNO-A2 526,280,881	2MASS 2nd Release 162,213,354	114,916,919	2h17m17s
USNO-A2 526,280,881	2MASS-working 431,628,881	189,330,477	2h56m2s

Table 2. Timing Tests for Cross Comparison

4. Generalizations and Conclusions

In this paper, we discuss a fast cross-identification service based on an $O(N)$ fast algorithm for the cross-identifications of large astronomical catalogs, where N is the number of sources involved.

It is straightforward to generalize the algorithm to incorporate probability of associations based on positional uncertainty, density, luminosity, *etc.* to perform clustering analysis based on the cross-reference data generated via the fast algorithm.

The algorithm presented has been successfully deployed within NASA's 2MASS project, has the performance and scalability to process the anticipated one billion sources in the final release of 2MASS, and is actively used within the scientific community. Finally, a version of the fast algorithm has also been implemented on parallel computer.

Acknowledgments. The authors are grateful to G. B. Berriman for his review of this paper.

GAIA and Virtual Observatories

W. O'Mullane
Astrophysics Division, Space Science Department of ESA, ESTEC, 2200 AG Noordwijk, The Netherlands.

X. Luri
Departament d'Astronomia i Meteorolgia, Universitat de Barcelona, Barcelona 08028, Spain

Abstract. GAIA[1] is proposed for ESA's fifth cornerstone mission which has a prospective launch date of 2009. The objectives of GAIA are manyfold, but the core objective is the discovery of the origin and formation of the Galaxy. To do this GAIA will combine information from astrometry, photometry, and radial velocity instruments using the proven principles of the Hipparcos mission. The astrometry will be complete to $V = 20$ magnitude with accuracies of 4 microarcsec at $V = 10$, 10 microarcsec at $V = 15$ and 0.2 milliarcsec $V = 20$. The radial velocity measurements will have accuracies of 1 km/s at V~10 and 10 km/s at V~17. There will be a specifically designed 4 broadband photometric filter and 11 medium band filter photometric system.

GAIA is estimated to observe around 1.3 billion objects about 150 times over its 5 year lifetime producing around 10TB of raw data. The current estimate of processing needed for GAIA is 10^{19} flops.

1. The Scientific Goals of GAIA

GAIA's main goal is to clarify the origin and history of our Galaxy, by providing tests of the various formation theories, and of star formation and evolution. The GAIA results will precisely identify relics of tidally-disrupted accretion debris, probe the distribution of dark matter, establish the luminosity function for pre-main sequence stars, detect and categorize rapid evolutionary stellar phases, place unprecedented constraints on the age, internal structure and evolution of all stellar types, establish a rigorous distance scale framework throughout the Galaxy and beyond, and classify star formation and kinematical and dynamical behaviour within the Local Group of galaxies. However, GAIA's scientific harvest will not be limited to these (very important by themselves) results, but will extend to many other astrophysical fields. Solar System studies will receive a massive impetus through the detection of many tens of thousands of new minor planets. Tens of thousands of brown dwarfs and white dwarfs will be identified.

---
[1]http://astro.estec.esa.nl/GAIA

About 100,000 extragalactic supernovae will be discovered and details passed to ground-based observers for follow-up observations. Many thousands of extra-solar planets will be discovered.

Furthermore, the results will also contribute in several fields of fundamental physics. GAIA will follow the bending of star light by the Sun and major planets, over the entire celestial sphere, and therefore directly observe the structure of space-time —the accuracy of its measurement of General Relativistic light bending may reveal the long-sought scalar correction to its tensor form. The PPN parameters and the solar quadrupole moment J_2, will be determined with unprecedented precision. New constraints on the rate of change of the gravitational constant, and on gravitational wave energy over a certain frequency range, will be obtained.

2. GAIA Processing

Processing for survey missions requires data to be accessed readily in both the time and spatial domains. Typically a set of global iterative processes will run over the different types of raw data to produce calibrated values using calibration information which will include values from other processes. A successful example (O'Mullane & Lindegren 1999) of this type of processing was implemented using Hipparcos Intermediate data, based on Java and the Objectivity OODBMS. The algorithms produce a mission chromaticity matrix, reference great circle harmonics, and astrometry updates. Each process has an effect on the other two and therefore has a global effect on the result, similar to the type of dependencies which will occur in the GAIA data. Each process can easily be run in a parallel fashion on multiple machines under the supervision of a coordinator. To achieve insulation of the algorithm from the storage of the data, a data driven approach for the processing is adopted. Processes accept data and process them in the order they are given. This will allow for a process on a machine to be given the data on that machine first. Furthermore it can be given data in the order it is on disk which will lead to more efficient processing. The access pattern for a set of algorithms, then, can be encapsulated in a class from which specific algorithms inherit. Another advantage of this approach is that algorithm writers are not burdened with needing intimate knowledge of the storage and distribution of data.

Driven by the processing requirements data access needs to be provided to a large amount of data in both spatial and temporal domains. Furthermore the data is constantly being updated by the processing. A multi-dimensional index system should suit these purposes and indeed two such systems already exist: the Hierarchical Triangular Mesh (Kunst et al. 1999) and the Hierarchical Equal Area isoLatitude PIXelisation[2]. Both schemes allow spatial splitting of the sky. At a second level something like a kd-tree could be used for indexing on other parameters. Currently the GAIA simulator uses HTM as a mechanism for reproducing simulations for a given part of the sky. A study contract has

---

[2] http://www.tac.dk:80/~healpix

been awarded to examine the GAIA data access and analysis problem further using simulated data as input.

3. GAIA as an Observatory

A mission which extends to such faint and complete magnitude levels as GAIA, *i.e.*, V~20 mag., may be regarded at one extreme as a survey experiment to which the principles of an 'observatory' mission do not apply, or as the 'ultimate' observatory in which every interested scientist can acquire any or all of the astronomical targets of interest. Procedures should maintain as much flexibility as possible for scientists to acquire data from GAIA as rapidly as possible, whilst ensuring that the data are calibrated, verified, and properly documented before becoming public. As such many of the challenges faced by a "virtual observatory" and a "GAIA observatory" will be identical. Indeed as may be seen from Section 2 many of the goals of the "virtual observatory" are similar to the goals of GAIA. In general they involve large statistical studies over practically the entire body of data. It is easy to postulate that an observing program/ solicitation of proposals is unnecessary for a "virtual observatory" since all the data would be available. But, availability of data must be balanced with the availability of computational resources and bandwidth - allowing arbitrarily complex requests to the "observatory" may result in computationaly intensive processes.

In the case of GAIA, the data could conceivably be made available to a small number of "Observers" for an initial period of time before general release. Such "Observers" would be selected on scientific merits of submitted proposals. After this initial period the GAIA archive could indeed become part of the "virtual observatory" providing a uniform and consistent data set across the entire sky underpinning other disjoint surveys and increasing their inter operability. At this point the whole complex issue of servicing complex requests on a large data set comes in to play. An initiative such as the "virtual observatory" should produce an interesting model for reconciling data volume and computational power.

4. Public Outreach and Education

Public outreach and education are already seen as important for GAIA[3]. GAIA is exceptionally well-suited for raising educational awareness and providing technical training. GAIA will provide opportunities and challenges at all levels, from the evolution of the Galaxy and the search for extra-solar planets, through applied gravitation, to the technical challenges in accessing large data sets. Every one of these is of direct and topical interest, and produces knowledge of very wide and continuing general applicability. Among many examples of GAIA science which are directly appropriate for general educational opportunities, GAIA will provide the first detailed knowledge of the content and evolution of our own Milky Way, an aspect of astronomy with immediate resonance for all people at all levels. Thus the basic level of interest will be high. Building on this,

---

[3]http://astro.estec.esa.nl/GAIA/Outreach/Outreach.html

GAIA will provide detailed knowledge of kinematics, allowing a natural forum for explanation of Newtonian and General Relativity gravitational theory, chaos theory, and orbits. This can be provided naturally at the pictorial level — a movie of the sky — through to the highly technical — metric mapping, gravitational distortions of space-time — appropriate to all ages and interests, and all levels of educational requirements. At a wider level, by providing a precise measure of the distribution of dark matter near the Sun, and throughout the Galaxy, GAIA will set the boundaries of our understanding of the nature of matter, luminous and dark. The direct links with particle physics and fundamental physics are well known, and of wide general appeal.

GAIA will provide educational opportunities at every level, from starting school through to University research training, with complementary personal home learning. In the future, introductory self-teaching interactive modules for internet learning may become the norm, with many trial developments appropriate to this scale already in operation. These educational opportunities will naturally involve astronomy, physics, technology, and information technology, and involve access by schools, universities, libraries, museums, and private individuals.

5. Conclusion

Many problems faced by a virtual observatory are also faced by GAIA. Indeed taking the analogy to the extreme GAIA may itself be seen as a virtual observatory because of its all-sky coverage. The task in reducing and cleaning the GAIA data will be arduous, and smart database/data mining techniques will be required to make this huge amount of high quality information available to a wide audience. A tool such as the International Virtual Observatory may ease this task.

References

Kunst P., et al. 1999, The Indexing of the SDSS Science Archive, Proceedings of ADASS 99.
O'Mullane W. & Lindegren L. 1999, An OO Framework for GAIA Data Processing. Baltic Astronomy, V.8, 57–72

Virtual Observatories of the Future
ASP Conference Series, Vol. 225, 2001
R.J. Brunner, S.G. Djorgovski, and A.S. Szalay, eds.

Mining the Virtual Sky

Patricio F. Ortiz
Centre de Données de Strasbourg

Francois Ochsenbein
Centre de Données de Strasbourg

Francoise Genova
Centre de Données de Strasbourg

Andreas Wicenec
ESO, Garching

Abstract. An effort was carried on at CDS in the context of the ESO/CDS datamining project as a first attempt towards making the information available in the **VizieR** service open to users with the purpose of mining. The problems and lessons learned are described.

1. Introduction

We describe the experience carried at Centre de Données de Strasbourg (CDS) in the context of the *ESO/CDS datamining project* (EC-dm) as one possible scenario of mining in the context of a virtual observatory.

The contents of the Virtual Sky are CDS are various and are provided by different servers. EC-dm did not use all of the resources available at CDS, but concentrated in the usage of the catalog service known as **VizieR** (Ochsenbein, 1998, Ochsenbein *et al.* 2000, Ortiz *et al.* 1998). In general, the kind of data we can find nowadays fall in any of the following categories: a) **Published Catalogs** listing: positions, magnitudes, fluxes, etc; b) **Compilation and cross correlation catalogs**, usually involving multi-wavelength information and listing similar quantities as normal catalogs. c) **Images**: digitized plates, and CCD's. d) **Spectra**: mostly from space missions like IUE. e) **Telescope/Mission's Archives**, HST, CFHT, *etc*. And f) **Bibliographic references**

CDS provides several tools (all available over the web) to explore and reach different parts of either local information or links to access servers elsewhere. Among these tools we have: a) **Simbad**, provides positional querying, name solving capabilities and bibliographic references among other things. b) **VizieR**, on-line database system which holds thousands of catalogs coming mostly from the literature. c) **Aladin** Java based Visualization tool which merges data from images and catalogs (including observing logs).

2. The CDS/ESO Datamining Project

One of the primary goals of the project is to provide the community with tools to cross correlate either user provided data or already available data against the contents of catalogs handled by **VizieR**.

The Datamining software provides three main tools: a) **Positional Cross Correlator**, the most basic kind of mining in astronomy, b) **Multi-dimensional Cross Correlator**: correlations which may or may not involve positions, and c) **Catalog Selection Tool**: search for patterns and words in catalogs' metadata. Help pages and Examples are provided for all operations. Examples are guided sessions.

3. The Positional Cross Correlator

Figure 1 shows the options a user has to perform a positional cross correlation.

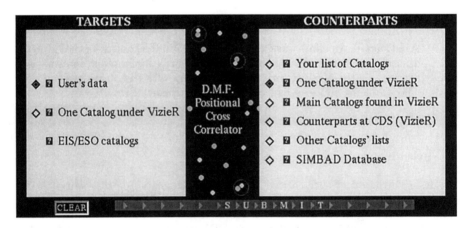

Figure 1. Positional Cross correlator selection panel

Positions listed in the **TARGET**'s side are used to search for matches in the **COUNTERPARTS**' side (**VizieR**) within a user-defined radius. Users can submit data (in the form of ASCII files) to the TARGET side containing at least positional information (RA + Dec). User favorite catalog lists —ASCII files as well— can be provided on the COUNTERPART side

The user has the opportunity to choose the same catalog used in the TARGET side. This "self correlation" allows in particular to check pairs of objects, like candidates for lensing, errors, *etc.*

The final result of a cross correlation is the result of an intermediate step (page) in which the user chooses which (and in which order) quantities are to be printed amongst the catalog variables. The results can be ordered according to maximum number of matches, proximity of closest match or kept in the same order as entered in the target side. Only target lines with matches are shown by default.

Mining the Virtual Sky 207

3.1. Preparation for Cross Correlation: Sky Indexing

One of the key issues in datamining catalogs of any size is I/O speed. Whatever extra step that could be avoided should be avoided. In order to speed up the positional cross correlation it was necessary to develop a sky indexing system and apply it to every catalog containing positional information.

The indexing consists of splitting the sky in tiles of about $30' \times 30'$, starting with the first tile in the South Celestial Pole. The whole sky contains ~ 162000 "tiles". The positional data and a pointer to **VizieR**'s internal denomination are then stored in the order given by the indexing, and a file is used to keep track of the number of elements contained in each tile.

These files are used when computing a cross correlation between any data and the corresponding catalog in **VizieR** (except huge surveys such as USNO which are handled separately). Based on the indexing information we build a table which contains for each of the $30' \times 30'$ tile the names of the tables with elements in such tile. This information is used when the user wants to find the list of all catalogs with potential counterparts to the TARGET list: only a fraction of the 2200 tables with positional information is consulted for possible counterparts.

3.2. Performance Issues

COUNTERPARTS → TARGETS ↓	Hipparcos (117955)	Tycho-2 (2539914)	Main Catalogs
Veron BL Lacs (492)	< 1 sec	< 1 sec	4 sec
Veron QSO's (13214)	< 1 sec	1 sec	17 sec
GCVS (31918)	1 sec	5 sec	—
Hipparcos (117955)	5 sec	12 sec	—

Table 1. Benchmarks for positional cross-correlation (Sun Ultra 400MHz, CPU time)

Table 1 lists the time currently needed to achieve the positional cross-correlation between widely used catalogs. Obviously, the time needed by the user to analyze the result is much longer ! As a comparison, consulting all tables in **VizieR** for positional matches around a list of 1000 target objects lasts about 2 minutes .

3.3. An Issue to be solved: Blind mining

One of the most delicate issues of positional cross correlation has to do with the way we handle the question: "*Which objects are located within X degrees from this direction?*" There is a need that future catalogs involving extended objects contain not just positional information but also information which would allow us to solve the problem more appropriately (*e.g.*, major/minor axes, position angle, or box size, *etc.*). The reason we cannot do this today is that for most catalogs this information is missing or not readily available, or the geometry of the object is complex; this means that we are obliged to treat objects as if they were point sources. Besides the question asked above one would like to know answer as well: "*Within which objects this direction fall?*" Typical examples would be to know whether a star lies near an extended galaxy or not (photometric properties like absorption will change), or whether a galaxy or QSO may be part of any Abell cluster. All of this, without the user having to use an unmanageable large search radius.

4. A Prerequisite for Multi-Dimensional Cross-Correlation: Unified Content Descriptors

There are serious problems to retrieve catalogs with similar column content, because of nomenclature diversity: for example, Johnson's V magnitude is named in 113 different ways, showing that column name can not be used as a unique descriptor of a column; and taking the units into account is not sufficient to solve the problem. In order to solve this problem, we developed a hierarchical classification of the columns contents based on the meta-information stored in **VizieR**: the column names, units, and short explanations (Ortiz 2000). In this scheme, each column is assigned a **UCD** (Unified Content Descriptor) allowing the recognition of columns with comparable contents.

The usage of UCDs allows to: a) find columns (other than RA and Dec) which can be used for cross correlation, b) find catalogs which contain parameters (columns) or interest. This information can hardly be found in the classical catalog's Meta-Data.

The **Catalog Selection Tool** uses UCDs to help users find catalogs based on their contents: the number of catalogs found by means of the UCDs usually exceeds by large the number of catalogs found by searches on their titles and/or author fields.

UCDs prove particularly useful in the exchange of information between different servers; it is hopeless to expect a unification of the column names in the tables created by different authors, but by attaching this piece of information the contents can be accurately interpreted.

References

Ochsenbein, F. 1998, ASP Conference Series, Vol 145, Astronomical Data Analysis Software and Systems VII, R. Albrecht, R.N. Hook, H.A. Bushouse, ed. (ASP, San Francisco, 1998) 387

Ortiz, P., Ochsenbein, F., Wicenec, A., Albrecht, M., 1998, in ASP Conference Series, Vol 172, Astronomical Data Analysis Software and Systems VIII., D.M. Mehringer, R.L. Plante, D.A. Roberts ed. (ASP, San Francisco) 379

Ochsenbein, F., Bauer, P., Marcout, J. 2000, A&AS, 143, 23–32

Ortiz, P. 2000, in Computer Physics Communication, Vol 127, 188

The Chandra Data Archive

Arnold Rots
Harvard-Smithsonian Center for Astrophysics, 60 Garden Street, Cambridge, MA 02138

Abstract. We provide a quick overview of the design and usage attributes of the *Chandra* Data Archive. Its functional design incorporates a single observation database that controls the complete life cycle of all observations, The data products in the archive are mainly FITS files. Data distribution is expected to take place mainly through a web interface and FTP. We describe the archive access design and discuss the FITS conventions that were adopted for the data products. The time scale adopted for the data products is Terrestrial Time (TT).

1. Introduction

The *Chandra* X-ray Observatory is the latest of NASA's great observatories, launched July 23, 1999. The data processing takes place in the *Chandra* X-ray Center (CXC) at CfA where also the *Chandra* Data Archive (CDA[1]) is located. Although the interfaces and services of the CDA are not yet up to the standards of more mature missions, we have been able to provide access to most public observations starting one month after launch. This paper gives a brief overview of the most relevant aspects of the CDA in the context of the NVO.

2. The Functional Design of the Chandra Data Archive

The *Chandra* Data Archive (CDA) is depository as well as source of information for all Chandra-related activities (Proposal submission and management, Mission Planning, Automated Processing, Data Retrieval). The observation catalog (ObsCat), embodied in the database *axafocat* plays a central role in this.

axafocat, with the help of related databases, described and controls observations through their complete life cycle. It:

- receives information from approved proposals,
- provides that information to Mission Planning,
- receives scheduling information from Mission Planning,
- provides information on scheduling, target data, and instrument configurations to Automated Processing,

---

[1]http://asc.harvard.edu/cda

- receives updates to that information from Automated Processing,
- informs the user what observations are being planned and what data are available,
- and aids in retrieving those data products.

3. The Data Products in the CDA

The CXC Automated Processing system generates data products in three main levels (Levels 0, 1, 2) and two sublevels (Levels 0.5, 1.5) of processing. All products are archived and divided into three categories: primary, secondary, and supporting products. The primary products contain all (science) Level-2 products and some crucial science Level-1 products. The secondary products contain the remaining science Level-1 and Level-1.5 products, Level-1 aspect products, and Level-1 orbit ephemeris. The remaining products are classified as supporting. Primary and secondary products are shipped to our users automatically. Individual supporting products are available upon request.

The basic data product (file) retrieval keys for observational data are:

- Time (or observation)
- Data product type (CONTENT)

(As far as retrieval by the user is concerned, the immediately visible keys are the usual observation-related ones, such as source, type of object, time, observing mode, RA, Dec, *etc.* (which largely translate into time), and standard data product packages.)

The vast majority of the data products that the user will encounter are FITS files; for these the metadata are extracted from the headers. In addition, the archive holds raw binary files, with accompanying log files for the metadata, and is equipped to deal with other standard formats, such as Postscript, PDF, JPEG, *etc.*

A FITS data product's type is identified by its CONTENT keyword. The list of recognized CONTENT values has, in principle, a flat structure, but for most querying activities a hierarchy is imposed by grouping them into instrument/subsystem and level-of-processing categories.

4. Data Distribution

Although we will continue the practice of sending proprietary data out to Guest Observers on physical media (4-mm DAT tape, 8mm Exabyte tape, CD-ROM), we expect that the bulk of our data transfers will happen through our web interface, for public as well as proprietary data.

The CDA Search and Retrieval interface is discussed in the next section. It will enable the community to download any set of CDA products, public or proprietary (provided one has the proper authorization), or stage the data for anonymous FTP transfer.

In the meantime, we have implemented a provisional web interface that allows retrieval of primary and secondary products for public observations that have been approved by the CXC Verification & Validation process. Distribution of proprietary data takes place through separate secure FTP and physical media.

5. Archive Access

The archive is built as a twin server system: one server for the RDBMS ("sql server"), the other holding the data products ("archive server"). An in-house designed and implemented 4GL product, *arc4gl*, provides ingest, browse, and retrieve services for the archive server. This interface is used by the Automated Processing pipelines and for operational personnel. For a more extensive description, see Chary & Zografou (1998) and Zografou *et al.* (1998).

We have been designing and implementing a Java-based public interface allowing access to public and proprietary data (with proper security features). This will provide the community with a flexible and sophisticated tool to access and retrieve information and data products from the Chandra Data Archive. Until the time that this interface becomes available, we are employing a provisional interface implemented as a *cgi* script that interfaces through *arc4gl*.

6. FITS Conventions

Starting with the HEASARC FITS Working Group guidelines and conventions, we have developed an extended set of conventions to which all FITS files produced by the CXC are supposed to adhere.

FITS files consist of one or more Header-Data Units (HDU), each containing a header and a data part. For all practical purposes, an HDU may be one of the following:

- Primary array
- Image extension
- Binary table extension

The first HDU in a FITS file must be a primary array; all subsequent ones (if any) are extensions. The length of the header as well as the data part of each HDU must be an integer (positive for headers, non-negative for data parts) multiple of 2880 bytes.

All FITS files are to be checksummed: the 32-bit one's complement checksum over each HDU (and therefore over the entire file) must be -0 (all ones). Hence, first order data product integrity may be ascertained by verifying the checksum.

The FITS headers are required to have a number of standard components:

- M: Mandatory FITS keywords for HDU type
- CC: Configuration control component
- T: Timing component

- O: Observation info component
- IC: Image coordinate system keywords (CTYPEn, *etc.*)
- TC: Table column specifications (TTYPE, *etc.*), coordinate system, and ranges (TLMIN/TLMAX)

There are different specifications for these components, depending on HDU type. The processing history of CXC data products is inserted at the end of the headers of principal HDUs through HISTORY keywords, formatted as:

```
         1         2         3         4         5         6         7         8
12345678901234567890123456789012345678901234567890123456789012345678901234567890
HISTORY llll   :ssssssssssssssssssssssssssssssssssssssssssssssssssssssssssASCnnnnn

llll    label in columns 10-13 :  TOOL | PARM | CONT
s..s    string in columns 17-72
nnnnn   Sequence number in columns 76-80, with leading zeroes
```

TOOL: The string contains the tool's name and version.
PARM: The string contains one or more *keyword=value* pairs
CONT: Continuation of the HISTORY keyword that has the previous sequence number.

These HISTORY keyword values will be used in the order specified by the sequence number, not necessarily in the order in which they appear.

7. Time and Ephemerides

Time is kept in FITS files as MET (Mission Elapsed Time): seconds since 1998-01-01T00:00:00 TT (Terrestrial Time). It is tied to absolute time by the keyword MJDREF (50814.0). At that epoch, the difference with UTC was: TT−UTC = 63.184 s; by launch: TT−UTC = 64.184 s.

We have adopted the FITS formats developed in the context of the RXTE mission for orbit ephemeris files and the JPL DE-200 and DE-405 solar and planetary ephemerides. The latter come with an ANSI C interface, using HEASARC's *cfitsio*, that is functionally equivalent to the JPL Fortran software. Consequently, the RXTE multi-mission software package for time conversions, barycenter corrections, and pulsar phase binning (including absolute timing and phase-resolved spectroscopy) can be used for *Chandra* data.

Acknowledgments. The CDA is supported by NASA contract NAS8-39073, the Chandra X-ray Center.

References

Chary, S. and Zografou, P. 1998, ASP Conference Series 145: Astronomical Data Analysis Software and Systems VII, 7, 408

Zografou, P., Chary, S., Duprie, K., Estes, A., Harbo, P. and Pak, K. 1998, ASP Conference Series 145: Astronomical Data Analysis Software and Systems VII, 7, 391

Virtual Observatories of the Future
ASP Conference Series, Vol. 225, 2001
R.J. Brunner, S.G. Djorgovski, and A.S. Szalay, eds.

The Space-Time Profile for ISAIA

Arnold Rots
Harvard-Smithsonian Center for Astrophysics, 60 Garden Street, Cambridge, MA 02138

Abstract. We consider the problem of a user searching for observational information in a given region of the space-time universe where the user's specification may not match the coordinate systems used by all archives being browsed. We propose a draft profile to characterize the coordinate conventions used by both resource and query, allowing general coordinate transformations between what the seeker is looking for and what the resource has to offer.

1. Introduction

If we are to attempt to provide universal access to a large variety of observations, we need to make certain that it is clear where and when the observations are made. Put differently, if I am looking for observations of a particular object at a given time, I would expect the search engine to do the necessary transformations from the coordinate system that I use to the ones used by all the different data depositories. A space physicist should not have to convert all lines of sight to geocentric (X,Y,Z) and an extragalactic astronomer should not have to look up the ephemeris of planetary observations.

In order to allow this free exchange, we need a standard that defines the information needed to facilitate such transformations easily. The way to achieve this is to attach to each observatory/telescope/mission data archive depository a profile that describes the coordinate systems and time scales used to define the telescope position as well as each observation. If coordinate and time information in data queries are defined through the same uniform profile, we will have the ability to conduct those queries in a consistent manner. In the context of the ISAIA initiative, we have proposed a space-time coordinate profile that specifies these characteristics for each observatory/telescope/mission; attaching the same profile to data mining queries will allow the search engine to perform the necessary transformations.

The proposal may sound esoteric but these profiles are at the core of our ability to achieve full data and information exchange.

2. The Profile Attributes

There are two profiles to be defined: the Resource Profile and the Query Profile. However, both can be built from the same attribute components, making the task of implementing the transformation tool easier. Some components are defined by telescope (or observatory, or mission), others by instrument. The Resource Profile consists of four major building blocks:

1. *The space-time coordinates of the telescope*
   This says where the telescope is as a function of time.

2. *The space-time coordinates of the observations*
   This says what space-time coordinates are attached to the observations made with this telescope. In most cases the telescope position will be the center of the observation's coordinate system, but there are exceptions such as heliographs and planetary mapping.

3. *The physical phenomenon observed*
   The most obvious ones are E-M radiation (photons), particles, magnetic field, and gas properties.

4. *The parameters that are measured*
   These are the observables.

The Resource Profile contains all four categories; the Query Profile only the last three.

The attributes in the four categories are:

```
1.      The space-time coordinates of the telescope
1.1         center
1.2         reference frame
1.3         type of coordinates
1.4         coordinate ephemeris (this ties space to time).

2.      The space-time coordinates of the observations
2.1         center
2.2         reference frame
2.3         time scale
2.4         time reference position
2.5         type of coordinates
2.6         coordinate bounds
2.7         resolution (by instrument)
2.8         field of view (by instrument)

3.      The physical phenomenon observed (by instrument)

4.      The parameters that are measured (by instrument)
4.1         type
4.2         bounds
4.3         resolution
```

3. The Profile Attribute Values

The list below provides more details on the full range of attribute values and options. For bounds, resolution, and field of view values or ranges of values need

to be provided in the profile. For 4.1 (measured parameter type) one or more of
the enumerated attribute values may be selected. For all other attributes one of
the listed options may be selected.

```
1 Observatory/Telescope Spatial Coordinates
  1.1 Center
    1.1.1 Geocenter
    1.1.2 Barycenter
    1.1.3 ...
  1.2 Reference frame
    1.2.1 ICRS
    1.2.2 FK5
    1.2.3 FK4
    1.2.4 ...
  1.3 Type
    1.3.1 (X,Y,Z)
    1.3.2 (Long,Lat,R)
  1.4 Coordinate ephemeris
      (\ie spatial coordinates as a function of time)
    1.4.1 Orbit ephemeris
    1.4.2 (Long,Lat,R) and host rotation ephemeris

2 Observational Space-Time Coordinates (Telescope)
  2.1 Center
    2.1.1 Telescope position
    2.1.2 Geocenter
    2.1.3 Barycenter
    2.1.4 Heliocenter
    2.1.5 ...
  2.2 Reference frame
    2.2.1 ICRS (\ie RA,Dec J2000)
    2.2.2 FK5 (\ie RA,Dec J2000)
    2.2.3 FK4 (\ie RA,Dec B1950)
    2.2.4 Galactic
    2.2.5 Ecliptic
    2.2.6 Solar disk
    2.2.7 Planetary surface
    2.2.8 Spacecraft frame
    2.2.9 ...
  2.3 Time scale
    2.3.1 TT
    2.3.2 TDB
    2.3.3 TCB
    2.3.4 TAI
    2.3.5 UTC
    2.3.6 LST
  2.4 Time reference position
    2.4.1 Telescope
    2.4.2 Geocenter
```

2.4.2 Barycenter
2.4.2 ...
2.5 Type
    2.5.1 (X,Y,Z,t)
    2.5.2 (Long,Lat,R,t)
    2.5.3 (theta,phi,t)
2.6 Coordinate bounds
    2.6.1 Single-valued bounds for individual coordinates
    2.6.2 Ranges of bounds for individual coordinates
2.7 Resolution (Instrument)
    2.7.1 Single-valued resolution for individual coordinates
    2.7.2 Single-valued resolution for coordinate combinations
    2.7.3 Ranges of resolution for individual coordinates
    2.7.4 Ranges of resolution for coordinate combinations
2.8 Fields of view (Instrument)
    2.8.1 Single-valued FOV for individual coordinates
    2.8.2 Single-valued FOV for coordinate combinations
    2.8.3 Ranges of FOV for individual coordinates
    2.8.4 Ranges of FOV for coordinates combinations

3 Physical Phenomenon Observed (Instrument)
    3.1 E-M radiation
    3.2 Particles
    3.2 Magnetic field
    3.3 Gas properties
    3.4 ...

4 Parameters Measured (Instrument)
    4.1 Type
        4.1.1 Flux density
        4.1.2 Polarization
        4.1.3 Direction
        4.1.4 Wavelength/frequency/energy
        4.1.5 Velocity
        4.1.6 Pressure
        4.1.7 Temperature
    4.2 Bounds
        4.2.1 Single-valued bounds for individual parameters
        4.2.2 Ranges of bounds for individual parameters
    4.3 Resolution
        4.3.1 Single-valued resolution for individual parameters
        4.3.2 Single-valued resolution for parameter combinations
        4.3.3 Ranges of resolution for individual parameters
        4.3.4 Ranges of resolution for parameter combinations

Acknowledgments. This work was supported by NASA contract NAS8-39073, the Chandra X-ray Center.

The Carl Sagan Solar-Stellar Observatory

A. Sanchez-Ibarra, & J. Saucedo-Morales
Area de Astronomia, CIFUS, Universidad de Sonora, Hermosillo Sonora, Mexico

Abstract. The Astronomy Area of the Center for Research in Physics at the University of Sonora, was established in 1990. The Area operates the only working solar observatory in Mexico, and a 41 cm stellar telescope which is mostly used for educational purposes. The main research topics carried out in the Area are studies of the solar coronal activity and extragalactic Astronomy. In 1996 a project for a new, remotely controlled, solar-stellar observatory was presented. The observatory will be built at "Cerro Azul", with an elevation of 2480 m above sea level. A summary of the scientific goals and of the advances of this project will be presented. Aside from the scientific aspects, this project expects to have a strong impact on educational and outreach programs. For these reasons the new observatory will be named in honor of the late astronomer Carl Sagan.

1. Introduction

The Astronomy Area has had two observatories in operation since 1990. The "Estacion de Observacion Solar", EOS, and the 41 cm reflector telescope at the "Centro Ecologico de Sonora" Observatory. At EOS, a heliostat with two 25 cm flat mirrors and a fixed C5 telescope, are used to carry out a continuous observation program to record and patrol solar activity in the continuum, as well as in the H_α and Ca lines[1]. The 41 cm telescope is mainly used for educational purposes, making an important contribution to the community, but it has also been used in small observational projects (comets Hyakutake and Hale-Boop, supernovae, *etc.*). However, both observatories are located within the city of Hermosillo, and are subject to light and dust pollution. For these reasons, a project to build a remote control observatory at "Cerro Azul" was presented in 1996. Cerro Azul is located at approximately 200 km north of Hermosillo and it has an altitude of 2480 m. It is an isolated mountain in a very dark place (ideal conditions for an astronomical site). An array of four telescopes for solar and stellar observation has been designed by David Lunt of Coronado Instrument Group[2] for this project. The array includes a 55 cm stellar telescope, three 16 cm Maksutov solar telescopes and a 16 cm telescope to be used as an auto-guider for

---

[1] http://cosmos.cifus.uson.mx/eosdata.htm

[2] http://www.coronadofilters.com/

the stellar work. The observatory building is a 5 m diameter, 8 m high cylinder with a shell dome built by Astro Haven[3]. Electric power will be supplied by solar energy. For this purpose an array of solar panels and accumulators will be used. Communications will be done through radio-modem from the mountain to a node of a Mexican telephone company located at 20 km from the observatory, and from there through optical fiber to observatory bases in Magdalena de Kino (at 36 km from the site) and at the University of Sonora campus in Hermosillo. Under good weather conditions, the observatory will be operating 24 hours in solar and stellar research programs, as well as in educational programs that will be available through the Web.

2. Site and Instrumentation

"Cerro Azul" is a mountain located at West longitude 110° 34', North latitude 30° 44', with an elevation of 2480 m above sea level, in the Sonora-Arizona desert, a well regarded region for an optical Astronomy site. It is 36 km from the city of Magdalena and 200 km from Hermosillo. At the top of the mountain, there is a flat area of 850 m by 150 m. No significant amount of light from any town or city can reach the mountain. The "Guillermo Haro" observatory at "La Mariquita", the MMT observatory and Kitt Peak National Observatory are visible from the summit of "Cerro Azul". Five telescopes will be installed at the Observatory:

- A 55 cm Cassegrain reflector with a surface RMS of the order of 1 nanometer, f/8.4 with focal length of 457 cm. A multi-device adapter will permit to install a CCD camera, a spectrograph and a photometer.

- Three 15 cm vacuum telescopes with Maksutov design for solar observation at lines H_α, Calcium K, and He I 1083 nm. The optics surface RMS will be of the order of 2–3 Å. Each telescope will have a CCD camera.

- A 15 cm telescope for auto-guiding.

2.1. Logistics and Operation

The "Carl Sagan" Observatory is designed to operate completely by remote control from Bases at Magdalena de Kino and Hermosillo. A maintenance visit to the mountain will be scheduled every three weeks. A prototype building has been almost finished to test the instruments, control software, solar energy supply system, meteorological station, and communications system. Once all the systems are working, the instruments will be translated to Cerro Azul. The prototype building will be used as the control and observation base in Hermosillo. Observations will be possible both from Hermosillo and Magdalena, and also through the Web.

---

[3]http://www.astrohaven.com/

3. Solar Observations

The main goal of the solar observation at the Carl Sagan Observatory will be to expand the capabilities of our current observatory on campus, EOS, and to get data to support our main subject of solar research: Coronal Holes and its relationship with other types of solar activity.

The reason for observing the He I line in this facility is related to the above purpose. In our new long term plan we will observe simultaneously the chromospheric network at the H_α and Calcium lines, as well as the evolution of Coronal Holes that will be observed at the He I line at the Carl Sagan Observatory. Our goal is to understand at high temporal-resolution the evolution of Coronal Holes and their relation with other types of activity. Our data will be combined with magnetograms of SOHO and NSO/KPNO, with soft X-ray Yohkoh, and with UV SOHO images.

Meanwhile EOS will continue recording the global solar activity level, the observations with the solar array at Carl Sagan Observatory will be focused to study the border limits of Coronal Holes, to study chromospheric network variations. The solar array will also be used to observe big flares and prominence ejections. We plan to put on the Web all these observations.

4. Stellar Observations

The main goal of the 55 cm stellar telescope will be to search and study supernovae. In particular, those of type Ia that have been shown to be so valuable as cosmological standard candles. The purpose of this work, is to be able to improve our understanding of SN Ia in the local universe.

Each night several hundreds galaxies will be observed in search for new supernovae. The galaxies will be observed at intervals of about a week to increase the chances of discovering supernovae before maximum. A typical integration time of the order of one minute will be used, which will allow us to detect supernovae of ~ 19.0 magnitude with good S/N.

A fraction of the time will be dedicated to secure data for known supernovae. Preference will be given to follow up SN Ia known to be in the pre-maximum phase of the light curve. The relatively large field (28'x28') of the CCD frames, will also be very useful to study variable stars and moving objects like asteroids and comets. Aside from the scientific value, several educational and outreach programs are expected to be developed as a result of this project.

5. Education and Outreach Programs

Education in Astronomy has been an important goal for the Astronomy Area. An extensive program that includes lectures, public observations, newspaper and magazine articles, TV and radio programs, and basic courses in Astronomy has been operational throughout the history of the Astronomy Area. To complement the scientific project, A fraction of the time will be scheduled for pilot educational programs at all levels at the Carl Sagan Observatory.

One of these plans contemplates, to put on line both the solar and stellar images moments after the observations are obtained. This will permit, both the public and the astronomic community, to make use of the data from the Carl Sagan Observatory.

Strong participation by students is expected in all phases of this project. In particular, for the reduction and analysis of the data, which will have to be done almost in real time. Furthermore, small observational projects not necessarily related to the two major goals of the observatory will also be conducted by students with the advise of one of the astronomers.

These programs will be supported with on line information about the observatory and relevant information about the data and supporting software. An interactive Astronomy basic course will also be included.

6. Conclusions

It is expected that the Carl Sagan Solar-Stellar Telescope will provide a high return in scientific and educational programs at a relatively modest cost. If successful it could become a model for other small observatories throughout the world.

Virtual Observatories of the Future
ASP Conference Series, Vol. 225, 2001
R.J. Brunner, S.G. Djorgovski, and A.S. Szalay, eds.

The Role of Existing Data Archive Centers in the International Virtual Observatory: PixelSets and Catalogs

David Schade
National Research Council of Canada, Herzberg Institute of Astrophysics, Canadian Astronomy Data Centre

Abstract. The International Virtual Observatory (IVO) is a vision of what could be, in terms of information management tools in astronomy. An important part of this vision is the role of large surveys in astronomical research and the need to be able to reap the scientific harvest of these surveys. Another facet of the IVO vision is the many data archive centers that manage multi-wavelength and multi-mission datasets and distribute these to the community. The activities of these groups, in many countries, have laid the groundwork for the IVO and made that vision conceivable. Their experience and expertise will be critical for the realization of the International Virtual Observatory.

1. Introduction

A major motivation of the International Virtual Observatory initiative is the urgent need for tools to access and exploit very large homogeneous surveys (*e.g.*, SDSS, 2MASS) and to provide powerful links with existing surveys (*e.g.*, IRAS, ROSAT, *etc.*). There is clearly a big scientific payoff to be gained from such linkage. The initial distributions of material from these large surveys will frequently be in the form of catalogs which represent a processed and compressed view of the information content of the survey. Full distributions of pixel data and the provision of effective interfaces to pixel data are more difficult than handling catalogs because of the much larger data volumes. Survey projects often create their own archiving and distribution systems. This makes a great deal of sense because the surveys need to construct internal working archives during the production phases of the survey. But it may mean that the survey catalog is not linked effectively with other datasets.

In contrast, existing archive sites host large, valuable, but largely heterogeneous collections of pixel data. These collections have been built up over a number of years from large numbers of individual observing programs where observing procedures vary enormously and science goals cover a very broad range of topics. Catalogs of sources and derived parameters frequently do not exist. These sites primarily deliver pixel data to their users and these users must carry out further processing in order to do science.

How will convergence be reached between these two regimes which are very different in terms of data production, handling, processing, and distribution?

Such convergence is necessary to produce the next generation of information management systems for astronomy.

The vision of the capabilities of the IVO project has evolved naturally from the experiences of data centers and their users worldwide and these data centers need to be an important component in the realization of the IVO because of their vision and because of their knowledge of how to do what needs to be done.

2. Large Homogeneous Survey Datasets

The astronomical community is now dedicating a larger fraction of its resources to very large surveys, for example the Sloan Digital Sky Survey (SDSS), and the Two-Micron All Sky Survey (2MASS), than it has in the past. Observatories such as NOAO, CFHT, and ESO have all recognized the value of survey programs and are increasing the fraction of telescope time given to large projects, in particular wide-field imaging surveys.

The scientific payoff from survey projects will be massive but the data volumes are large and the data handling, processing, and analysis requirements are onerous. The ability to perform analysis across many surveys and at multiple wavelengths will multiply the scientific impact of this type of approach to research. The opportunities to do excellent science will be far more numerous than can be exploited by the limited number of individuals in the groups that plan and execute the survey projects and who may have privileged access to data for a period of time. For all of these reasons, an effective archiving and distribution system is fundamental to the scientific success of these projects.

Initial distributions of data from surveys will be in the form of catalogs. The larger data volumes represented by the pixelsets means that these will be distributed in a more limited fashion. One of the first goals of the IVO might legitimately be to construct tools that allow powerful cross-survey access to the catalogs from a number of large surveys, that is, catalog-level data mining. It is a matter of debate whether the IVO can or should provide the means of reprocessing pixelsets (which might be termed pixel-level data mining) at some limited number of host locations.

3. Data Archive Centers: Pixelsets and Catalogs

The datasets residing in data archive centers are typically far less homogeneous than those originating in surveys. Furthermore, these sites host primarily pixelsets rather than catalogs of derived scientific quantities. These centers have evolved over a number of years and their content, goals, and capabilities have evolved along with them. Their focus has been on developing efficient tools that enable their users to query catalogs of observations and to retrieve specified pixel datasets. These tools remain fairly primitive and queries are based on technical parameters (*e.g.*, instrument, filter) rather than scientific ones. These data centers are widely used to access data for research but also to access auxiliary datasets (*e.g.*, NED or the Digital Sky Survey) that facilitate research and form an important component of the research infrastructure.

How do these archive centers fit into the overall scheme represented by the IVO?

Data archive centers need to constantly re-invent themselves. Technological change rapidly expands the range of what is possible. Archive centers must anticipate the capabilities that their users will require if they are to continue to deliver an important service. The vision represented by IVO, that is, a vision of vastly more powerful data and information management capabilities for astronomical research, follows from this need for archive centers to be constantly evolving.

As one example of this evolution, the Canadian Astronomy Data Centre (CADC) began development in 1998 of a system for scientific manipulation of fully processed datasets (for example, the Canadian Network for Observational Cosmology Surveys and the Canada-France Redshift Survey) and this work has led to a fairly complete design for data warehousing of scientific data which accommodates quantities of many different types, from catalogs of derived parameters all the way through to the pixel data themselves.

Development of a data warehousing system is a change in course for CADC that was motivated by our recognition that data archives need to be much more than pixel collections. Archive centers need to produce and to distribute more refined data products, such as catalogs of derived parameters and collections of fully-processed data. This process of adding value to data holdings is the future of data archive facilities.

One example of our approach to populating the data warehouse from our own pixel collections is the WFPC2 source catalogs pipeline that has been running at CADC for over one year. This pipeline reads raw images from CD-ROM and ultimately produces a catalog containing fully-calibrated photometry, galaxy surface photometry and structural parameters, and astrometry. We have committed to do the same thing for CFHT and other mosaic camera surveys. The resulting catalogs can be digested by the IVO but we retain the linkage with the original pixels and the ability to re-process the pixels in new and interesting ways.

So CADC recognizes the need to move into the business of producing and distributing catalogs. However, the continuing need to handle pixel data is also clearly recognized. The function of surveys is to provide a broad view of the populations of objects in the universe and to provide the means to construct sub-samples of highly interesting objects for further study at a more detailed level. The more detailed studies will be done with new pointed observations but also by using the broad range of instrumentation and observational parameters available in archives of pointed observations. Surveys do not render pointed observations obsolete. In addition, the pixelsets of the surveys themselves need to be archived and be available for the purpose of verification and re-processing.

It may appear that the large, homogeneous surveys will produce the cream of data mining results whereas the processing of pixelsets, at CADC for example, forces us to work very hard for our scientific results. But it is only hard to process existing pixelsets because they reside in "bad" archives. They have been ingested without sufficient metadata to describe them fully. They have been ingested into archives whose structure is insufficient to permit data mining. And a proper archive structure (a "virtual observatory" in CADC internal jargon) would make pixelset processing and data mining automatic. Although we cannot change the way existing archiving is done in a single generation of facilities, we are trying to incorporate what we have learned into our design work for the Gemini science archive with the goal, at minimum, of not creating insurmountable obstacles to effective integration into the IVO.

4. Conclusions

Large, homogeneous surveys are a key component of the IVO concept with good reason. These surveys and the development of effective linkage between them will produce great scientific rewards. A large investment has been made, by many countries, in data archive centers and these centers will play an important role in the IVO by bringing their technical and scientific expertise into the project. It is the experience of these archive centers over the past decade that has made it possible to conceive of a system with the capabilities of the IVO.

The first steps toward realizing a system like the IVO have been taken in many locations. We describe some of the work done at the Canadian Astronomy Data Centre. The way forward for data archive centers is a way that leads to convergence with the IVO goals of providing researchers with vastly more powerful tools with which to pursue an understanding of the universe.

Virtual Observatories of the Future
ASP Conference Series, Vol. 225, 2001
R.J. Brunner, S.G. Djorgovski, and A.S. Szalay, eds.

Solar Web: A Web Tool For searching in Web-based Solar Databases

Isabelle Scholl

Institut d'Astrophysique Spatiale, CNRS/Université Paris XI, Bâtiment 121, F-91405 ORSAY, France

Abstract. Today, scientists interested in solar data have access to a large variety of data collections (coming from space or ground-based observatories), but only available with different and heterogeneous software. Moreover, there is no common definition for keywords describing the data. Consequently, this makes the pre-analysis period long and difficult.
This paper presents the early design and architecture of the 'Solar Web Project'. Its aim is to provide scientists with one unique interface for browsing worldwide solar databases. The idea is to make interoperable heterogeneous databases and their access systems. It is under development at MEDOC (Multi-Experiment Data and Operations Centre), located at the Institut d'Astrophysique Spatiale (Orsay, France) which is one of the European Archives for SOHO and TRACE data. Technologies like XML and JAVA are being used for designing and developing this tool.

1. Introduction

In the frame of the SOHO mission, ESA has approved 3 European long-term archives for the SOHO data. One of these centers is MEDOC (Scholl 1999). This facility has also an operational role in SOHO operations and works as an analysis center for various solar data. MEDOC functions go well beyond the SOHO archive. Actually, the MEDOC archive is not dedicated to SOHO only: TRACE data are already included in the archive, the THEMIS catalog will be available within the MEDOC one, and it is already planned to include future solar mission data.

The development of the SOHO archive is a joint effort between ESA/NASA and MEDOC teams. After designing the catalog in common, the development was shared among the two teams. The ESA/NASA team was in charge of the catalog data ingestion programs while, at MEDOC, we designed and developed a system to let the user interactively browse the catalog using dynamic multi-instrument queries and retrieving data over the web (Scholl 1998).

The MEDOC development was made with open concepts in mind: this tool should be generic in order to be able to easily manage other solar observations than SOHO. Later on, we enhanced its basic capabilities dedicated to SOHO data in order to support multi-mission data. This new tool is currently available from the MEDOC web site[1] and provides users with SOHO and TRACE data. But even if some new mission data will be hosted at MEDOC, it is not desirable to co-locate there all kinds of solar data, according to the principle that data should reside close to the relevant scientific experts.

2. The Solar Web project

For several years, thanks to technology progress, more and more powerful instruments are built. They generate a huge volume of data that covers time over periods longer than ever. Moreover, the observational mode of these observatories (ground-based or from space) are often organized around multi-mission concurrent programs. But scientists involved in this kind of collaboration encounter two main problems:
• One difficulty arises from the tools available to identify the total data related to a collaborative observation. Currently, when scientists need to compare several observations coming from different observatories, they have to use several query systems in order to retrieve these files from different data centers. Some of these tools may be available from the web, others use only XWindow-based software.
• Another problem is linked to the difficulty of interpreting data coming from several instruments that generally have different characteristics and heterogenous formats and semantics. This is a general problem in Solar Physics. Solar scientists and more recently other scientists involved in Space Weather programs are facing large amounts of data and as many tools to identify them. In this context, the SOHO archive access system is yet another stand-alone tool.

The need to provide users with a unique and centralized tool for browsing simultaneously several catalogs is obvious. Moreover such a tool that would present, in a homogeneous way, data coming from various sources would certainly ease scientists' day-to-day work.

The software available at MEDOC already offers users this capability for SOHO, TRACE and soon for THEMIS observations, but it is limited to access the local database. Consequently, the idea is to extend the current system capabilities to access simultaneously remote archive systems, to merge and show all remote and local query results on one single web display. Later on, users should be able also to request data from one unique web page.

This project is not an easy one: from a technical point of view, there are no major difficulties, excepted that very new technologies will be used. But from an organizational point of view, the most difficult point is to have several data centers agreeing on some basic definitions. If no tools similar to this are readily available (a few projects already exist in Astronomy and Planetary Science) it is because:
• Data centers must first collaborate to produce a standard for exchanging data between their applications. Basic concepts such as vocabulary and protocol must be adopted by all data centers interested in sharing data they host with others. In the frame of the Reference Model for an 'Open Archival System' (CCSDS

---

[1] http://www.medoc-ias.u-psud.fr/archive/

1999), a model for archive interoperability is introduced and can be followed as a starting point for this project.
• Until recently, no technologies were available to implement such a distributed web software.

3. System Design

Before developing any software to access various distributed databases, it is essential to define a common model based on a standardized vocabulary. The goal is to describe a 'Generic Solar Observation' obtained from a Ground-Based or Space Observatory. Such an observation can be defined with:
• General parameters such as:
  o the observatory that made the observation
  o the instrument used
  o the start date/time and the end date/time
• Observational technical parameters:
  o the pointing: coordinates, coordinate system used and the field of view
  o the spatial resolution (pixel size)
  o the spectral resolution and the spectral window size
  o the wavelength used and the targeted spectral lines
  o the integration time (that gives indications on the type of observation)
  o the observation type (spectrum, magnetogram, image, full disk, ...)
• Scientific parameters:
  o the scientific objective
  o the targeted object
  o links to referenced events (*e.g.*, NOAA Active Regions Catalog)
• Physical localization of the data:
  o an archive location (the data center where files can be retrieved)
  o a file name (the file containing observation results) at this location
  o a thumbnail (in an image format like GIF).

This list represents only a few keywords but these are sufficient for the purpose of selecting the data, an activity which is very different from using the data. By using such a model, existing data archives will be able to communicate in a standard way, even if some of them would probably not have all this information or maybe should convert their own keywords to this model.

4. System Architecture

As shown in Figure 1, remote systems can work collaboratively to exchange the common keywords listed above: users are connected through a web browser to the MEDOC web server (or any other collaborative 'Solar Web' Server). This process accommodates the 'Solar Web Application Server' that submits queries to:

1. The local catalog using the JDBC API.

2. Any registered remote databases that offer a direct JDBC access.

3. Any registered remote web servers through existing cgi scripts that query its local databases using an HTTP GET method access.

4. Any other 'Solar Web Application Servers' which query their local databases using a specific protocol.

Remark: no developments are needed on remote site for items 2 and 3.

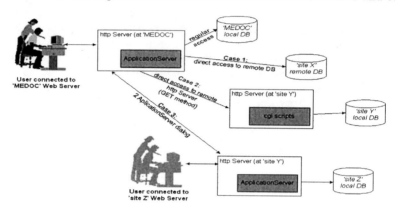

Figure 1. Collaborative System Architecture.

Exchanges between the main server and remote applications are made using either a native application protocol to simplify development or XML with a standardization objective.

Technologies used for this development are: JAVA and JDBC driver, XML, the HTTP GET method and servlets. Security is a major concern in this project. Permitting direct remote accesses is always a potential security risk for any organization. A secure protocol for communications between remote servers is currently under study.

The software is based on a 3-tier architecture. It has been designed open to new kinds of data. The client (except for the dialog with the server) and the query generator (the module in charge of a specific database) are the only business specific components. These two modules will be available as an API of the generic software for any extensions.

5. The Future of Solar Web

'Solar Web' is currently a stand alone development effort from MEDOC. Discussions are in progress with potential partners such as: the MDI team (Michelson Doppler Imager on SOHO from the Hansen Experimental Physics Laboratory, Stanford University) and Bass2000 (Themis data, at Tarbes, France). The software specification has already been defined and its development is in progress[2]. We are open to suggestions for new collaborations.

---

[2]http://www.medoc-ias.u-psud.fr/archive/solarweb/

References

Scholl, I., et al. 1998, in ASP Conference Series, 172, Astronomical Data Analysis Software and Systems VIII, Ed. D.M. Mehringer, R.L. Plante, D.A. Roberts, 253

Scholl, I. 1999, in Proceedings 8th SOHO Workshop, (ESA Publ. SP-446) (Paris 1999), 611

CCSDS, 1999, CCSDS 650.0-R-1, Red Book, Issue 1, May 1999

Multi-Threaded Decomposition of Queries: The SDSS Model

Aniruddha R. Thakar, Peter Z. Kunszt and Alexander S. Szalay

The Johns Hopkins University, Center for Astrophysical Sciences, 3701 San Martin Drive, Baltimore MD 21218-2695

Abstract. Data mining in the individual member archives of a virtual observatory must make optimal use of system resources and data architecture. As described below, the Sloan Digital Sky Survey's Science Archive uses a distributed multi-threaded query agent and engine to achieve parallel scalable decomposition of queries that takes maximum advantage of both SMP and Beowulf cluster architectures.

1. Introduction

The Sloan Digital Sky Survey (SDSS) is a multi-institution project to map 10,000 square degrees of the northern sky in unprecedented detail. The SDSS Science Archive is the science database that will serve the survey data up to scientists. The Science Archive (SX) is expected to be several Terabytes (Tb) in size when completed, and it is among the first to tackle the challenge of data mining in multi-Tb scientific archives. The SX is designed to meet this challenge by using an extensively multi-threaded server architecture and the ability to decompose queries into several query nodes that can be executed in parallel.

The SX incorporates the following main features:

- a commercial object-oriented DBMS (Objectivity) as the data warehouse layer
- a client/server interface to the data warehouse
- simple proprietary communication protocol between client and server
- queries formulated in SQL-compatible query language (SXQL)
- high degree of parallelism in design
- scalable performance for SMP and cluster environments

In addition to facilitating efficient data mining in a multi-Tb archive, the SX must also provide fast network access to the data for a user community spread across several institutions in the US and the world. To achieve this, the SX will be replicated at each of the SDSS institutions with a local archive and server running at each site. Here we describe in detail the multi-threaded design of the server and the parallel execution of queries that takes maximum advantage of a multi-processor CPU architecture and a distributed data environment to deliver optimal performance both for SMP-type hardware like the SGI Origin 2000 that currently hosts the SDSS master archive, as well as Beowulf-type Linux clusters like the one that will host the JHU local archive.

2. Server Multi-Threading

The SX server (query agent) is designed to meet the following requirements:

- support multiple simultaneous user sessions
- allow multiple simultaneous queries per user session
- execute each query execution tree in parallel
- distribute query among database partitions if applicable

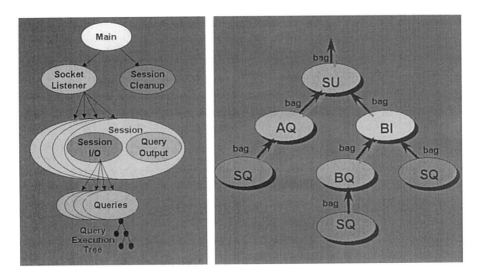

Figure 1. Multi-threading of the SX server (left) and the query engine (right, SQ=Scoped Query, BQ=Bag Query, AQ=Association Query, SU=Set Union, BI=Bag Intersection).

The main server thread splits into two threads (Figure 1) — the socket listener thread to listen on the client socket and accept new user logins, and the session cleanup thread to monitor closed sessions and delete them. The socket listener thread spawns two new threads for each successfully authenticated user session — the session I/O thread which handles the communication between the client and the server, and the query output thread to handle query output. The session I/O thread spawns a new query execution tree thread for each query entered by the user. The query execution then proceeds in parallel as described under Section 3, with each node of the tree executed as a separate thread. If the data is distributed among multiple data "partitions", some of these threads are run on slave servers as described in Section 4.

SX Queries SX queries are formulated in SXQL (SX Query Language). SXQL is a simple SQL-like language that implements the basic subset of clauses and functions necessary to formulate queries for an object database. There is no attempt to be OQL-compliant, although we have borrowed some concepts from OQL (Object Query Language). SXQL recognizes the SELECT-FROM-WHERE statement of standard SQL, which can be nested and combined using the UNION, INTERSECT and EXCEPT set operators. It further allows specification of association links in the SELECT, FROM and WHERE sub-clauses. Associations are links to other objects, and can be either to-one (unary) or to-many (n-ary) links. The query predicate in the WHERE clause can be built using the usual arithmetic and boolean operators, but in addition can also contain references to associations. SXQL also contains a number of macros and keywords specific to astronomical queries (*e.g.*, RA(), DEC(), J2000). SXQL also recognizes a proximity query syntax, which allows the user to search for all objects that are close to a given object in space. The SXQL query is parsed and then intersected with a sky and flux index to determine the scope of the search in terms of the number of database nodes that need to be scanned. In addition to generating the database node list for the query, the intersection with the index allows the rough cost of the query to be determined *a priori*. This information is made available to the user so they can decide whether to continue the query.

3. Query Execution Tree

SX queries are executed by first building a Query Execution Tree (QET) whose nodes are one or more of the following query primitives:

- Scoped Query- This specifies a subset of the database in terms of the database nodes to scan as well as a specific type of object to which the query predicate is to be applied.

- Bag Query - The given predicate is to be applied to each object in the given bag.

- Association Query - Selects all the objects linked to each object in the given bag through to-one or to-many associations, and applies the predicate to the selected objects.

- Proximity Query - Finds all objects spatially close to each object in the given bag. This is a special kind of Scoped Query where the node list is the neighborhood.

- Set/Bag Union - Perform the set union operation on the two bags given as operands. If duplicates are allowed in the resulting bag, it is a set union, else it is a bag union.

- Set/Bag Intersection - Perform the set intersection operation on the two given bags.

- Set/Bag Difference - The resulting bag is the difference of the two given bags.

Scoped Queries are the only ones that operate directly on the database and do not need a bag operand. Therefore they are the leaf nodes of the QET. The other types of queries form the interior nodes (see Figure 1).

Each node of the QET is executed as a separate thread to deliver the highest possible degree of parallelism and performance. Results from query nodes are pushed up the tree using stacks as soon as they are generated, ensuring that even for very time-consuming queries, output is seen as soon as possible. In the case of blocking operations like set intersection and difference, one of the two operand nodes must finish before results can be pushed up the tree. Some of the query nodes may be executed partly or entirely on remote processors running slave servers if the data they require resides on remote data partitions (see "Distributed Query Processing" below).

4. Distributed Query Processing

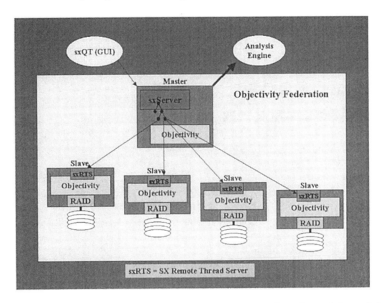

Figure 2.   Distributed processing of queries in a distributed data federation.

Federation is the Objectivity term for a distributed database, where portions of the database are located physically on different systems. Each remotely located section of the database is called a partition. The SX server parcels out the processing for remote partition searches to slave servers running on the host processors local to the partitions (Figure 2). Query nodes that require searches on remote partitions are sent to the slaves, which are also multi-threaded servers with the ability to run multiple query nodes simultaneously. The slaves are known as Remote Thread Servers, and are lightweight versions of the master server. The master server is responsible for collecting all the output from the slave servers and routing it back to either the client (GUI) or an Analysis Engine sitting on a different socket.

The ROSAT X-ray Database from All-Sky Survey and Pointed Observations

Wolfgang Voges, Thomas Boller, Jakob Englhauser, Michael Freyberg, & Rodrigo Supper on behalf of the MPE-ROSAT team

Max-Planck-Institut für extraterrestrische Physik, D-85741 Garching, Germany

Abstract. During the 8.5 years of operations ROSAT performed the first all-sky survey in soft X-rays and more than 9200 pointings with the Position Sensitive Proportional Counter (PSPC) and High Resolution Imager (HRI). All of the original data and derived products as there are various source catalogs and thousands of X-ray images have been released to the public. These products are briefly described in this paper.

1. Introduction

The ROSAT All-Sky Survey and pointed observations have led to a dramatic increase of our knowledge of the X-ray sky. The total number of publications based on ROSAT is larger than 3,000; they concern almost all astrophysical fields from comets to cosmology.

ROSAT was launched on June 1, 1990. Four concentric parabolic-hyperbolic mirror pairs form the ROSAT Wolter Type-I telescope with a focal length of 2.4 m. Its mirror surfaces have a residual roughness of less than 3 Å, which is responsible for the excellent contrast of the ROSAT telescope. Another important feature of ROSAT is the low intrinsic background of the PSPC.

The total number of new X-ray sources discovered by ROSAT during the all-sky survey and pointed observations is larger than 150,000 (see, *e.g.*, Figure 1 for their distribution), which is more than a factor of 20 larger compared with the number of X-ray sources known before ROSAT. The source lists were produced automatically by the Standard Analysis Software System (SASS). For each X-ray source the following properties are calculated: detection likelihood, source position, source and local background count-rates, exposure time, hardness ratios, source extent and corresponding likelihood. To produce reliable source catalogs, an automated as well as a visual screening procedure of individual sources was applied to some of the data sets.

2. The Catalogs and the X-ray Background Maps

The results from our analysis of the soft X-ray background as well as the detected sources from the ROSAT All-Sky Survey (covering 99.9% of the sky) and from the First ROSAT Source Catalog of pointed observations (covering 15.2% of the sky) are summarized in Table 1 and explained in the following.

ROSAT ALL-SKY SURVEY Bright Sources
Aitoff Projection
Galactic II Coordinate System

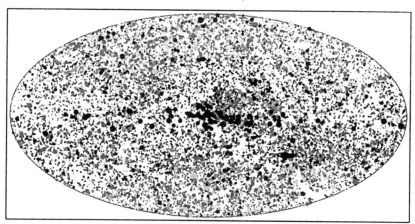

Figure 1. The RASS-BSC sources above a count-rate of 0.05 cts s^{-1} in the energy band 0.1–2.4 keV. The original color coded hardness ratio is translated into a grey scale (light grey: spectrally soft and black: spectrally hard sources). The size of the symbols scales with the logarithm of the count-rate.

The first **ROSAT-1RXP** catalog of pointed observations with the PSPC contains more than 74,000 entries from 2,917 pointed observations performed over 3 years. Compared to the RASS-BSC this catalog is less reliable with respect to source parameters like extent, count-rate and position (see Figure 2).

The X-ray source catalog **WGACAT** is also constructed from 3 years of ROSAT pointed observations and contains about 69,000 sources. The main difference to the ROSAT-1RXP catalog is the use of a different source detection algorithm. The source counts for WGACAT are calculated for three energy bands: low-band (0.1–0.4 keV), mid-band (0.4–0.9 keV), high-band (0.9–2.0 keV).

The first ROSAT source catalog of pointed observations with the high resolution imager (covering 1.2% of the sky) (**ROSHRICAT**) contains arcsecond positions and count rates of detected sources from more than 2,000 public ROSAT HRI observations, including more than 5,700 bright sources (S/N > 4).

ROSAT all-sky survey maps of the diffuse X-ray background are produced in up to 7 energy bands, and have various angular resolutions between 12' and 40' with and without point-source exclusion.

The **RASS-BSC-1RXS** catalog is a visually screened derivative of the RASS II catalog. By applying a threshold to the count-rate at 0.05 cts/s (0.1–2.4 keV) a list of approximately 25,000 entries was generated. A careful visual inspection and the setting of a threshold on the source-likelihood (≥ 15) and

THE FIRST ROSAT SOURCE CATALOGUE Sources
Aitoff Projection
Galactic II Coordinate System

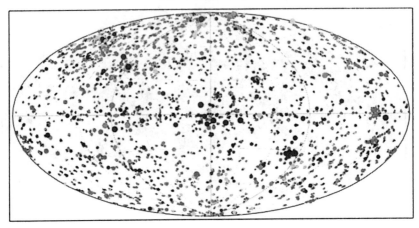

Figure 2. The ROSAT-1RXP sources (see also Figure 1).

counts (≥ 15), resulted in a final bright source list of 18,811 entries (see Figure 1).

The **RASS-FSC-1RXS** catalog is an automatically screened derivative of the RASS II catalog. Setting a threshold on the source-likelihood (≥ 7) and counts (≥ 6), resulted in a final faint source list of 105.924 entries.

The foundation for the RASS-BSC and the RASS-FSC, the **RASS II** catalog is based on a source detection likelihood threshold of 7 and contains 145,060 entries.

The ROSAT variable source catalog **ROSAT-VSC** was generated by correlating the RASS I and RASS II catalogs created from the RASS data and the ROSAT-1RXP. Additionally, the Photon Event Files from the RASS and from public ROSAT pointings were used to determine upper limit values. From the Pointed versus Survey detection **1,062** sources exceed a factor of variability above 3, whereas from the Pointed versus Pointed detections **1,451** sources were found to show variability above a factor of 3.

3. The ROSAT All-Sky Survey Photons

The ROSAT all-sky survey reprocessing (RASS3) has been recently completed and the data-products have been released to the scientific community. (ROSAT-News of 22-MAR-2000).

There are 1378 RASS3 fields each $6°.4 \times 6°.4$ covering the whole sky. Neighboring fields are overlapping by at least 0.23 degrees. Each field can be identified by an equatorial latitude zone number (1 to 33) and an equatorial longitude segment number (1 to 64, depending on the zone number).

Catalog or Map	Number of Sources	Detection Likelihood	Sky Coverage
ROSAT PSPC Pointing (ROSAT-1RXP)[a]	74,407	≥ 10	15.2%
WGACAT Pointing[b]	68,907	opt. χ^2	14.0%
ROSHRICAT HRI Pointing[c]	5,700	$S/N \geq 4$	1.2%
40' RASS Diffuse X-ray bkg maps[d]	3 bands		100.0%
12' RASS Diffuse X-ray bkg maps[e]	7 bands		100.0%
RASS Bright Source (RASS-BSC-1RXS)[f]	18,811	≥ 15	99.9%
RASS Faint Source (RASS-FSC-1RXS)[g]	105,924	≥ 7	99.9%
ROSAT Variable Source (ROSAT-VSC)[h]	$\sim 2,000$	≥ 10	15.2%

Table 1. ROSAT catalogs and maps

[a]Voges et al. (1996)
[b]White et al. (1994)
[c]ROSAT Collaboration (1998)
[d]Snowden et al. (1995)
[e]Snowden et al. (1997)
[f]Voges et al. (1999)
[g]Voges et al. (2000)
[h]Voges et al. (1998)

Accordingly there are 1378 directories in the ROSAT Data Archive each containing a README file and 10 standard FITS files. The names of the data directories have been adapted to match other ROSAT archival data. A 6-digit number (ROR number) is followed by 'p' for the detector used (PSPC). The ROR number consists of '93' as the first two digits ('9' for the ROR category 'other and survey data','3' stands for RASS3), digits 3 and 4 correspond to the zone number, 5 and 6 to the segment number. RASS FITS data files start with the 2 characters "rs" to indicate survey data. There are several ways to retrieve the ROSAT all-sky survey data; they are listed in the help file to be found on our web page[1].

References

Snowden, S.L., M.J. Freyberg, P.P. Plucinsky, J.H.M.M. Schmitt, J. Trümper, W. Voges, R.J. Edgar, D. McCammon and W.T. Sanders 1995, First Maps of the Soft X-Ray Diffuse Background from the ROSAT XRT/PSPC All-Sky Survey, ApJ, 454, 643

Snowden, S.L., R. Egger, M.J. Freyberg, D. McCammon, P.P. Plucinsky, W.T. Sanders, J.H.M.M. Schmitt, J. Trümper & W. Voges 1997,: ROSAT Survey Diffuse X-Ray Background Maps. II., ApJ, 485, 125

Voges, W., Boller, Th., Dennerl, K., Englhauser, J., Gruber, F., Paul, W., Pietsch, W., Trümper, J., Zimmermann, H.-U.: Identification of the

---

[1]http://wave.xray.mpe.mpg.de/rosat/survey

ROSAT All-Sky Survey Sources[2], in: Proc. of the Conference Röntgenstrahlung from the Universe, 1996, MPE Report 263, 637

Voges, W., et al. 1999, The ROSAT All-Sky Survey bright source catalogue[3], A&A, 349, 389

Voges, W., et al. 2000, The ROSAT All-Sky Survey faint source catalogue[4], IAUC 7432

Voges, W., Boller, Th., Šimić, D. 2000, A&A, in preparation

White, N.E., Giommi, P., Angelini, L., 1994, WGACAT[5], IAU Circ. 6100

---

[2] ftp://ftp.xray.mpe.mpg.de/rosat/catalogues/1rxp

[3] http://wave.xray.mpe.mpg.de/rosat/catalogues/rass-bsc/

[4] http://wave.xray.mpe.mpg.de/rosat/catalogues/rass-fsc/

[5] ftp://ftp.xray.mpe.mpg.de/rosat/catalogues/1wga

Part 3
Computer Science & Statistics for a Virtual Observatory

Computer Technology Forecast

Jim Gray

Microsoft Research, 301 Howard St #830, San Francisco, CA 94105

Abstract. This is a brief overview of the computer, storage network, and software trends that will enable Virtual Observatories.

1. Introduction

I was asked, as a computer scientist, to give a sense of what computer technologies the VOF can design for over the next decade. In designing the VOF we need to think in terms of how much storage, bandwidth, and processing power we will have five and ten years from now—because that is the equipment we will actually be using.

The good news is that processing, storage, and networking are improving at an exponential pace. In the limit, they are infinitely fast and capacious, and cost nothing. Unfortunately, most of us do not live in the limit; we live in the present so computers will not be free anytime soon. The bad news is that people are getting more expensive, and indeed the management and programming costs routinely exceed hardware costs.

If everything gets better at the same rate, then noting really changes, but some ratios are changing. The people:computer ratio is changing so that we now try to minimize people costs. Although computers are getting faster, the speed of light is not changing (so if you want to get data from the other coast, you must wait 60 milliseconds (request-response round-trip time)). In addition, WAN costs have not improved much in the last decade. This is a surprise: the cost of WAN networking is much reduced. But, it seems costs and prices are not correlated in the telecommunications industry. Perhaps this will change in the future, but if you live in the present, you want to put the computation "close" to the data to minimize costs–rather than moving huge datasets to the computation. There is a recurring theme in this presentation: processing, storage, and bandwidth are all getting exponentially faster, cheaper, and bigger, but access times are not improving nearly as much, if at all. So the process:access ratios are changing, and forcing us to change our processing strategy by making things more sequential, and by using both pipeline and partition parallelism.

Astronomy is in a paradoxical state, there is not exponential growth in glass, but there is exponential growth in CCDs. So astronomy seems to be tracking Moore's law, doubling data volumes every two years as more and better CCDs are put behind the glass and as new instruments come online.

Computing involves progress in many areas, data visualization, human-computer interfaces (vision, speech, robotics), mobile applications (wireless and batteries), micro-electron-mechanical-systems (MEMS), and the more prosaic areas of networking, processing, and data storage and access. This talk focuses primarily on these classic areas (processing, storage, networking). Even within these classic areas, I am going to go light on topics such as object-oriented programming, databases, data visualization, data mining, and the open source movement. Rather in the 20 minutes I have, I want to just quickly give you a sense of the base architecture issues of server-side processors, memory, storage, and networking. I also want to covey the essential role that parallelism must play in our designs, and point out that we need to place processing next to the data. This may mean that we replicate data near the processors, and it argues against a design that has processors in one place and data stores in another.

2. How much information is there, and how much astronomy information is there?

To, start, lets talk about how much information there is—and how astronomers fit into the scheme of things. It appears that all the astronomy data adds up to a few petabytes today. Large archives are tens of terabytes (TB). The Sloan will be 40 TB raw and 3 TB cooked—but that is five years off. The radio telescopes generate a few terabytes per day, but at present we are not saving most of that data.

How does this relate to the rest of the world? The public Internet is about 30 TB toady (Sept 2000). The Library of Congress in ASCII is about 20 TB (20 million books at 1MB/book). If you include all the film, all the music, all the photos, and images of all the books, then LC is a few tens of Petabytes. Each year the disk industry builds a few exabytes of disk. Mike Lesk (following Alan Newell) points out that ALL the recorded information is a few tens of exabytes. He points out that we can now record everything on disk or tape, and that the precious resource is human attention. People will never see most data. Computers will read, analyze, and summarize this data for us. Human attention is the scarce resource in this new world[1].

So, to put things in perspective, astronomy data sets are huge, but not so huge that they are impossible. They are Wal-Mart size, not impossible size. They come in units of 10TB and sum to a few PB (petabytes). They can all be put online (see below) for a few million dollars as a federated virtual observatory. Each member of the federation can contribute its part to the overall corpus. Some members of the federation must serve the unifying role of coordinating access, much as Yahoo! does for the current Internet.

I think astronomy data is the best playpen for us computer scientists to experiment with data analysis and summarization techniques. You astronomers are unique: You share your data. That is surprisingly rare (for a variety of reasons data in physics, in geology, in medicine, in finance, and in the Internet is encumbered by privacy and by commercial interests). Your data has rich correlations, many of which you have discovered. And you have so much data, and it has such high dimensionality that you need help from computers and computer scientists and statisticians (and you know you need help). So, it seems

---

[1] http://www.lesk.com/mlesk/ksg97/ksg.html

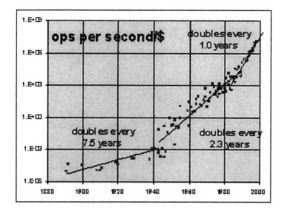

Figure 1. Operations/s/$ doubling times.

you astronomers are willing to tolerate us computer scientists. I have formed personal relationships with some of you, and I really like the community. And, lastly, as each of you knows, Astronomy is just such an amazing study—you are always on the threshold of discovering whole new realms.

3. Moore's, Gilder's, Metcalf's, and Amdahl's Laws

We computer scientists observe phenomena and coin "laws" to explain them. We know, and you know, that these laws are made to be broken (they reflect the current technology universe.) Memory chip capacity has been doubling every 18 months since 1965—so that now 64 Megabit chips are common and 256Mb chips are available in small quantities. This doubling is called Moore's law. It gives a hundred- fold improvement every 10 years.

Moore's law applies more generally to microprocessor clock speeds (they have been doubling every 18 months for a long time), disk capacity (doubling every 12 months since 1990), and network bandwidth (doubling every 8 months according to George Gilder (and others) (Gilder 1997, 2000).

An interesting thing about these laws, is that in the next doubling period, you install as much capacity as in the history of time ($\int_0^t = \int_t^{t+1}$). That is amazing. Among other things it means that if you can afford to store the data for one time period (a year) then storing it forever only doubles the hardware cost. If you compute something today, the computation in 3t years will run 8 times faster and be 8 times cheaper (so do not bother to store it, just re-compute it if you do not expect to access it in the interim.) I will come back to some of these rules-of-thumb later.

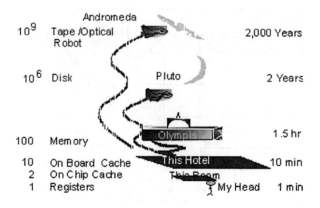

Figure 2. How far away is your data?

3.1. Processors

Figure 1 shows how processing has gotten less expensive since 1880. Electro mechanical devices performance/price improved at a rate of 2x per 7.5 years. With the advent of vacuum tubes and then transistors, the doubling time dropped to 2.3 years. With the advent of integrated circuits, the doubling time dropped to one year! That is the cost of doing an n-bit operation has been dropping in half each year. I can cite lots of evidence for this (see, for example, progress on sorting at Sort Benchmark[2]). About 50% of this speedup comes from faster devices (disks, cpus, memories, *etc.*). The other 50% comes from parallelism In 1980 processors executed a few million instructions per second, while current processors deliver about 500 to 1000 million instructions per second. That's "only" a 100 fold to 1000-fold improvement. But doubling for 20 years is a growth of 220   1,000,000. So, where did the other factor of 1,000 come from? Well, it came from parallelism. The fastest sort uses 2,000 processors! Each of those processors is moving 64-bit words rather and 16-bit words in parallel. Each of those processors is heavily pipelined with much parallelism inside and outside the processor. It also uses about 5,000 disks. That's parallelism. You will find that I harp on this a lot, here and forever. Parallelism is your friend. Tom Sterling's Beowulf clusters are your friends. The 8,000 cpus at Hotmail are your friends, all running in parallel. The 6,000 processors at Goggle are your friends, all running in parallel. You could not do Hotmail, or Goggle, or Yahoo, or MSN or AOL on an SMP—because people are waiting. You can do all your astronomy on an SMP if you are patient, but it will go a LOT faster, and a lot cheaper if you use commodity clusters of inexpensive disks and processors.

The processor consumes programs and data from memory and sends answers (data) to memory. In a perfect world, whenever a processor wanted a piece of

---

[2]http://research.microsoft.com/barc/SortBenchmark

data, that datum would be sitting there instantly available. We do not live in that perfect world. Often, the processor does not know what data it will need until the last minute. For example if it is following a linked list, then the next data is not know until the current node of the list is examined. So, how far away is the data in, if it is a surprise? I think the picture in Figure 2 is a good way of thinking about this question.

Processors live by clock cycles: today a clock cycle is a nano-second (for a 1GHz processor). In that time, a modern processor can execute 4 instructions in parallel, if all the data is available (a quad issue processor). But, what if the data is not in the processor registers? Then the processor must look in the processor cache (typically a fraction of a megabyte of data near the processor (e.g. on the processor chip/package)). If the data is there, great, that just costs 2 clocks. If the data is not there, the processor has to go to the level 2 cache (a few megabytes) that are typically 10 clocks away. If it is not there, the processor has to go to the L3 cache (typically main memory of a few Gigabytes). RAM data is 100 or more clocks away. If this is a big SMP, some memory can be even further away (say 400 clocks in a NUMA). If the data is not there, it must be on disk. Disk data is 10,000,000 clocks away. And if it is not there, then it must be on tape that is a trillion clocks away. To put this in human terms, our clocks run in minutes. If I ask you a question, you say: "Just a minute, you compute and you tell me the answer." If you do not know, you ask someone in this room (cache) and it is two minutes. If you have to go to cache it is like going somewhere on your campus. If you have to go to main memory, it is an hour's drive there and back. If you have to go to disk, it is 2 years (like going to the Pluto!), and if you have to go to disk it is like going to Andromeda (there is at least one joke associated with that).

If your program is doing list processing on disk, your program will execute 100 instructions per second—and you will be back with the 1940's class computers. If you want a modern computer you need to have the data ready for the processor. That means that you have to have good program locality (no surprises in the data stream). You have to go sequentially through the data so that the system can prefetch it and pipeline it to the processor. If you program in this streaming style, you get a 4 billion instructions per second rather than 100 instructions per second. In practice, people are achieving about one clock per instruction (there are some surprises in the data, but there is also lots of regularity.)

In the future we expect to see each "chip" contain several processors. In the mean time, there is a tendency to put many processors on a shared memory (Shared Memory Processors or SMP). Figure 3 shows this. You can get "hot" uniprocessors and you can get fairly high-performance 2-way 4-way, 8-way and even 64-way multiprocessors. You will notice that an N-way multiprocessor costs much more than N times more than N uni-processors, and that the N-way multiprocessor is typically not much more powerful than a N/2 way system. For example an 8-way Intel is about 3 times slower than a 64 way Sun E10000 (so 40 processors are wasted), and an 8 way Intel is only 4x faster than an Intel uniprocessor, so half the processors are wasted (on a standard transaction processing benchmark.)

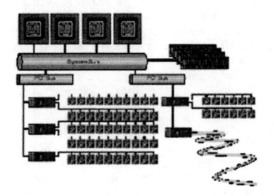

Figure 3. System Balance: what is the right ratio of cpus:memory:disk:networks? Amdahl's law suggests that a 1 million instructions per second processor needs 1 MB of RAM and 1 Mb/s of IO. These laws are still approximately true sand suggest that a billion instructions per second processor needs a Gigabyte of RAM, about 25-50 disk drives, and a Gigabit Ethernet connection. See Gray et al. (2000) for more information.

3.2. Memory and the Memory Hierarchy

So, now we come to the memory hierarchy. You might think that Moore's law makes memory boring: it just doubles in size every 18 months. That's true. Memory is even more boring than that. It comes in various speeds: slow, slower, slower still, very slow, very very slow, and so on. Typical memory is a few hundred nanoseconds to access (when it is packaged and mapped). This is 10x faster than the memory of 1960. So, main memory speeds have improved about 10x in 40 years. That compares poorly to the return on Treasury bills (they grow at 5%). As I said before, if everything gets faster at the same rate, then the ratios do not change and noting really changes. One ratio that is changing is the speed and bandwidth to memory. Processor speeds are rising much faster than memory speeds, and processor bandwidth is rising much faster than memory bandwidth. The net of this is that it does not make a lot of sense to put two infinitely fast processors on a finite speed memory. Computer architects are struggling with this problem, (put the processors next to the memory); but today there is not much progress. My simple advice is to stick with low degrees of SMP and work hard on program locality so that you have streamed access to the memory.

In the past, main memory was a precious resource. In the future terabyte memories will be economic (a few tens of thousands of dollars). This will simplify some programming tasks (right now a TB of RAM costs a $1M to $100M depending on the vendor.)

Memory addresses are a key to the way we program. We are in a transition from 32-bit addressing to 64-bit addressing. Since memory doubles every 18 months, the memory needs an extra address bit every 18 months. Right now 4 gigabytes of memory costs about $5,000. Notice that 4GB ($= 2^{32}$) needs 32 bit addresses, exhausting the 32-bit pointers used in most of our software. The MIPS, Alpha, PowerPC, Sparc and (someday) Itanium processors all support 64-bit addressing, as does IRIX, VMS, HPUX, Solaris, and AIX, but Linux, SCO, Windows, Netware, MacOS, and MVS do not yet have a 64bit memory option. Also most compilers and software systems have not upgraded to 64-bit addressing. This transition will give you (and me) considerable pain over then next 5 years. There is now a crash program underway to overhaul our software to be 64-bit friendly (the Microsoft Windows2000 and Linux solutions are undergoing beta test today—but that is just the first layer of the software stack). Now for the good news, $32/1.5 = 20$. So it will be 20 years before we run out of 64-bit addresses and have to do this all over again. If I am lucky I will get to do this three times in my life (I had fun going from 16/24 bits to 32 bits).

Of course, the main good news is that large and inexpensive main memory lets you write your programs more simply: but remember Figure 2. This memory is getting relatively slower with every processor generation. In 10 years it will look about 100x slower in relative terms.

This naturally segues to disks and file systems and database systems. Again, many file systems have been late to convert to 64-bit addressing, so for them the largest file is 2GB (Linux, NFS, and Informix are examples). But most file systems and database systems are 64-bit (e.g. Windows, Solaris, AIX, IRIX, HPUX, Oracle, DB2, SQL Server, *etc.*).

Disks are relatively inexpensive $100 to $1000 each. Well-balanced systems have many disks (about one for every 10 MIPS, or about 40 disks per cpu with today's ratios. Managing that many disks might take some effort, so most file systems aggregate many disks into one logical volume. The resulting volume looks like one big disk (of about a terabyte).

Each disk has a mean time to failure of 50 years, but when you have 50 of them, the group has a one-year mean time to failure. To eliminate these problems, the hardware or software duplicates the data on the disks (either duplexing it or using a parity scheme). The trend is to go towards duplexing because it is simpler and better utilizes the disk arms. The net of this is that 2 terabytes of "raw" disk becomes 1 terabyte of "cooked" (fault tolerant, file system mapped, ...) storage.

Just as with RAM, disks hate surprises. A single disk can deliver about 30 MBps of data if the disk is read sequentially. If you read the same disk in random order (8KB pages), it will deliver about 120 reads per second and so about 1 MBps. This is thirty times slower than the sequential access pattern. This sequential:random bandwidth ratio is increasing. Therefore is increasingly important that large computations lay the data out on disk in a sequential order, and scan it sequentially if possible. Of course if you are looking at just one bit of data (if you just want to find a particular galaxy) you need not read all the data, you can just look the page address up in an index and then go directly to

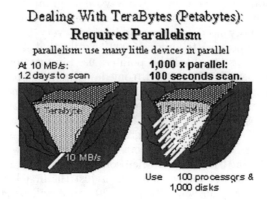

Figure 4. Parallelism gives 10^3x speedups.

the page that holds that galaxy. But, as a rule of thumb, if you are going to read 1% or more of the pages, you should read all the pages sequentially.

Like memories, disks have gotten faster: in the last 5 years the sequential access speed has gone from 5MBps to 30 MBps, and from 70 random accesses per second to 125 random accesses per second. These 6x and 2x improvements are modest gains compared to the 30x increase in capacity (from 2GB to 60GB). Disk technologies now in the laboratory allow disks with current form factors to have 1 TB capacity, 200 random accesses per second, and 100 MBps bandwidth. Notice that it will take a LONG time to scan such a disk. We may end up storing the data 3 times on 3 disks for fault tolerance. This will allow us to scan the logical disk in 1,000 seconds, and will allow any single disk to fail without degrading performance.

So, here comes another pitch for parallelism. Suppose you have a Terabyte store and you read it sequentially. How long will it take? Well 1TB/(30MBps) = 30,000 seconds or about 8 hours. But that terabyte is probably stored on 40 disks (each 50 GB but each paired with it's mirror). So, if you read in parallel, the scan will take 833 seconds or 13 minutes rather than 8 hours. Sequential access gives a 30x speedup; parallel access gives a 40x speedup (or 10,000x speedup if you use 10,000 streams).

I have lunatic-fringe ideas about disk architecture. So, to be fair, I should start by saying that the conventional view is that you should buy disk cabinets from an established vendor like IBM, Compaq, EMC, or others. These cabinets house hundreds of disks, and about 10TB today. They then connect to your processors via a system area network (storage area network often built from fiber-channel). This is how the TerraServer[3] is built. But, this approach is not inexpensive, and I think that the future may make this approach less attractive.

---

[3] http://terraserver.microsoft.com/

Figure 5. Inexpensive Terabytes. A no-name PC includes a 700 MHz processor, 256 MB RAM, Gigabit Ethernet, and housing for 8 disks. This costs $1k for the computer and $2k for the disks and controller. The net cost is $7k per TB. The disks are hot-pluggable. They can be used for data interchange or for backup, eliminating the need for tape.

First, these storage boxes are expensive: the empty box can cost $500k and a populated box can cost $1M per terabyte, compared to $0.7k to $15k for the disks themselves (wow! what a markup!). Second, the fiber channel interconnect does not offer bandwidth comparable to several hundred disks (each fiber channel link is guaranteed not to exceed 100MBps).

I advocate that you buy your disks with your processors, packaged together in Beowulf style. I have been building these storage bricks for our day-to-day work on the TerraServer, SDSS, and some other projects. The basic plan is to stuff a no-name PC with inexpensive disks. The folks at the Internet Archive[4] and many web-centric stores are taking a similar approach. The Archive has a 30TB store that cost them about 7k$/TB (including the processing and networking). The basic idea is that one buys inexpensive disks (right now 60 GB disks cost 200$ in quantity 1) and put 8 of them in an inexpensive server, as in Figure 5.

Now we come to tape. Let me start by saying that I hate tape and tape hates me. My experience with tape is that it is slow, unreliable, and the software to manipulate it is full of bugs (because the software is not used very much.) So, I try to avoid using tape—and I caution my friends and family to avoid it. Replace your tapes with disk.

But, the simple fact is that offline tape is cheaper than disk. Again, taking the TerraServer example, we back it up to DLT7000 tapes, each is 40 GB and each costs us $60. So the backup of a terabyte costs about 25*$60 = $1,500. In practice, we get some fragmentation so it costs more like $2,000. Now $2,000 is less than the $7,000 in Figure 5, and half the cost of the raw disk price. But when one includes the cost of a tape robot ($10k to $1M), then the price for nearline

---

[4]http://www.archive.org/

tape rises to $10k/TB, which exceeds the price for online disk. Technology trends will exacerbate this trend.

Tape access times are measured in minutes (not milliseconds). Tape bandwidths are 5x worse than disks. So tapes are a relatively slow and inconvenient way to store data that you want to read. I can scan all the data on the disks in my system in an hour or two. It takes a week to scan the data in a tape robot.

You no doubt use tape for archive and for data interchange, and are probably wondering how I expect to perform those functions with disks. For data interchange, I want to use the network to move small (multi- Gigabyte) objects. For larger objects, I propose to ship disks as in the right side of Figure 5.

As for archive, it takes a LONG time to restore a petabyte database from tape. With current tapes (6MBps), it will take 5.4 tape years (166M tape seconds) to restore a petabyte. Parallelism could speed this up by 1,000x if you had 1,000x more tape readers. But, still it is an unacceptably long time at an unacceptably high price. So, I want to store all the data online at two places. This means that each place stores the data twice (on 2 mirrored disks), and then we do site mirroring. Each observatory will count on others in the federation to back up its data, and conversely, the observatory will store some part of the federation's backup data at its site. This has to be part of the VO architecture. There is an initial 4x cost multiplier to do this reliable storage, but it comes with 4x more read bandwidth (since you can read all four copies in parallel.) Hence, if you need the read bandwidth, this redundancy may come for free if disk capacities are so large that disk bandwidth is the only limit to how much you put on a drive.

Having two (or more) online copies of the same data offers some interesting possibilities. Recall that the data should be laid out sequentially. The two copies can have different sequential orders: one could be spectral, the other spatial or some other variant.

Summarizing the storage discussion: storage access times and bandwidth are the scarce resources. Storage capacities are doubling every year or 18 months and are predicted to continue this trend for at least the next 5 years. You can optimize storage access by using sequential (predictable) access patterns so that pipeline parallelism is possible. You can leverage inexpensive processors and disks by partitioning your task into many independent parallel streams that can execute in parallel.

3.3. Communications

The trends in data communications are exaggerated forms of the trends in processing and storage. At the technology level, the speed of light is not changing, but there are both software and hardware revolutions in our ability to deliver bandwidth. For short distances (tens of meters) one can use point-to-point copper and coax to send signals at up to 10Gbps. For longer distances, optical networks can carry signals at about 10 Gbps per wavelength, and can carry 500 different wavelengths, for a total of about 1 Tbps per fiber. These fibers can come in bundles of a thousand, so there is lots of bandwidth to go around. In 1995, George Gilder coined the law of the telecosm: deployed bandwidth grows 300% each year. So far, his prediction has held true, and some people are beginning to believe him (Gilder 1997, 2000).

In the local area, gigabit Ethernet is about to become ubiquitous and inexpensive. This delivers 100 MBps to each processor and each device. If that is not enough then you can get multiples of these links. The cost of such links is about $1,000 now (NIC+cable+port at a switch). This is expensive compared to about $100 for a switched 100Mbps Ethernet, but the link price for GbE will drop to $100 over the next 2 years. Indeed, 10GbE is now emerging from the laboratories.

In parallel to this, most of the hardware vendors (IBM, Intel, Sun, Compaq, Dell, ...) are working on a system area network called Infiniband[5]. This is a 10Gbps processor-to-processor link that is not as flexible as GbE, but has the virtue that the NIC is built onto the chip. Future machines will likely be a combination of 10GbE and Infiniband interconnects. (This will likely displace ATM and FiberChannel and MyriNet and GigaNet and ServerNet and ...), mostly because these latter technologies are low volume and hence relatively expensive.

The sad fact is that local networks have been software limited: the software costs of sending a message have dominated the hardware latencies. The basic problem has been that tcp/ip and the sockets interface were designed for a wide area network with long latencies, high error rates, and low bandwidth. The local area network (LAN) performance suffered with this common software stack. Over the last 5 years, the software stacks have improved enormously, mostly driven by the need to improve web server performance and partly driven by the use of tcp/ip in parallel programming environments. Now most operating systems are willing to offload much of the tcp/ip protocol to the NIC (checksums, DMA, interrupt moderation, buffering) so the software is much less of a bottleneck.

As a demonstration of this, colleagues at ISI-east (in Arlington Virginia), U. Washington (Seattle, Washington), and Microsoft (Redmond Washington) have been showing off high-speed desktop-to-desktop WAN bandwidths in the 100MBps (megabyte per second) range. This is not so impressive in a LAN, but they have also demonstrated it coast-to-coast using commodity workstations and out-of-the-box Windows2000 software. The Unix community is working along similar lines. Indeed, the Windows2000 demo interoperates with a Linux box at similar speeds.

These are "hero" benchmarks that employ very high-speed links (2.5 Gbps), expensive switches (Juniper) costing hundreds of thousands of dollars each, and of gurus (who are priceless). The challenge now is to close this guru gap: to allow ordinary scientists and high-school students to get this same high-performance. The empirical fact is that most scientists are getting less than 100KBps to their desktop, 1,000 times less than the "advertised" performance. Companies like Microsoft are working on this problem for their product—one goal is for Windows2001 to deliver this performance with no user tuning. I believe other companies are also working to close the guru gap. The USG has given several

---

[5] http://www.infiniband.org

million dollars to the Pittsburgh Supercomputer Center to build an open source auto-tuner for Linux[6].

You can expect that links in the wide area network will be able to move data at about 1GBps (10 Gbps). This means that it will be fairly easy to move astronomy datasets of less than a terabyte from place to place within an hour. In that same timeframe, disks will be about 1TB and will have bandwidth of about 200MBps, so the technologies will be reasonably well matched (you will use parallel disks).

Now, the bad news: Networking costs have been dropping fast. Public network prices have not improved much in the last 10 years. The retail cost of a Megabit link has remained about the same since 1990. While the price of the technology has dropped about 1,000x the cost of the right-of-way, trucks, cones, and people has increased in that period. One might expect therefore, that you could buy a Gigabit link in 2000 for the price of a Megabit link on 1990. But, sadly, that is not the case. Someone is making a lot of money on these artificially high prices. The hope is that competition will bring prices in line with costs over the next 5 years, but that hope failed over the last 5 years[7].

3.4. Software

If you have watched the evolution of computing, it is easy to believe that there have been exponential improvements in hardware and linear improvements in software. After all: there is not that much difference between Java and Fortran.

In some ways I agree with this view, in some ways I disagree. If you want to compute GCD (greatest common devisor using the Chinese Remainder Theorem) then there has not been much progress in software. On the other hand if you want to sort, then there has been huge progress (50% of the sorting speedup is due to parallelism). If you want to do a database application, then there has been huge progress in the tools space. And if you want to build a web site, there has been huge progress in tools.

The VOF effort is more like a database problem or a web site problem than it is like GCD. The traditional Fortran/Cobol/Algol/C/Pascal/Ada/C++/Smalltalk/Java community has made substantial progress on traditional programming. I use these tools almost every day. But the real quantum leap has been to raise the level of abstraction so that you are programming at a meta-level. PERL-Unix operates at this meta-level. More generally, scripting languages like JavaScript, VBscript combined with powerful libraries seem to offer huge productivity gains over traditional approaches. John Ousterhout has been making this point for a long time[8]. By now, I think most of us get it.

The traditional programming languages (including the scripting languages) run tasks sequentially. They do not make parallel execution easy to specify or to execute. Throughout this talk, I have tried to emphasize the need to execute things in parallel, if you want to be able to process large quantities of data

---

[6]http://www.web100.org/

[7]http://www.research.att.com/ amo/doc/history.communications0.pdf

[8]http://www.scriptics.com/people/john.ousterhout/

Computer Technology Forecast 253

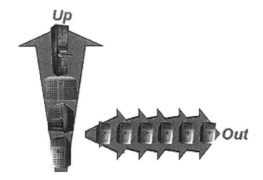

Figure 6. By parallel database techniques on many inexpensive processors and disks to search the astronomy archives in parallel, we hope to give Astronomers automatic parallelism. Searching Terabyte datasets in a few minutes.

quickly. The scientific community has tried to recast the Fortran model in a parallel style, first with High Performance Fortran (HPF) and more recently with PVM and MPI[9]. These are laudable efforts, and MPI is the workhorse of Beowulf clusters. But, MPI requires great skill on the part of the programmer. Meanwhile, the database community has figured out ways to automatically extract parallelism from SQL queries and execute those queries in parallel. Companies like Wal-Mart have huge clusters of machines (thousands of processors, 10s of thousands of disks) storing and searching the Wal-Mart inventory database.

Don Slutz and I have been working with Alex Szalay to apply these techniques to the Sloan Digital Sky Survey data. I think it is fair to say that we have made great progress. Alex is able to read data in parallel at about 200 MBps per node using an SQL interface. A 10-node cluster costing about $200k should be able to read data at 4 GBps and be able to answer most questions within 20 minutes. If we can put a Graphical query interface on the front of this, and a good data visualization system on the output, we should be able to allow astronomers to do data analysis without having to write programs. Rather the astronomers would just write the equations that we then execute in parallel against all the relevant objects (Szalay 2000). We hope that this approach will bring the benefits of parallelism to the VOF community.

We expect that groups will set up servers on the Internet that allow that group's data to be searched. Astronomers will be able to send queries to the server and get back answers either immediately (if it is a simple question) or within an hour (if the question requires scanning the entire archive). If an astronomer is interested in a particular part of the dataset, she can copy that part

---

[9]http://www.mpi-forum.org/

to a local parallel server (costing $1k/Terabyte of disk including the processors to manage and search that terabyte).

There have also been great advances in data mining technology for massive data sets—mostly being used in the commercial sector. Wall-mart, Amazon, AOL, and Microsoft are each investing heavily in these technologies to help their marketing and their Internet business. Similar, even more confidential, efforts are underway in the financial community and in the security agencies. Much of this work is proprietary—and so not generally understood. On the surface, it seems that these tools ought to be useful to Astronomy, since they are designed to find patterns in huge datasets of very high dimension with considerable noise in the data. It is also possible that advances in the Astronomy community could help these commercial interests. This is all quite speculative, but the public nature of the Astronomy datasets makes them an excellent testbed for new data mining algorithms.

How will the members of the Virtual Observatory exchange data? How will anyone find anything? How can the archives cooperate? The current answer is FITS (Flexible Image Transport System[10]) files. This is actually an OK answer for some questions. Certainly FITS is a good way of representing astronomy data sets for data interchange. The more trendy approach would be to use XML as the data representation. That replaces ASCII with UNICODE and replaces all binary numbers with their Unicode representation. The result will be a bigger file (which compresses nicely) but the result will not be semantically any different from FITS. There are many tools that work with XML and XML is more widely known than FITS—both of these are important points, and probably justify a FITS-XML marriage.

In order for the archives to cooperate and in order for astronomers to compare data from multiple archives, the data in each archive needs to have a clear description of what it means. The units, coordinate system, precision, accuracy needs to be specified if other scientists are to be able to use the data. It is a difficult and thankless task to construct this meta-data. This fact means that a virtual observatory is not possible unless the scientists are willing to perform this difficult and thankless task. It is hard for me to guess the outcome. But, I must assume that the astronomers went to great pains to gather the data, so they will exert the effort to preserve it, and make it available to future generations.

Let's assume that the metadata has been built for a few important data sets. Then what? How will you or I find the data and access it? The best current story seems to be a combination of OpenGIS and SOAP (sorry to be talking in acronyms—but that is how it goes with us geeks). OpenGIS[11] is defining standard ways to access geo-spatial data (GIS stands for Geographical Information System.) They have defined a standard meta-data framework for spatial data and are doing an XML representation now. This also includes logic for accessing the data: extracting data from an archive, and adding new data to it. It seems likely that there are strong ties between this spatial data and the

---

[10] http://fits.gsfc.nasa.gov/

[11] http://www.opengis.org/index.htm

spatial data in the astronomy community. Some combination of the FITS work with OpenGIS will probably emerge within the next 5 years.

The next step is to have a worldwide catalog of the metadata for all the astronomy data sets. Rather than bore you with more acronyms, I will just say that the commercial sector is cooperating to define such a "yellow pages" and "green pages" for Internet services. The Astronomy community can piggyback on this effort (XML is a core technology for these directory services). The Astronomy community already has a good start on this with the Vizier project[12].

SOAP (Simple Object Access Protocol[13]) is a way to call subroutines anywhere in the Internet—from any language to any language. It maps your subroutine call into an XML message, and sends it to the server. It maps the answer onto XML and sends it back to your program. SOAP works for Fortran, Cobol, C++, Java, JavaScript, VB, and C#, and your favorite programming language (it is agnostic about languages). It uses web servers like Apache or IIS as the ORB, replacing the need for a CORBA or DCOM ORB.

As Tony Hoare observes "algorithms are procedures plus data." XML is just data, SOAP is a way to externalize (publish) your procedures. The Astronomy Archives need to define the procedures they want to publish and support. These procedures would return the data you (the client) want from the archive. There is not time to go into it here, but we are doing something like this for the TerraServer as a textbook case of how to externalize information from a web services to clients. This is a big part of Microsoft's .NET initiative.

4. Conclusion

I have tried to covey three main messages:

1. Technology progress in processors, storage, and networks is keeping pace with the growth of Astronomy datasets.

2. Access times are not improving so rapidly, hence we must use parallelism to ameliorate this gap. With parallelism, we will be able to scan the archives within an hour. This parallelism should be automatic. I believe database technology can give this automatic parallelism.

3. The Virtual Observatory will be a federation of database systems. As such, the federation needs to define interoperability standards. FITS is a good start, but the community could benefit by looking at progress in the commercial sector in using directories, portals, XML, and SOAP for data interchange.

---

[12] http://vizier.u-strasbg.fr/doc/astrores.htx
[13] http://www.w3.org/TR/SOAP

References

Chung, L., & Gray, J. 2000, MSR-TR-2000-55[14]
Devlin, B., Gray, J., Laing, B., & Spix, G. 2000, MSR-TR-2000-85[15]
Gilder, G. 1997, Forbes, April 7, 1997[16]
Gilder, G. 2000, Free Press, ISBN: 068480930
Gray, J., & Shernoy, P. 2000, ICDE 2000[17]
Szalay, A., Kunszt, P., Thakar, A., Gray, J., Slutz, D., & Brunner, R.J. 2000, ACM SIGMOD Conference 2000: 451-462[18]

---

[14] http://research.microsoft.com/scripts/pubs/view.asp?TR_ID=MSR-TR-2000-55

[15] http://research.microsoft.com/scripts/pubs/view.asp?TR_ID=MSR-TR-99-85

[16] http://www.forbes.com/asap/97/0407/090.htm

[17] http://research.microsoft.com/scripts/pubs/view.asp?TR_ID=MSR-TR-99-100

[18] http://research.microsoft.com/scripts/pubs/view.asp?TR_ID=MSR-TR-99-30

Astronomy Image Collections

Reagan W. Moore

San Diego Supercomputer Center, La Jolla, CA 92093

Abstract. At the San Diego Supercomputer Center, image collections are being supported for multiple disciplines, including astronomy, high energy physics, earth systems science, and the humanities. Based upon the experience learned through implementation of these distributed data collections, the key infrastructure elements needed for managing image collections can be identified. The associated infrastructure includes persistent archives for managing technology evolution, data handling systems for collection-based access to data, collection management systems for organizing information catalogs, digital library services for manipulating data sets, and data grids for federating multiple collections. The infrastructure components can be characterized as interoperability systems for digital object management, information management, and knowledge management. The architecture for building a distributed astronomy image collection is illustrated with a summary of the infrastructure used to support the 2-Micron All Sky Survey image collection.

1. Data Management Architecture

A collection-based data management system has the following software infrastructure layers:

- Data Grid—for federation of access to multiple data collections and digital libraries

- Digital library—to provide services for discovering, manipulating, and presenting data from collections

- Knowledge repository—to describe relationships between concepts used to organize the collection

- Information repository—to provide support for querying extensible attributes associated with the images

- Data handling system—to provide persistent identifiers for collection-based access to data sets

- Persistent archive—to provide collection-based storage of data sets, with the ability to handle evolution of the software infrastructure.

The essential infrastructure component is the data handling system. It is possible to use data handling systems to assemble distributed data collections, integrate digital libraries with archival storage systems, federate multiple collections into a data grid, and create persistent archives. An example that encompasses all of these cases is the 2-Micron All Sky Survey image archive.

2. 2-MASS All Sky Survey Image Collection

The 2MASS survey is an astronomy project led by the University of Massachusetts and Caltech to assemble a catalog of all stellar objects that are visible at the 2-micron wavelength. The goal of the project is to provide a catalog that lists attributes of each object, such as brightness and location. The final catalog can contain as many as 2 billion stars and 200 million galaxies. Of interest to astronomers is the ability to analyze the images of all the galaxies. This is a massive data analysis problem since there will be a total of 5 million images comprising 10 terabytes of data.

A collaboration between IPAC at Caltech and the NPACI program at SDSC is building an image catalog of all of the 2MASS observations. A digital library is being created at Caltech that records which image contains each object. All images that are put into the archive are first aggregated into physical containers (or files) to minimize the number of discrete digital objects that the archival storage system must manage. Also, the images are sorted into 147,000 containers to co-locate all images from the same area in the sky. The image collection is then replicated between archives at Caltech and SDSC to provide disaster recovery. The SDSC Storage Resource Broker is used as the data handling system to provide access to the archives at Caltech and SDSC where the images are replicated.

The usage model builds upon the following access procedures:

- Astronomers access the catalog at Caltech to identify galaxy types of interest.

- Digital library procedures determine which images need to be retrieved.

- The data handling system maps the image to the appropriate container, retrieves the container from the archive, and caches the container on a disk.

- The desired image is then read from the disk and returned to the user through the digital library.

Since the images are accessed through the data handling system, the desired images can be retrieved from either archive depending upon load or availability. If the container that holds the image has already been migrated to disk cache, the data handling system can immediately retrieve the image from disk avoiding the access latency inherent in reading from the archive tape system. If a fire destroys one of the archives, the data handling system automatically defaults to the alternate storage system. If data is migrated to alternate storage systems, the persistent identifier for the image remains the same and the data handling system adds the location and access protocol metadata for the new storage system. The system incorporates persistent archive technology, data handling systems, collection management tools, and digital library services in order to support analysis of galactic images.

Given the ability to access a 10-terabyte collection that contains a complete sky survey at the 2-micron wavelength, it is then possible to do data intensive analyses on terascale computer platforms. The data rates that are supported by a teraflops-capable computer will be over 100 Megabytes per second. It will be possible to read the entire collection in 30 hours. During this time period, up to ten billion operations can be done on each of the five million images. Effectively, the entire survey can be analyzed in a "typical" computation on a teraflops computer.

3. Infrastructure Components

The development of infrastructure to support the creation of distributed image collections must recognize that information repositories and knowledge bases are also needed. One can differentiate between infrastructure components that provide:

- Data storage of digital objects that are either simulation output or remote sensing data. The digital objects are representations of reality, generated either through a hardware remote sensing device or by execution of an application.

- Information repositories that store attributes about the digital objects. The attributes are typically stored as metadata in a catalog or database.

- Knowledge bases that characterize relationships between sets of metadata. An example is rule-based ontology mapping that provides the ability to correlate information stored in multiple metadata catalogs.

A image digital library will need to support ingestion of digital objects, querying of metadata catalogs to identify objects of interest, and integration of responses from multiple information repositories. Fortunately, a rapid convergence of information management technology and data handling systems is occurring for the support of scientific data collections. The goal is to provide mechanisms for the publication of scientific data for use by an entire research community. The approach used at the San Diego Supercomputer Center is to organize distributed data sets through creation of a logical collection. The ownership of the data sets is assigned to the collection, and a data handling system is used to create, move, copy, replicate, and read collection data sets. Since all accesses to the collection data sets are done through the data handling system, it then becomes possible to put the data sets under strict management control, and implement features such as access control lists, usage audit trails, replica management, and persistent identifiers.

Effectively, a distributed collection can be created in which the local resources remain under the control of the local site, but the data sets are managed by the global logical collection. Researchers authenticate themselves to the collection, and the collection in turn authenticates itself to the distributed storage systems on which the data sets reside. The collection manages the access control lists for each data set independently of the local site. The local resources are effectively encapsulated into a collection service, removing the need for researchers to have user accounts at each site where the data sets are stored.

The data handling system serves as an interoperability mechanism for managing storage systems. Instead of directly storing digital objects in an archive or file system, the interposition of a data handling system allows the creation of a collection that spans multiple storage systems. It is then possible to automate the creation of a replica in an archival storage system, cache a copy of a digital object onto a local disk, and support the remote manipulation of the digital object. The creation of data handling systems for collection-based access to published scientific data sets makes it possible to automate all data management tasks. In turn, this makes it possible to support data mining against collections of data sets, including comparisons between simulation and measurement, and statistical analysis of the properties of multiple data sets. Data set handling systems can be characterized as interoperability mechanisms that integrate local data resources into global resources. The interoperability mechanisms include

- inter-domain authentication,

- transparent protocol conversion for access to all storage systems,

- global persistent identifiers that are location and protocol independent,

- replica management for cached and archived copies,

- container technology to optimize archival storage performance and co-locate small data sets, and

- tools for uniform collection management across file systems, databases, and archives.

4. Data Handling Infrastructure

The data management infrastructure is based upon technology from multiple communities that are developing archival storage systems, XML database management systems, digital library services, distributed computing environments, and persistent archives. The combination of these systems is resulting in the ability to describe, manage, access, and build very large distributed scientific data collections. Several key factors are driving the technology convergence:

- Development of an appropriate information exchange protocol and information tagging model. The ability to tag the information content makes it possible to directly manipulate information. The eXtensible Markup Language (XML) provides a common information model for tagging data set context and provenance. Document Type Definitions (and related organizational methods such as X Schema) provide a way to organize the tagged attributes. Currently, each discipline is developing their own markup language (set of attributes) for describing their domain-specific information. The library community has developed some generic attribute sets such as the Dublin core to describe provenance information. The combination of the Dublin core metadata and discipline specific metadata can be used to describe scientific data sets.

- Differentiation between the physical organization of a collection (conceptually the table structures used to store attributes in object-relational databases) and the logical organization of a collection (the schema). If both contexts are published, it becomes possible to automate the generation of the SQL commands used to query relational databases. For XML-based collections, the emergence of XML Matching and Structuring languages makes it possible to construct queries based upon specification of attributes within XML DTDs. Thus attribute-based identification of data sets no longer requires the ability to generate SQL or XQL commands from within an application.

- Differentiation of the organization and access mechanisms for a logical collection from the organization and access mechanisms required by a particular storage system. Conceptually, data handling systems store data in storage systems rather than storage devices. By keeping the collection context independent of the physical storage devices, and providing interoperability mechanisms for data movement between storage systems, logical data set collections can be created across any type of storage system. Existing data collections can be transparently incorporated into the logical collection. The only requirement is that the logical collection be given access control permissions for the local data sets. The data handling system becomes the unifying middleware for access to distributed data sets.

- Differentiation of the management of information repositories from the storage of metadata into a catalog. Information management systems provide the ability to manage databases. It is then possible to migrate metadata catalogs between database instantiations, extend the schema used to organize the catalogs, and export metadata as XML or HTML formatted files.

The ability to manipulate data sets through collection-based access mechanisms enables the federation of data collections and the creation of persistent archives. Federation is enabled by publishing the schema used to organize a collection as an XML DTD. Information discovery can then be done through queries based upon the semi-structured representation of the collection attributes provided by the XML DTD. Distributed queries across multiple collections can be accomplished by mapping between the multiple DTDs, either through use of rule-based ontology mapping, or token-based attribute mapping.

Persistent archives can be enabled by archiving the context that defines both the physical and logical collection organization along with the data sets that comprise the collection. The collection context can then be used to recreate the collection on new database technology through an instantiation program.

This makes it possible to migrate a collection forward in time onto new technology. The collection description is instantiated on the new technology, while the data sets remain on the physical storage resource. The collection instantiation program is updated as database technology evolves, while the archived data remains under the control of the data handling system. As the archive technology evolves, new drivers are added to the data handling system to interoperate with the new data access protocols.

The implementation of information management technology needs to build upon the information models and manipulation abilities that are coming from the Digital Library community, and the remote data access and procedure execution support that are coming from the distributed computing community. The Data Access Working Group of the Grid Forum is promoting the development of standard implementation practices for the construction of data grids. Data grids are inherently distributed systems that tie together data, compute, and visualization resources. Researchers rely on the data grid to support all aspects of information management and data manipulation. An end-to-end system provides support for:

- Knowledge discovery—ability to identify relationships between digital objects stored in different discipline collections

- Information discovery—ability to query across multiple information repositories to identify data sets of interest

- Data handling—ability to read data from a remote site for use within an application

- Remote processing—ability to filter or subset a data set before transmission over the network

- Publication—ability to add data sets to collections for use by other researchers

- Analysis—ability to use data in scientific simulations, or for data mining, or for creation of new data collections

These services are implemented as middleware that hide the complexity of the diverse distributed heterogeneous resources that comprise data and compute grids. The services provide four key functionalities or transparencies that simplify the complexity of accessing distributed heterogeneous systems.

- Name transparency—Unique names for data sets are needed to guarantee a specific data set can be found and retrieved. However, it is not possible to know the unique name of every data set that can be accessed within a data grid (potentially billions of objects). Attribute based access is used so that any data set can be identified either by data handling system attributes, or Dublin core provenance attributes, or discipline specified attributes.

- Location transparency—Given the identification of a desired data set, a data handling system manages interactions with the possibly remote data set. The actual location of the data set can be maintained as part of the data handling system attributes. This makes it possible to automate remote data access. When data sets are replicated across multiple sites, attribute-based access is essential to allow the data handling system to retrieve the "closest" copy.

- Protocol transparency—Data grids provide access to heterogeneous data resources, including file systems, databases, and archives. The data handling system can use attributes stored in the collection catalog to determine the particular access protocol required to retrieve the desired data set. For heterogeneous systems, servers can be installed on each storage resource to automate the protocol conversion. Then an application can access objects stored in a database or in an archive through a uniform user interface.

- Time transparency—At least five mechanisms can be used to minimize retrieval time for distributed objects: data caching, data replication, data aggregation, parallel I/O, and remote data filtering. Each of these mechanisms can be automated as part of the data handling system. Data caching can be automated by having the data handling system pull data from the remote archive to a local data cache. Data replication across multiple storage resources can be used to minimize wide area network traffic. Data aggregation through the use of containers can be used to minimize access latency to archives or remote storage systems. Parallel I/O can be used to minimize the time needed to transfer a large data set. Remote data filtering can be used to minimize the amount of data that must be moved. This latter capability requires the ability to support remote procedure execution at the storage resource.

5. Summary

The development of a distributed image collection for accessing multiple astronomy all-sky surveys is greatly simplified by differentiating between data handling systems, information management systems, and knowledge management systems. Each component is effectively an interoperability mechanism for linking different implementations of the infrastructure. By focusing on the interoperability between currently available systems for data storage, information storage, and knowledge storage, it becomes possible to assemble a national virtual observatory that links current archives, information catalogs, and ontologies into a coherent system.

References

Moore, R., C. Baru, A. Rajasekar, B. Ludascher, R. Marciano, M. Wan, W. Schroeder, and A. Gupta, "Collection-Based Persistent Digital Archives — Part 1", D-Lib Magazine, March 2000[1]

---

[1] http://www.dlib.org/

Foster, I., Kesselman, C., "The Grid: Blueprint for a New Computing Infrastructure," Chapter 5, "Data Intensive Computing," Morgan Kaufmann, San Francisco, 1999.

Baru, C., R, Moore, A. Rajasekar, M. Wan, "The SDSC Storage Resource Broker," Proc. CASCON'98 Conference, Nov.30-Dec.3, 1998, Toronto, Canada.

NPACI Data Intensive Computing Environment thrust area[2]

Moore, R., C. Baru, P. Bourne, M. Ellisman, S. Karin, A. Rajasekar, S. Young, "Information Based Computing," Proceedings of the Workshop on Research Directions for the Next Generation Internet, May, 1997.

---

[2] http://www.npaci.edu/DICE/

Computational AstroStatistics: Fast Algorithms and Efficient Statistics for Density Estimation in Large Astronomical Datasets

R.C. Nichol,

Department of Physics, Carnegie Mellon University, 5000 Forbes Avenue, Pittsburgh, PA 15213

A.J. Connolly,

Department of Physics & Astronomy, University of Pittsburgh, 3941 O'Hara Street, Pittsburgh, PA 15260

A.W. Moore, J. Schneider,

Robotics Institute & the Computer Science Department, Carnegie Mellon University, 5000 Forbes Avenue, Pittsburgh, PA 15213

C. Genovese, L. Wasserman,

Department of Statistics, Carnegie Mellon University, 5000 Forbes Avenue, Pittsburgh, PA 15213

Abstract. We present initial results on the use of Mixture Models for density estimation in large astronomical databases. We provide herein both the theoretical and experimental background for using a mixture model of Gaussians based on the Expectation Maximization (EM) Algorithm. Applying these analyses to simulated data sets we show that the EM algorithm — using the both the AIC & BIC penalized likelihood to score the fit — can out-perform the best kernel density estimate of the distribution while requiring no "fine-tuning" of the input algorithm parameters. We find that EM can accurately recover the underlying density distribution from point processes thus providing an efficient adaptive smoothing method for astronomical source catalogs. To demonstrate the general application of this statistic to astrophysical problems we consider two cases of density estimation; the clustering of galaxies in redshift space and the clustering of stars in color space. From these data we show that EM provides an adaptive smoothing of the distribution of galaxies in redshift space (describing accurately both the small and large-scale features within the data) and a means of identifying outliers in multi-dimensional color-color space (*e.g.*, for the identification of high redshift QSOs). Automated tools such as those based on the EM algorithm will be needed in the analysis of the next generation of astronomical catalogs (2MASS, FIRST, PLANCK, SDSS) and ultimately in the development of the National Virtual Observatory.

1. Introduction

With recent technological advances in wide field survey astronomy it has now become possible to map the distribution of galaxies and stars within the local and distant Universe across a wide full spectral range (from X-rays through to the radio) *e.g.*, FIRST, MAP, Chandra, HST, ROSAT, SDSS, 2dF, Planck. In isolation, the scientific returns from these new data sets will be enormous, combined however, these data will represent a new paradigm for how we undertake astrophysical research. They present the opportunity to create a "Virtual Observatory" that will enable a user to seamlessly analyze and interact with a multi-frequency digital map of the sky. While the wealth of information contained within these new surveys is clear, the questions we must address is how to we efficiently analyze these massive and intrinsically multidimensional data sets. Standard statistical approaches do not easily scale to the regime of 10^8 data points and 100's of dimensions.

Therefore, we have formed a collaboration of computer scientists, statisticians and astrophysics to address this problem through the development of fast and efficient statistical algorithms that scale to many dimensions and large datasets like those found in astronomy. In this volume, we present outlines of our on-going research including the use of tree algorithms for fast n-point correlation functions, non-parametric density estimations (Genovese *et al.* 2001, these proceedings) and data visualization (Welling *et al.* 2001, these proceedings). In this paper, we present initial results from the application of Mixture Models to the general problem of density estimation in astrophysics; the reader is referred to Connolly *et al.* (2000, in preparation) for the full details of the theory, the algorithm and our results.

2. Outline of the Algorithm

Let $X^n = (X_1, \ldots, X_n)$ represent the data. Each X_i is a d-dimensional vector giving, for example, the location of the ith galaxy. We assume that X_i takes values in a set A where A is a patch of the sky from which we have observed. We regard X_i as having been drawn from a probability distribution with probability density function f. This means that $f \geq 0$, $\int_A f(x)dx = 1$ and the proportion of galaxies in a region B is given by $Pr(X_i \in B) = \int_B f(x)dx$. In other words, f is the the normalized galaxy density and the proportion of galaxies in a region B is just the integral of f over B. Our goal is to estimate f from the observed data X^n.

The density f is assumed to be of the form

$$f(x;\theta_k) = p_0 U(x) + \sum_{j=1}^{k} p_j \phi(x;\mu_j,\Sigma_j) \qquad (9)$$

where $\phi(x;\mu,\Sigma)$ denotes a d-dimensional Gaussian with mean μ and covariance Σ:

$$\phi(x;\mu,\Sigma) = \frac{1}{(2\pi)^{d/2}|\Sigma|^{1/2}} \exp\left\{-\frac{1}{2}(x-\mu)^T \Sigma^{-1}(x-\mu)\right\},$$

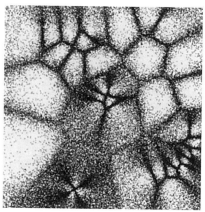

Figure 1. A comparison of the fixed kernel and our EM density estimation. Top left panel is the simulated Voronoi density distribution while top right shows 100,000 data points draw from this distribution.

and $U(\cdot)$ is a uniform density over A i.e., $U(x) = 1/V$ for all $x \in A$, where V is the volume of A. The unknown parameters in this model are k (the number of Gaussians) and $\theta_k = (p, \mu, \Sigma)$ where $p = (p_0, \ldots, p_k)$, $\mu = (\mu_1, \ldots, \mu_k)$ and $\Sigma = (\Sigma_1, \ldots, \Sigma_k)$. Here, $p_j \geq 0$ for all j and $\sum_{j=0}^{k} p_j = 1$. This model is called a mixture of Gaussians (with a uniform component). The parameter k controls the complexity of the density f. Larger values of k allow us to approximate very complex densities f but also entail estimating many more parameters. It is important to emphasize that we are not assuming that the true density f is exactly of the form (Equation 1). It suffices that f can be approximated by such a distribution of Gaussians. For large enough k, nearly any density can be approximated by such a distribution (see Roeder & Wasserman 1997).

2.1. The EM Algorithm

How do we find the maximum likelihood estimate of θ_k i.e., $\hat{\theta}$? (We assume for the moment that k is fixed and known.) The usual method for finding $\hat{\theta}$ is the EM (expectation maximization) algorithm. We can regard each data point X_i as arising from one of the k components of the mixture. Let $G = (G_1, \ldots, G_k)$ where $G_i = j$ means that X_i came from the jth component of the mixture. We do not observe G so G is called a latent (or hidden) variable. Let ℓ_c be the log-likelihood if we knew the latent labels G. The function ℓ_c is called the complete-data log-likelihood. The EM algorithm proceeds as follows. We begin with starting values θ_0. We compute $Q_0 = E(\ell_c; \theta_0)$, the expectation of ℓ_c, treating G as random, with the parameters fixed at θ_0. Next we maximize Q_0 over θ to obtain a new estimate θ_1. We then compute $Q_1 = E(\ell_c; \theta_1)$ and continue this process. Thus we obtain a sequence of estimates $\theta_0, \theta_1, \ldots$ which are known to converge to a local maximum of the likelihood. See Connolly et al. (2000) for more details.

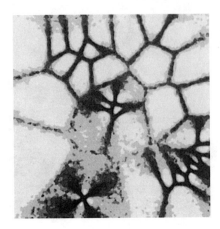 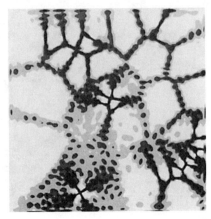

Figure 2. The resulting density estimates from the data in Figure 1: fixed kernel (left) and EM+AIC (left)

Above, we have assumed that k is known. One approach of choosing k from the data is to sequentially test a series of hypotheses i.e. test the density estimation using k versus the one with $k+1$ and repeat for various k. The usual test for comparing such hypotheses is called the "likelihood ratio test" which compares the value of the maximized log-likelihood under the two hypotheses. This approach is infeasible for large data sets where k might be huge. Also, this requires knowing the distribution of the likelihood ratio statistic which is not known. As an alternative (see Connolly et al. 2000 for the explanation), we use two common penalized log-likelihoods to test these two hypotheses, namely Akaike Information Criterion (AIC) and Bayesian Information Criterion (BIC) which take the general form of $\ell(\hat{\theta}_k) - \lambda_k R_k$ where R_k is the number of parameters in the kth model. For AIC $\lambda_k = 1$ while $\lambda_k = \log n/2$ gives the BIC criterion. See Connolly et al. (2000) for a full explanation.

Given the definition of the EM algorithm and the criteria for determining the number of components to the model using AIC and BIC we must now address how do we apply this formalism to massive multi-dimensional astronomical datasets. In its conventional implementation, each iteration of EM visits every data point pair, meaning kR evaluations of a M-dimensional Gaussian, where R is the number of data-points and k is the number of components in the mixture. It thus needs $O(M^2 kR)$ arithmetic operations per iteration. For data sets with 100s of attributes and millions of records such a scaling will clearly limit the application of the EM technique. We must therefore develop an algorithmic approach that reduces the cost and number of operations required to construct the mixture model. To do this, we have used Multi-resolution KD-trees to gain impressive speed-ups for the numerous range-searches involved in computing the fits of these k Gaussians to the data.

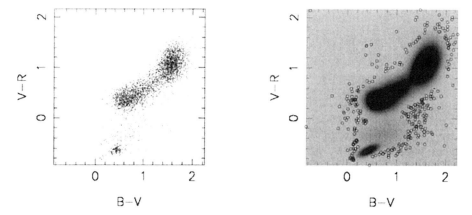

Figure 3. The application of our algorithm to density estimation in color space. The left panel shows the distribution of $B - V$ and $V - R$ colors of stellar sources while the right hand panel shows the density distribution, as a grayscale, of these sources derived from our algorithm (using AIC). The open circles are the 5% of the data that is least likely to be drawn from this distribution.

3. Applications to Astrophysical Problems

To test the sensitivity of the Mixture Model density estimation to the hierarchical clustering in the universe, and thus determine if it is better than present, more traditional, astronomical methods of density estimations, we have tested our algorithm using simulated data generated from a Voronoi Tessellation since this mimics the observed distribution of filaments and sheets of galaxies in the universe. In Figure 1, we present the underlying Voronoi density map we have constructed; this is the "truth" in our simulation. From this distribution we derive a set of 100,000 data points to represent a mock 2-dimensional galaxy catalog.

We have applied the EM algorithm (with both the AIC and BIC criteria) and a standard fixed kernel density estimator to these point-like data sets in order to reproduce the original density field. The latter involved finely binning the data and smoothing the subsequent grid with a binned Gaussian filter of fixed bandwidth which was chosen **by hand** to minimize the Kullback-Leibler distance $(D(f,g) = \int f(x) \log f(x)/g(x) dx)$ between the resulting smoothed map and the true underlying density distribution. Clearly, we have taken the optimal situation for the fixed kernel estimator since we have selected it's bandwidth to ensure as close a representation of the true density distribution as possible. This would not be the case in reality.

In Figure 2, we show the reconstructed density field using the fixed kernel and EM algorithm (AIC). As we would expect the fixed kernel technique provides an accurate representation of the overall density field. The kernel density map suffers, however, when we consider features that are thinner than the width of the kernel. Such filamentary structures are over-smoothed and have their significance reduced. In contrast the EM algorithm attempts to adapt to the size of the structures it is trying to reconstruct. The right panel shows that where narrow filamentary structures exist the algorithm places a large number of small Gaussians. For extended, low frequency, components larger Gaussians are applied. For the fixed kernel estimator, we have a measured KL divergence of 0.074 between the final smoothed map and the true underlying density map (remember, this is the smallest KL measurement by design). For the EM AIC density map we measure a KL divergence of 0.067 which is lower than the best fixed kernel KL score thus immediately illustrating the power of the EM methodology. We have not afforded the same prior knowledge to the EM measurement — *i.e.*, hand-tune it so as to minimize the KL divergence — yet we have beaten the kernel estimator.

We have extended the analysis above to real astronomical data *e.g.*, the the distribution of 6298 stellar sources in the B-V and V-R color space (with R<22) taken from a 1 square degree multicolor photometric survey of Szokoly *et al.* (2001). In this case, the mixture model density distribution naturally provides a probability density map that a stellar object drawn at random from the observed distribution of stars would have a particular set of B-V — V-R colors. We can now assign a probability to each star in the original data that describes the likelihood that it arises from the overall distribution of the stellar locus. We can then rank order all sources based on the likelihood that they were drawn from the parent distribution. The right panel of Figure 3 shows the colors of the 5% of sources with the lowest probabilities. These sources lie preferentially away from the stellar locus and thus can be used to find high redshift quasars as outliers to color-space. As we increase the cut in probability the colors of the selected sources move progressively closer to the stellar locus.

The advantage of the EM approach over standard color selection techniques is that we identify objects based on the probability that they lie away from the stellar locus (*i.e.*, we do not need to make orthogonal cuts in color space as the probability contours will trace accurately the true distribution of colors). While for two dimensions this is a relatively trivial statement as it is straightforward to identify regions in color-color space that lie away from the stellar locus (without being restricted to orthogonal cuts in color-color space) this is not the case when we move to higher dimensional data. For four and more colors we lose the ability to visualize the data with out projecting it down on to a lower dimensionality subspace (*i.e.*, we can only display easily 3 dimensional data). In practice we are, therefore, limited to defining cuts in these subspaces which may not map to the true multidimensional nature of the data. The EM algorithm does not suffer from these disadvantages as a probability density distribution can be defined in an arbitrary number of dimensions. It, therefore, provides a more natural description of both the general distribution of the data and for the identification of outlier points from high dimensionality data sets. With the new generation of multi-frequency surveys we expect that the need for algorithms that scale to a large number of dimensions will become more apparent.

References

Szokoly, G., *et al.* , ApJ, 2001, in preparation

Roeder, K., Wasserman, L., 1997, Journal of the American Statistical Association, 92, 894

Statistical Methodology for the National Virtual Observatory

G. Jogesh Babu

Department of Statistics, 319 Thomas Building, Pennsylvania State University, University Park, PA 16802

E. D. Feigelson

Department of Astronomy & Astrophysics, 518 Davey Laboratory, Pennsylvania State University, University Park, PA 16802

Abstract. Large-scale NVO multi-wavelength surveys present a variety of challenging statistical problems. Multivariate clustering and analysis methods, such as principal components analysis, can be very helpful to astronomers and are briefly described. Newer statistical methods such as projection pursuit, multivariate splines, and visualization tools such as XGobi are introduced. However, multivariate databases from astronomical surveys present significant challenges to the statistical community. These include treatments of heteroscedastic measurement errors, censoring and truncation due to flux limits, and parameter estimation for nonlinear astrophysical models.

1. Introduction

Large astronomical surveys from new high-throughput detectors and observatories are powerful motivators for more effective statistical techniques. Astronomical observatories now frequently generate gigabytes of information every day, with terabyte-size raw databases which produce reduced catalogs of $10^6 - -10^9$ objects. The National Virtual Observatory may become an important institution where these databases and collected and used for astronomical study. These catalogs or raw databases, which may include up to dozens of observational properties of each object, often contain heterogeneous populations which must be isolated prior to detailed analysis.

Quite an amount of work has also been directed to the automated analysis and classification of objects or images, particularly the discrimination of stars from galaxies on optical band photographic plates and CCD images Each object is characterized by a number of properties (*e.g.*, moments of its spatial distribution, surface brightness, total brightness, concentration, asymmetry), which are then passed through a supervised classification procedure. Methods include multivariate clustering, Bayesian decision theory, neural networks, k-means partitioning, CART (Classification and Regression Trees) and multi-resolution methods (White 1997; Bijaoui, Rué, & Savalle 1997). Such procedures are crucial to the creation of the largest astronomical databases with 1–2 billion objects derived from digitization of all-sky photographic surveys. But after images are characterized and other preliminary reduction of raw data is performed, the intermediate scientific product of astronomical surveys is frequently a large table with rows representing individual stars, galaxies, sources or locations and columns representing observed or inferred properties. Statistical characterization of such databases is the domain of *multivariate analysis*. We therefore concentrate on multivariate statistical methodology in the following sections.

2. Multivariate Analysis and Clustering

A sample obtained from one or more multi-wavelength surveys often will not constitute a single type of astronomical object. A multivariate database should be viewed as vectors in p-space which can have any form of structure, not just planar correlations parallel to the axes. Apparent relationships between the variables may elucidate astrophysical processes, or may arise from heterogeneity of the sample. It is thus important to search for groupings in p-space using multivariate clustering or classification algorithms. Dozens of such methods have been proposed. Unfortunately, most are procedural algorithms without formal statistical justification (*i.e.*, no probabilistic measures of merit) and there is little mathematical guidance which produces 'better' clusters.

Hierarchical clustering methods produces small clusters within larger clusters. One such procedure, 'percolation' or the 'friends-of-friends' algorithm is a favorite among astronomers. In statistical parlance, it is called *single linkage clustering* and can be easily obtained by successively removing the longest branches of the unique *minimal spanning tree* connecting the n points in p-space. Single linkage produces long stringy clusters. This may be appropriate for galaxy clustering studies, but researchers in other fields usually prefer *average or complete linkage* algorithms which produce more compact clusters. The many varieties of hierarchical clustering arise due to the choice of the metric (*e.g.*, should the 'distance' between objects be Euclidean or squared?), weighting (*e.g.*, how is the average location of a cluster defined?), and criteria for merging clusters (*e.g.*, should the total variance or internal group variance be minimized?).

An alternative method with some mathematical foundation is *k-means partitioning*. It finds the combination of k groups that minimizes intragroup variance. However, it is necessary to specify the number k of groups in advance.

For each homogeneous class, the study of pairwise relationships between variables provides valuable insights into the data structure. The sample covariance matrix S contains information for this approach, and lies at the root of many methods of multivariate analysis developed during the 1930–60s. The method most widely used in astronomy is *principal components analysis* (PCA). A quick search of astronomical literature database yielded hundreds of papers in astronomy using PCA. The 1st principal component is $e_1^T X$ where e_k is the eigenvector of S corresponding to the kth largest eigenvalue. This is equivalent to finding by the direction in p-space where the data are most elongated using least-squares to minimize the variance. The second component finds the elongation direction after the first component is removed, and so forth.

PCA is also used in analyzing the covariance function of a timeseries data. However, caution should be exercised in using PCA. The two timeseries, the so-called *Ramp Function* and the *Brownian Bridge* (*i.e.*, tied-down Brownian Motion, tied-down at 1), both have identical covariance function. So PCA cannot distinguish the data from these two timeseries. Ramp function X is obtained by choosing a random point ω on the unit interval $[0, 1]$ and defining

$$X(t) = \begin{cases} t & \text{if } 0 \leq t \leq \omega \\ t - 1 & \text{if } \omega \leq t \leq 1. \end{cases}$$

Note that the stochastic process generated by the Ramp function is highly regulated, while the Brownian Bridge is a white noise Gaussian process. Without additional model assumptions such as Gaussian structure PCA is not of much use. PCA also fails for non-linear relations (*e.g.*, X Gaussian and $Y = X^2$).

3. Application to Gamma-ray Bursts Data

As only a few gamma-ray burst (GRB) sources have astronomical counterparts at other wavebands, empirical studies of GRBs have been largely restricted to the analysis of their bulk properties such as fluence and spectral hardness, and evolution of these properties within a burst event. While bursts exhibit a vast range of complex temporal behaviors, their bulk properties appear simpler and amenable to straightforward statistical analyses. Examination of whether GRB bulk properties comprise a homogeneous population or are divided into distinct classes may lead to astrophysical insights of their origins.

It can be dangerous to look for correlations prior to classifying (or establishing the homogeneity of) the population. While the anticorrelation between hardness ratio and burst duration seen in full samples may be the manifestation of a single astrophysical process, it may alternatively reflect differences between distinct processes. Most multivariate analyses thus begin with a study of homogeneity and classification, and then investigate the variance-covariance structure (*i.e.*, correlations) within each class. We recently performed such an analysis (Mukherjee *et al.* 1998)

Two multivariate clustering procedures are used on a sample of 797 bursts from the Third BATSE Catalog: a nonparametric average linkage hierarchical agglomerative clustering procedure validated with Wilks' Λ^* and other MANOVA tests; and a parametric maximum likelihood model-based clustering procedure assuming multinormal populations calculated with the EM Algorithm and validated with the Bayesian Information Criterion. The two methods yield very similar results. We find the BATSE GRB population consists of three classes with the following Duration/Fluence/Spectrum bulk properties: Class I with long/bright/soft bursts, Class II with short/faint/hard bursts, and Class III with intermediate/intermediate/soft bursts. Class IV consists of a single point later found to have erroneous processing.

These are frames from the 'grand tour' movie of the 5-dimensional dataset provided by the XGobi software where each cluster is 'brushed' with a different symbol.

Statistical Methodology 275

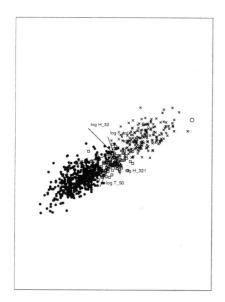
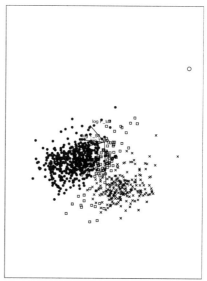

Figure 1. Two snapshots from the XGobi grand tour of the 5-dimensional database with bursts brushed according to the nonparametric Average Linkage clustering results: Class I (•), Class II (×), Class III (□) and Class IV (○).

4. New Methodology Needed for Astronomy

Many astronomical surveys are not amenable to traditional multivariate analysis and classification, and present serious needs for methodological advances by statisticians. Several major difficulties are outlined here.

Fluxes or other measured quantities are subject to *heteroscedastic measurement errors with known variances*. That is, each variable of each object has an associated measurement of the variable uncertainty, and these uncertainties can differ for each object. Surprisingly, statistical methodology is not well developed for such situations. For example, there does not seem to be a clustering algorithm that weights points by their known measurement errors. Astronomers also need density estimation, k-sample goodness-of-fit tests, spatial point processes, time series and multivariate analyses for such datasets.

Some objects may be undetected at one or many wavebands, leading to upper limits or *censoring* in one or many variables. A branch of statistics known as survival analysis, used principally for biomedical and industrial reliability applications, has been developed for censored datasets. A suite of survival methods is now widely used in astronomy (Feigelson 1992). However, most survival statistics apply only to univariate problems; Cox regression, the principal multivariate technique, permits censoring only in the single dependent variable. But a full multivariate survival analysis is not yet available.

Astronomical surveys nearly always suffer *truncation* in one or more variables due to sensitivity limits of the telescopes. Left-truncation is frequently present in any variable because faint objects are undetected. This can create spurious structure in the variance-covariance matrix and makes the sample distribution a biased estimate of the underlying population. As with censoring, little statistical attention has been directed towards such multivariate datasets, except for linear regression problems in econometrics (Maddala 1983).

Finally, following the traditions of celestial mechanics of previous centuries, modern astronomers often seek exploratory structural regression relations between variables, and wish to constrain *parameters of nonlinear astrophysical models*. Multivariate methodology was largely developed to assist social sciences and industry where such modeling does not arise. Classical regression techniques fail in the presence of heteroscedastic measurement errors, censoring and truncation. Often the model is so complex, particularly if survey selection effects are included within it, that the results are available only through Monte Carlo simulation.

While these issues have yet to be adequately addressed by statisticians, some recent advances can have significant benefits to astronomers. A number of approaches have emerged to facilitate both linear and nonlinear modeling of multivariate datasets. *Projection pursuit* regression uses local linear fits and sigmoidal smoothers to model nonlinear behavior (Huber 1985; Friedman 1987). *Multivariate Adaptive Regression Splines* (MARS) and a variety of similar methods fit the data with multidimensional splines (Friedman 1991). These methods are based on reasonable, but not unique, procedures for sparingly choosing the number of parameters that avoid overfitting the data.

Further, astronomers can greatly benefit from visualization tools that permit powerful exploration of complex multivariate datasets. **XGobi** provides a 2-dimensional *grand tour* of the database by displaying various projections of the data, with flexible interactive choice of variables, color brushing and projection pursuit options. **ExplorN**, ViSta, CViz, IVEE and other packages give d-dimensional grand tour, saturation brushing, parallel coordinate plots and other functionalities.

Finally, we note that this brief paper omits many topics in statistics with potential importance for astronomy. These include nonparametric methods, Bayesian approaches, wavelet analysis, bootstrap resampling, and many aspects of traditional multivariate analysis. The statistical methodology for understanding databases is vast and constantly growing.

5. Astrostatistics References and Codes

Many statistical methods are briefly reviewed in an astronomical context by Babu & Feigelson (1996), and multivariate methods are more thoroughly described (with Fortran codes) by Murtagh & Heck (1987). The conference proceedings by Feigelson & Babu (1992), Babu & Feigelson (1997) also have several articles containing use of multivariate analysis in astronomy. The monographs by Ripley (1996) and by Johnson & Wichern (1992) would be useful resources for astronomers.

While commercial statistical packages are the most powerful tools for implementing statistical procedures, a considerable amount of software is in the public domain on the World Wide Web. An informative essay on statistical software was written by Wegman (1997)[1]. Information on commercial statistical software packages such as SAS, SPSS and S-PLUS is available on-line[2]. Significant archives of on-line public domain statistical software reside at StatLib[3] and the *Guide to Available Mathematical Software*[4]. StatLib provides many state-of-the-art codes useful to astronomers such as XGobi, loess and MARS. Finally, we maintains a Web metasite called *StatCodes*[5] that gives links to on-line source codes and packages of multivariate, clustering, visualization and other statistical methods for astronomical research.

References

Babu, G. J., & Feigelson, E. D. 1996, Astrostatistics (London: Chapman & Hall)

Babu, G. J., & Feigelson, E. D. 1997, Statistical Challenges in Modern Astronomy II (New York: Springer-Verlag)

Bijaoui. A., Rué, F, & Savalle, R. 1997, in Statistical Challenges in Modern Astronomy II, ed. G. J. Babu & E. D. Feigelson (New York: Springer-Verlag), 173

Feigelson, E. D. 1992, in Statistical Challenges in Modern Astronomy, ed. E. D. Feigelson & G. J. Babu (New York: Springer-Verlag), 221.

Feigelson, E. D., & Babu, G. J. 1992, Statistical Challenges in Modern Astronomy (New York: Springer-Verlag)

Friedman, J. H. 1987, J. Amer. Stat. Assn., 82, 239

Friedman, J. H. 1991, Annals of Statistics, 19, 1

Huber, P. J. 1985, Annals of Statistics, 13, 435

Johnson, R. A., & Wichern, D. W. 1992, Applied Multivariate Statistical Analysis, 3rd edition (NJ: Prentice Hall)

Maddala, G. S. 1983, Limited-dependent and quantitative variables in econometrics (Cambridge: Cambridge University Press)

Mukherjee, S., Feigelson, E. D., Babu, G. J., Murtagh, F., Fraley, C., and Raftery, A. 1998, ApJ, 508, 314

Murtagh, F., & Heck, A. 1987, Multivariate Data Analysis (Dordrecht: Reidel)

Ripley, B. D. 1996, Pattern Recognition and Neural Networks (Cambridge: Cambridge University Press)

White, R. L. 1997, in Statistical Challenges in Modern Astronomy III, ed. G. J. Babu & E. D. Feigelson (New York: Springer-Verlag), 135

---

[1] http://www.galaxy.gmu.edu/papers/astr1.html

[2] http://www.stat.cornell.edu/compsites.html

[3] http://lib.stat.cmu.edu

[4] http://gams.nist.gov

[5] http://www.astro.psu.edu/statcodes

Wegman, E. J., Carr, D. B., King, R. D., Miller, J. J., Poston, W. L., Solka, J. L., & Wallin, J. 1997, in Statistical Challenges in Modern Astronomy II, ed. G. J. Babu & E. D. Feigelson (New York: Springer-Verlag), 185

Virtual Observatories of the Future
ASP Conference Series, Vol. 225, 2001
R.J. Brunner, S.G. Djorgovski, and A.S. Szalay, eds.

Nonparametric Density Estimation: A Brief and Selective Review

C.R. Genovese & L. Wasserman
Department of Statistics, Carnegie Mellon University, Pittsburgh, PA 15213

R.C. Nichol
Department of Physics, Carnegie Mellon University, Pittsburgh, PA 15213

A.J. Connolly
Department of Physics & Astronomy, University of Pittsburgh, 3941 O'Hara Street. Pittsburgh, PA 15260

J. Schneider & A.W. Moore
Robotics Institute & School of Computer Science, Carnegie Mellon University, Pittsburgh, PA 15213

Abstract. Nonparametric statistical methods have great potential for astrophysical data analysis. They provide a way to make inferences about complex structures from massive data sets without overly restrictive assumptions or intractable computations. Here, we review selected techniques for the problem of density estimation and draw some lessons for data analysts.

1. Introduction

In this age of high-precision cosmology, massive, complex data sets are the norm. With so much data, it becomes possible to make very accurate statistical inferences with the right techniques. At the same time, the complexity of the data and of the underlying structures being inferred make it hazardous to assume a simplistic functional form for the parameters being estimated. Fortunately, recent developments in statistical theory and practice offer powerful methods for inferring complex structure without overly restrictive assumptions. These methods are often called "nonparametric" methods; we describe them in more detail and contrast them with "parametric" methods below.

This paper reviews some of these developments for a specific sub-problem: density estimation. Density estimation is interesting to us both as a prototype for other such problems and as problem that we face frequently in our work. We consider two important methods for density estimation: kernel estimation (with a data-driven bandwidth selector) and wavelet estimation.

1.1. Take-Home Points

- The choice of kernel shape is relatively unimportant.
- The choice of bandwidth in kernel estimations is crucial.
- Nonlinearity in wavelet estimators is essential to performance.

1.2. A Handy Statistics-Physics Notation Translator

- Expectation: $\mathsf{E}Y = \langle Y \rangle$
- Variance: $\mathsf{Var} Y = \langle (Y - \langle Y \rangle)^2 \rangle$
- Estimators: $\hat{}$ denotes estimator, i.e., $\hat{f}$ estimates f.
- Distributions: $Y_1, \ldots, Y_n \sim f$ denotes that Y_i has distribution (density, cdf, etc) f; an "iid" over the $\sim$ means independent and identically distributed.

2. Parametric Versus Nonparametric Inference

Statistical inference is concerned with characterizing the structure of a random system from the observed outputs of that system. A statistical model is a list of assumptions that indicates how the structure of the system determines the probability distribution of the observed outputs.

The most basic statistical model (for independent data) is as follows:

$$Y_1, \ldots, Y_n \stackrel{\text{iid}}{\sim} f_{\theta_*}$$

where $\theta_* \in \Theta$ for some specified set Θ. Here, Θ is called the *parameter space* of the model, and points $\theta \in \Theta$ are possible value for the model parameter. Both can be multi-dimensional or even infinite-dimensional, as we will see. The family of distributions f_θ for $\theta \in \Theta$ determines the possible joint distributions of the data. The task of statistical inference is to use the data to get as much information as possible about θ_*. Since θ_* cannot usually be determined exactly, uncertainty will remain; the goal is to devise methods for reducing this uncertainty as much as possible.

Classical or *parametric* inference is the case when the parameter space Θ is a finite-dimensional set. For example, we might assume in the model above that f_θ is a Gaussian(μ, σ^2) distribution. That is, $f_\theta(y) = \frac{1}{\sigma\sqrt{2\pi}} \exp(-(y-\mu)^2/2\sigma^2)$ where $\theta = (\mu, \sigma)$ and $\Theta = (-\infty, \infty) \times (0, \infty)$.

Parametric inference has a well-developed statistical theory. Effective procedures for estimating θ_* are known and their performance properties are well-understood. For example, a common approach is the Maximum Likelihood Estimator: $\hat{\theta} = \arg\max_{\theta \in \Theta} \mathcal{L}(\theta)$. The function $\mathcal{L}(\theta)$,s called the likelihood function, gives (roughly speaking) the probability of observing the observed data for each possible value of the parameter. In the independent data model above, we have $\mathcal{L}(\theta) = \prod_{i=1}^{n} f_\theta(Y_i)$, and we know that $\hat{\theta}$ has approximately a Gaussian($\theta_*, \frac{1}{n} I^{-1}(\hat{\theta})$) where $I_{jk}(\theta) = -\mathsf{E}_\theta \left(\frac{\partial^2}{\partial \theta_j \partial \theta_k} \log f_\theta(Y) \right)$. This estimator is "statistically efficient" in its use of data: $\mathsf{E}(\hat{\theta} - \theta_*) = O(1/n)$ and gives an easy 95% confidence interval $\hat{\theta} \pm 1.96/\sqrt{nI(\hat{\theta})}$.

So, with powerful statistical methods in hand, all seems well. Unfortunately, for most interesting systems, the structures to be estimated are more complicated than the simple models above allow. Thus, if the Y_is are luminosities and θ is the luminosity distribution, we may not believe a priori any prespecified form that we can write down. (Our goal might be to estimate θ as a function of other variables such as redshift and spectral type; this easily fits within the above framework.) Similarly, if the Y_is are the positions of galaxies and θ is the galaxy density, from which we would like to identify clusters as a function of redshift, then the parameter we seek — θ — represents a fairly complictated object, namely the distribution of galaxies in the universe. Restricting ourselves to a finite-dimensional family of parametrics forms is likely to exclude interesting and important structure, potentially dooming the model to substantial bias.

Put another way, the family of joint distributions for the observed data are often too complex to specify with confidence in a simple mathematical form. This is particularly true with data of many dimensions. In recent years, this realization has prompted an explosion of techniques that allow for a more general structure for Θ. These techniques are usually called "nonparametric", which is an ill-chosen teerm because it offers the false hope of conclusions without assumptions. Nonetheless, ...

Nonparametric inference deals with the case in which the parameter space Θ is infinite dimensional. In other words, each $\theta \in \Theta$ is a function. For example, in the independence model above, we might take $\Theta = \{\text{probabilitydensities } g : \int (g'')^2 < C\}$ for some constant $C > 0$ and $f_\theta = \theta$. This parameter space includes a wide variety of functions, including those too complicated to write down in a neat expression. Herein lies both the power and the challenge of nonparametric methods. We are freed of making overly restrictive assumptions on the form of the unknown function, but this Θ also includes many functions that would likely be considered implausible as well.

3. Density Estimation

In this paper, we focus on the problem of nonparametric density estimation. This corresponds to the above independent data model with $f_\theta = \theta$ and with a parameter space Θ consisting of functions that are probability density functions — functions $g \geq 0$ with $\int g = 1$. (The density f determines the probability that the data Y takes values in any set: $P\{a < Y < b\} = \int_a^b f(y)\,dy$, for any a, b.) Our goal is to estimate the distribution of the data points θ.

One common approach to finding effective procedures for density estimation is to minimize a global measure of deviation between the estimate and the truth. At any given point y, define the pointwise Mean Squared Error $\text{MSE}_y = \mathsf{E}(\hat{f}(y) - f(y))^2$. Then, our global measure is the Mean Integrated Squared Error (MISE) given by $\text{MISE} = \int dy \text{MSE}_y$.

Below, we consider two classes of procedures along these lines: Kernel Estimators and Wavelet Estimators.

4. Kernel Estimators

Data $Y_1, \ldots, Y_n \overset{iid}{\sim} f \in \mathcal{F}$, where $\mathcal{F}$ is some a priori class of densities.

$$\hat{f}(y) = \frac{1}{n}\sum_{i=1}^{n} \frac{1}{h} K\left(\frac{y-Y_i}{h}\right) = \int \frac{1}{h} K\left(\frac{y-t}{h}\right) dF_n(t)$$

Key Points:

- Choice of kernel function K not important.
- Choice of bandwidth h crucial.

Choose h to minimize Mean Integrated Squared Error (MISE)

$$\text{MISE} = \text{EISE} = \mathsf{E}\int (\hat{f} - f)^2 = \text{Bias}^2 + \text{Variance} = \frac{c_1 h^4}{4} + \frac{c_2}{nh}.$$

Optimal $h = cn^{-1/5}$ balances bias and variance; MISE $= O(n^{-4/5})$. In high dimension d, get "curse of dimensionality" MISE $= O(n^{-4/(d+4)})$.

5. Cross Validation

Idea: Estimate MISE from data using a successive "leave-one-out" and minimize over h. Note ISE $= \int \hat{f}^2 - 2\int f\hat{f} +$ constant, and we can approximate the middle term. Empirical rule:

$$\hat{h} = \arg\min_h \int \hat{f}^2 - \frac{2}{n}\sum_{i=1}^{n} \hat{f}_{-i}(Y_i)$$

where $\hat{f}_{-i}$ is the kernel estimate without the ith data point. Stone's theorem: ISE$(\hat{h}) \approx$ ISE(h_{optimal}) for large sample sizes. See books by Silverman and Scott for more information.

6. Truncation Bias

Note that kernel is $\hat{f}(y) = \int h^{-1} K((y-t)/h) dF_n(t)$ where F_n puts mass $1/n$ at every data point. With no truncation, F_n is the nonparametric maximum likelihood estimate. With truncation, use Efron-Petrosian to find nonparametric maximum likelihood estimate $\hat{F}_n$ then use $\hat{f}(y) = \int h^{-1} K((y-t)/h) d\hat{F}_n(t)$. Work in progress: optimal smoothing.

7. Wavelets

A wavelet basis is a collection of localized wiggles related to each other by dilation and translation. Wavelet bases give sparse representations (most coefficients nearly 0) for a wide range of function shapes.

Take $f(y) \approx \sum_k \alpha_{j_0 k}\phi_{j_0,k}(y) + \sum_k \sum_j \beta_{jk}\psi_{j,k}(y)$ where ϕ is father wavelet, ψ is mother wavelet, $\phi_{jk}(y) = 2^{j/2}\phi(2^j y - k)$, and $\psi_{jk}(y) = 2^{j/2}\psi(2^j y - k)$. To estimate coefficients: apply wavelet transform to data and then apply nonlinear thresholding.

- Good for large data sets, inhomogeneous functions.

- The benefits follow from the nonlinear estimation (thresholding). Don't use wavelets as basis in a linear procedure.

- Sparse representations are the key to performance. Other bases may be better for estimating certain types of functions.

See the work of Donoho and Johnstone for more information.

References

Donoho, D., et al. 1995, Wavelet shrinkage: Asymptopia? *Journal of the Royal Statistical Society, B*, 301–369

Efron, B., & Petrosian, V. 1999, Nonparametric methods for doubly truncated data, *Journal of the American Statistical Association*, 94, 824–834

Ogden, T. 1997, *Essential Wavelets*

Scott, D. 1992, *Multivariate Density Estimation*

Silverman, B. 1986, *Density Estimation*

Visualization of Large Multi-Dimensional Datasets

Joel Welling

Department of Statistics, Carnegie Mellon University, & Pittsburgh Supercomputing Center, Pittsburgh, PA 15213

Mark Derthick

Human Computer Interaction Institute, Carnegie Mellon University, Pittsburgh, PA 15213

Abstract. Visualization techniques are well developed for many problem domains, but these systems break down for datasets which are very large or multidimensional. Techniques for data which is discrete rather than continuous are also less well studied. Astronomy datasets like the Sloan Digital Sky Survey are very much in this category. We propose the extension of information visualization techniques to these very large record-oriented datasets. Specifically, we describe the possible adaptation of the *Visage* information visualization tool to terabyte astronomy datasets.

1. Introduction

How can huge data sources (gigabytes up to terabytes) be quickly and easily analyzed? There is no *off-the-shelf* technology for this. There are devastating computational and statistical difficulties; manual analysis of such data sources is now passing from being simply tedious into a new, fundamentally impossible realm where the data sources are just too large to be assimilated by humans. The only alternative is to provide extensive computer support for the process of discovery.

The focus of this workshop is the challenge posed by the next generation of large astronomical sky surveys. Specifically, we concentrate on the Sloan Digital Sky Survey (SDSS), which will create over one terabyte of reduced data over the next 5 years: How does one navigate such a huge, multi-dimensional, dataset? The techniques of information visualization and visual discovery may be extended to such datasets, allowing the scientist to interactively explore and understand her results in real time.

This work is the product of a collaboration at Carnegie Mellon University and the University of Pittsburgh, involving astronomers, experts in traditional scientific visualization and interactive information visualization, and computer and computational scientists. Condensed data representations from the data mining and machine learning community (Nichol *et al.* 2001, these proceedings) make it possible to explore these huge datasets interactively.

Visualization of Large Multi-Dimensional Datasets   285

2. Visualization Challenges of the Virtual Observatory

Over the next ten years, we will witness a revolution in how astrophysical research is performed. This is primarily due to the large number of new sky surveys presently underway (or completed) that are designed to map the Universe to higher sensitivity and resolution than ever previously envisaged. We are quickly approaching the prospect of a Virtual Observatory, where one can digitally reconstruct the whole sky. These surveys, and the virtual observatory, present scientists with a "gold mine" that the next generation of astrophysicists will spend their whole careers exploring.

Two cornerstones of the Virtual Observatory are the 2 Micron All Sky Survey (2MASS) and the optical Sloan Digital Sky Survey (SDSS). The 2MASS survey (which is 91% complete of which 50% has been publicly released) is a near infrared imaging survey covering the full sky in three passbands (from 1–2.2 microns). The SDSS is an imaging and spectroscopic survey that will cover one quarter of the sky at five different wavelengths. Together these two surveys will detect over 200 million objects (galaxies and stars) and from these detections positions, fluxes, shapes, textures and bitmaps will be extracted. In addition to the scientific information will be bookkeeping information that describes the observations themselves, *e.g.*, whether the sky was cloudy or there were problems with the instrumentation. We must also understand these possible systematic uncertainties present within the data. The total 2MASS and SDSS surveys are expected to acquire 500 GB of cataloged attributes and 1 TB of postage stamp images (cutout images around each detected object) over the next 5 years of operation.

For each object detected hundreds of attributes will be recorded. The size and large dimensionality of these new data sets means that simple visualization and analysis techniques cannot be applied directly, because they do not scale effectively. The questions then become: How do we quickly determine the important dimensions within such a data set? Which dimensions tell us about how galaxies or stars form and how matter is distributed throughout the individual galaxies or, the Universe as a whole? New techniques developed for 2MASS and SDSS will be applicable to the analyses of all large observational data sets (such as the all-sky surveys of GALEX, ROSAT, MAP and PLANCK) and for the visualization of other large physical and biological experiments.

3. The Breakdown of Visual Discovery

Traditional astronomy has relied on the study of small numbers of rare objects. With hundreds of millions of objects in the Virtual Observatory, "rare" objects will themselves number in the millions. Statistical methods become necessary to group objects into classes for study.

This situation is opposite to that for which traditional visualization methods are best suited. Large dataset size alone kills interactivity. The record-oriented nature of the dataset makes it difficult to assimilate, because the brain is evolved to understand continuous systems that exist in three dimensions plus time. If too broad a view of the data is taken, important details can be literally too small to see. Too narrow a view is also disastrous. It becomes very easy to be distracted by structures that *look* important.

Supercomputer simulations routinely produce very large datasets. Standard methods of visualizing such data typically include the generation of one or more animations (assuming the system evolves in time), or interactive visualization of tiny subsets of the data. In most problem domains the researcher's physical intuition applies, and can help her to assimilate the evolution of the system.

Large computing facilities are developing interactive visualization tools for these terabyte-sized datasets. Direct application of traditional visualization typically requires a supercomputer, and even with one available interactivity is very difficult to attain. Current rendering hardware can draw at most about 10 million polygons per second (though this number is rising), so interactive drawing of hundreds of millions of objects is simply impossible.

4. Desirable Capabilities

Here are illustrative questions that astrophysicists will want to ask of the data:

- A range of counting queries such as:
  - How many elliptical galaxies, with a redshift above 0.3, are there?
  - What is the mean and variance of ellipticity among radio galaxies within clusters of galaxies compared to outside clusters of galaxies?
  - How does the distribution of galaxy colors observed on 1 May 2001 compare to those seen on 1 June 2002?

- Sophisticated statistical queries that require the clustering, classification, regression, filtering and newer probabilistic inference techniques.

- Visualization requests to answer questions such as:
  - Give me a smoothed map of all X-ray detected galaxies.
  - Display all emission-line galaxies in detected clusters of galaxies; are they in the center or on the outskirts?

5. XGobi, a Simple Tool for Visualizing Multidimensional Data

XGobi, one of the most widely used information visualization tools, was written at Lucent Technologies[1] (Swayne, Cook, & Buja 1998). This tool interactively produces arbitrary 3D projections of n-dimensional data. For example, Figure 1 shows the correspondence between data groups which differ in one projection but are cluster together in another. XGobi deals only with scalar data, and is designed for relatively small datasets.

XGobi supports a method called brushing, whereby the user selects data items in one window and the selection is propagated to the corresponding items

---

[1]XGobi is freely available from http://www.research.att.com/areas/stat/xgobi/.

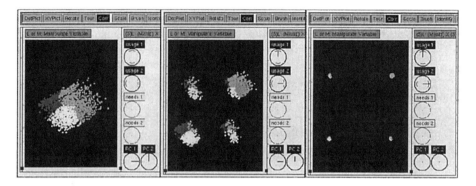

Figure 1. The information visualization tool XGobi

in all other windows. This has proven to be a useful technique for mentally integrating multiple views.

6. Visage, an Information Exploration Tool

At Carnegie Mellon University, in collaboration with Maya Design Group, we have already developed an interactive data exploration system called Visage (Kolojejchick, Roth, & Lucas 1997). It is effective for analyzing high dimensional data, but it uses only 2D visualizations and is limited to small discrete datasets. Below, we describe Visage's features, while in the next subsection, we discuss our plans to generalize these features and expand the capacity of Visage to handle massive datasets. We call this new system TeraVisage.

Visage presents data as graphemes (visual elements) organized by frames (lightweight nestable windows that impose some visualization discipline on the graphemes presented within them). Within a frame, the data are presented by graphemes such as bars, text labels, marks and gauges. The visual properties of the graphemes (*e.g.*, the color or size of a plot point) encode attributes of the objects they represent (*e.g.*, a galaxy). While each grapheme stands for one object, an object may be represented by many graphemes of varying appearance in different frames.

Some of the basic operations provided by Visage are:

- Drag-and-drop objects from one frame to another,

- Navigate (drill-down) from an object along a relation creates graphemes for the related objects,

- Aggregate (roll-up) a set of selected objects to create a new object with properties computed from its members,

- Brush an object or a set of objects in a choice of colors, as described in Section 5,

- Dynamically query by an attribute to render invisible all objects within a frame whose values for that attribute fall outside a range selected by a slider widget.

The results of these operations depend only on the underlying data object, and are uniformly applicable in any type of Visage frame.

6.1. Going to Large Datasets

In this subsection, we discuss the methods by which we propose to adapt Visage to very large datasets. Interactive visualization requires interactive speeds; the challenge is to maintain these speeds during interrogation of a large (500GB) database. Our plan is to maintain a hierarchy of representations and subsets, with a hierarchy of access speeds.

Figure 2 presents our vision of how this would play out for exploring an astronomical dataset. The data comes from a simulation of the coming merger of our Milky Way galaxy with the Andromeda galaxy[2].

The dataset we examine here corresponds to the last frame of the simulation. In the course of the collision large "tidal tails" of stars, dust, and gas have been drawn from both galaxies. At the moment in question the bulk of the galactic matter has formed a roughly elliptical collision remnant, but two tidal tails remain. The matter in one tail is to be selected and examined.

In the left frame labeled "All Star Groups", the astronomer has displayed 7 attributes of all the star groups in the two galaxies. The 3D visualization shows spatial position (x, y, and z). The histograms additionally show the distribution of distance from the center of the galaxy, magnitude of velocity, x component of velocity, and original galaxy each star group belonged to. On the galaxy histogram the astronomer has brushed the Milky Way star groups purple and the Andromeda ones yellow. The relative distributions can be seen in the other histograms and the density plot. It is apparent that star groups that originated in the Milky Way dominate the streamers. The astronomer selects a group of stars from one of the streamers using a bounding box. These stars are brightened in the 3D density plot. The brushing and selection operations should take about one second.

The astronomer then creates a new frame with a density plot showing the x, y, and z components of the velocities and drags (copies) the selected star groups there. She labels the new frame "+Z Tail" (right), and copies the histograms as well. Appropriate condensed representations should allow this operation to be carried out in about 10 seconds. Being interested in the relationship between the distance from the center and the magnitude of the velocity, she creates a 2D scatter plot of those variables. By moving the distance slider, she controls which stars remain fully visible and which are grayed out. Feedback from slider changes should occur in roughly 100ms.

---

[2]Welling carried out the simulation using conditions specified by John Dubinski and code by Lars Hernquist and others (Mihos,C. 1998, A.A.S, #176); further information is available from http://www.cita.utoronto.ca/~dubinski/index.html. An animation of the merger can be found at http://www.psc.edu/~welling/big-merger.mpg.

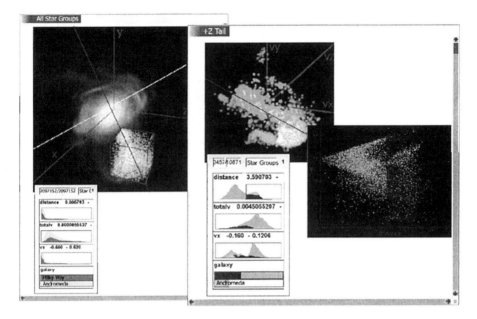

Figure 2. Hypothetical TeraVisage operation

It is easy to see how to extend most of the graphemes of Visage to large datasets. Sets of individual marks become density distributions in some space, and are selected by selecting regions in the space and/or using DQ filters. Gauges remain gauges, but now represent statistical summaries of subsets of the data. Text labels remain, but become labels for sub-aggregates found in traversing kd-trees or in more sophisticated models. Many confusing issues remain, however.

7. Interactive Visualization Through Condensed Representations

In addition to facilitating machine learning and database operations, appropriate condensed representations can accelerate such tasks as volume rendering by allowing data items to be grouped and drawn together if they are similar to within specified error bounds. A large dataset can easily have more data elements than there are pixels on the display, and there is no sense rendering at a level of detail which will be invisible to the user. These hierarchical rendering methods are well studied in computer graphics and map well to condensed representations for knowledge discovery, for example, see Laur & Hanrahan (1991).

References

Kolojejchick, J. A., Roth, S. F., & Lucas, P. 1997, in Computer Graphics and Applications, Vol. 17 No 4, *Information Appliances and Tools in Visage*, 32[3]

Laur & Hanrahan 1991 in Computer Graphics, Vol. 25 No. 4, *Hierarchical Splatting: A Progressive Refinement Algorithm for Volume Rendering*, 285

Swayne, D. F., Cook, D., & Buja, A. 1998, in Journal of Computational and Graphical Statistics 7 (1), *XGobi: Interactive Dynamic Data Visualization in the X Window System*

---

[3]http://www.cs.cmu.edu/ sage/PDF/Appliances.pdf

Large-Scale Visualization of Digital Sky Surveys

Joseph C. Jacob

Jet Propulsion Laboratory, California Institute of Technology, M/S 126-234, 4800 Oak Grove Drive, Pasadena, CA 91109-8099

Laura E. Husman

Jet Propulsion Laboratory, California Institute of Technology, M/S 126-234, 4800 Oak Grove Drive, Pasadena, CA 91109-8099

Abstract. Designed and developed within JPL/Caltech's supercomputing and visualization environment, the virtual observatory software described in this paper permits high performance visualization and analysis of large sky images and catalogs. Functionality includes parallel generation of large sky image mosaics and synthetic maps from catalogs, techniques for interactive exploration of large images on workstation or special high resolution displays, image enhancement features, catalog viewing, catalog to image relation, and run-time compositing of multi-resolution datasets. The tools and technologies being developed are intended to be components of the National (and eventually Global) Virtual Observatory effort, which will federate and provide seamless access to a wide variety of geographically distributed, multi-terabyte datasets.

1. Introduction

Modern sky surveys have archived many terabytes of catalog and image data at various resolutions and wavelengths. The great advances that have been made in remote sensing pose a tremendous challenge to the state of the art of information technology. Sheer volumes of scientific data are of little value without an effective ability to analyze them quickly and thoroughly, and to distill the essence of scientific knowledge from them.

The *Digital Sky* project[1] is an effort to design and prototype an interactive database search, analysis and visualization system spanning the catalog and image archives of multiple sky surveys. The work reported in this paper is the product of the *Digital Sky Virtual Observatory* task, which provides high-performance visualization, supercomputer level analysis (where appropriate), and large-scale data management for the *Digital Sky* project. The objective of this work is to prototype a "virtual observatory" that allows multi-spectral, multi-survey viewing and analysis of sky data, both images and catalogs, at the terabyte level.

---

[1] http://www.digital-sky.org

The datasets that we are currently using to develop and test this new technology are at one arc minute resolution, Infrared Astronomical Satellite (IRAS)[2] in the infrared, and at one arc second, Digital Palomar Observatory Sky Survey (DPOSS)[3] in the visible and Two Micron All Sky Survey (2MASS)[4] in the near infrared. DPOSS and 2MASS each have multiple terabyte archives, making them ideal test surveys for development of technologies essential for a National Virtual Observatory[5]

2. High Performance Information Technology Environment

The supercomputing and visualization environment at JPL and Caltech is well-suited for the large scale visualization and interactive analysis tool described in this paper. The hardware includes the following: SGI Onyx2 with 128 processors (peak speed of 76 Gflops), 32 GB of memory, and 6 InfiniteReality2 graphics pipes; HP V2500 with 128 processors, 128 GB of memory, and 1.8 TB of disk space; two 300 TB mass storage systems; 3×2 "PowerWall" display; and $4\times$ Gigabit Ethernet and OC-12 communications interconnecting all major assets. The PowerWall display is a 3×2 matrix of 1280×1024 displays that are synchronized by the software to provide an effective display resolution of 3840×2048.

3. Large Image Mosaicing

The virtual observatory software that we are developing can be used both to generate large image mosaics from sky image patches and to view these large images. The mosaicing software is fully automated and can be run in parallel across 1,2,4, ..., 128 processors on the Onyx2. The input image patches may be in the common FITS format at any resolution, in any coordinate system and projection, and having any data type supported by FITS. The software translates these into the user-selected resolution and coordinate system of the output mosaic. The World Coordinate System library (Greisen & Calabretta 1999) is used to convert the input FITS images into a common output coordinate system and projection. The galactic, ecliptic, J2000, and B1950 coordinate systems are supported.

To date, the following sky mosaics have been constructed using this software: all-sky IRAS mosaic at 1 arc minute resolution, constructed from 430 individual image patches with 4 bands each; $10° \times 10°$ 2MASS mosaic at 1 arc second resolution constructed from over 1000 individual images with 3 bands each; $2° \times 2°$ DPOSS mosaic at 1 arc second resolution, constructed from 100 individual images with 3 bands each; $14° \times 13°$ DPOSS mosaic constructed from two full DPOSS plates. In addition, the 2MASS and smaller DPOSS mosaic

---

[2] http://www.ipac.caltech.edu/ipac/iras/iras.html

[3] http://astro.caltech.edu/~rrg/science/dposs_public.html

[4] http://www.ipac.caltech.edu/2mass

[5] http://astro.caltech.edu/nvoconf/white_paper.ps

listed above were registered to each other and combined to produce a 6 band multi-spectral mosaic that includes both the visible and near infrared.

4. Large Image Exploration

4.1. Large-Scale Image Input

The size of a mosaic that can be constructed and viewed is limited only by the available disk space. A full sky mosaic at 1 arc second resolution in one spectral band approaches one terabyte in size. For high performance viewing of these large mosaics, the data is stored in a custom hierarchical, tiled format with two defining characteristics. The data is stored as a series of 512×512 pixel "tiles," and at multiple resolutions, including full resolution, half resolution, quarter resolution, all the way down to the resolution where the entire image fits in a single tile. The tiled nature of the data storage format permits any subset of the data to be quickly referenced and extracted for viewing without requiring that all of the data be read into memory. It also permits the image to be quickly extracted and served for viewing at any arbitrary resolution by resampling with a kernel no larger than 2×2 pixels.

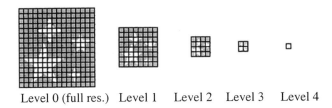

Level 0 (full res.)　Level 1　Level 2　Level 3　Level 4

Figure 1. Hierarchical, tiled data format. Each level represents as 512×512 pixel tiles the entire image at a resolution which is half the resolution of the previous level.

In summary, the tiled nature of the data permits rapid panning to any arbitrary location and the hierarchical resolution storage permits rapid zooming to any arbitrary zoom level. Also, intelligent data caching keeps recently visited tiles in memory for rapid retrieval. In addition, the software includes a lookahead cache that loads additional tiles just outside the bounds of what is visible on the display and at neighboring resolution layers for improved performance. The disk storage penalty paid for this rapid panning and zooming capability is about one-third the size of the full resolution image.

4.2. Large-Scale Image Output

Just as the image viewing software scales on the input side, the software also scales on the output side, allowing displays ranging from single screen workstation displays up to large PowerWall displays (described in Section 2). Single pipe and multi-pipe support is provided. The software can be configured to use any rectangular subset of the available display screens. The supported display configurations are specified in a configuration file and are easily selectable at run-time with a command line option or by setting an environment variable.

High performance on a 3×2 PowerWall display (3840×2048 effective resolution) has been demonstrated.

4.3. Image and Catalog Navigation and Viewing

The software that has been developed permits high performance visualization of both sky image and catalog data. A number of features are provided to enable efficient navigation of the potentially huge images. Mouse and keyboard interfaces for smooth, variable speed panning and zooming are provided. A *Global Map View* of the dataset is a popup window that shows the entire dataset with a box highlighting the region that is currently visible on the display. Users may jump to any location in the image by clicking at the location in the Global Map View. Another way to quickly jump to any location in the image is to specify that location in either pixel or sky coordinates in the *Coordinate Selection* window. Galactic, Ecliptic, and Celestial coordinate systems are supported. The Coordinate Selection window may also be used to retrieve the pixel or sky coordinate of any pixel that is selected by clicking with the mouse on the image.

The image mosaicing software described in Section 3 is capable of constructing mosaics having many spectral bands. The image viewing software permits any subset of three bands to be mapped to red, green, and blue at run-time. This allows the user to do such things as view both visible and infrared bands from different sky surveys simultaneously and then switch back to all visible or all infrared.

Keyboard and graphical user interfaces for brightness and contrast adjustments are provided, and may be applied to all three video channels (red, green, and blue), or to any channel individually. For example, this feature allows viewing of both Andromeda and the center of the Milky Way in our IRAS mosaic, as illustrated in Figure 2

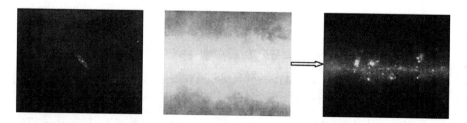

Figure 2. In the full sky IRAS image, the brightness and contrast settings for optimal viewing of the Andromeda galaxy (left) causes the center of the Milky Way to saturate (center). The brightness and contrast can then be adjusted to enhance the structure at the center of the Milky Way (right).

The catalogs may be viewed in ASCII text in a scrollable window and as image overlays, as illustrated in Figure 3, or as synthetic maps. With overlays which are vectors drawn over the image, the shape, size, and color of the overlay objects may be set according to the values in one or more columns of the catalog. With synthetic maps, the catalog positions and values are used to generated a pixelated image. For instance, the Hipparcos catalog was used to generate a synthetic map of the sky with 118,218 stars down to 14th magnitude.

The image and catalog viewing capabilities are tightly coupled, allowing easy relation of a location in the images to catalog entries for celestial objects in the proximity and vice versa. For instance, the user may select a region of the sky in an image and see the catalog entries for those objects in that region highlighted in both the image and in the catalog window, as illustrated in Figure 3. Alternatively, the user may select a catalog entry in the scrollable list and see that object highlighted in the image or jump to the position of that object in the image. This demonstrates both image to catalog and catalog to image relations.

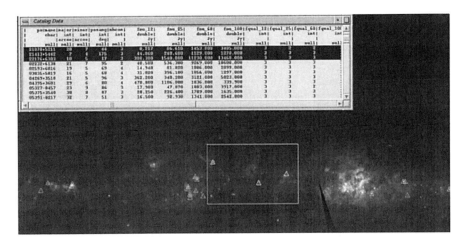

Figure 3. Image to catalog relation. The catalog shown here was obtained from a search of the IRAS catalog via the IRSA CatScan website: http://irsa.ipac.caltech.edu/applications/CatScan/.

5. Automatic Dataset Compositing

The image viewing software described in this paper has support for fully automated **run-time** compositing of multiple datasets, correctly positioned based on pixel resolution and latitude and longitude at a corner. Any number of datasets may be composited, although panning and zooming performance is degraded slightly with each one added. This capability allows the user to do such things as view high resolution insets of particular celestial objects or regions of the sky overlaid on top of lower resolution imagery of the whole sky. As an example, refer to Figure 4 which shows four composited datasets, all-sky IRAS and Hipparcos synthetic map at 1 arcminute resolution, and high resolution insets of 2MASS and DPOSS at 1 arcsecond. The user can start at a zoom setting that permits synoptical viewing of the whole sky, smoothly zoom in to full resolution of IRAS, and seamlessly continue zooming in to the full resolution of the high resolution datasets.

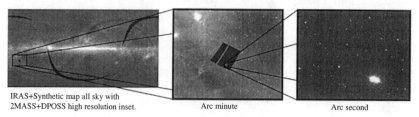

Figure 4. Automatic dataset compositing.

6. Conclusion

In this paper, the new technology we are developing for generation of large sky image mosaics and viewing of the large images and sky catalogs was described. Many capabilities are included for large image and catalog navigation and exploration, including both smooth and discontinuous pan and zoom over the whole sky and at any resolution. The catalogs may be viewed as ASCII tables, image overlays, or synthetic sky maps. Multiple image datasets of differing resolutions may be composited at run-time and viewed simultaneously, enabling novel multi-spectral views of the sky. The image and catalog viewing capabilities are tightly coupled, allowing easy relation from image to catalog and vice versa. The software is scalable both on the input side, permitting arbitrary sized images, and on the output side, permitting use of high resolution PowerWall displays.

Acknowledgments. We are grateful to our Digital Sky colleagues Robert Brunner, George Djorgovski, John Good, Ashish Mahabal, and Steve Odewahn. This work is sponsored by NASA Code SR, JPL Space and Earth Science Programs Directorate, Space Science Applications of Information Technology Program (formerly Interactive Analysis Environments Program).

The research described in this paper was carried out at the Jet Propulsion Laboratory, California Institute of Technology, under a contract with the National Aeronautics and Space Administration.

References

Greisen, E. W. & Calabretta, M., 1999, A&A, submitted[6]

---

[6]http://www.atnf.csiro.au/computing/software/wcslib.html

High Speed Interconnects and Parallel Software Libraries: Enabling Technologies for the NVO

Jeremy Kepner, & Janice McMahon

MIT Lincoln Laboratory, Lexington, MA 02420

1. Introduction

The National Virtual Observatory (NVO) will directly or indirectly touch upon all steps in the process of transforming raw observational data into "meaningful" results. These steps include:

(1) Acquisition and storage of raw data.

(2) Data reduction (*i.e.*, translating raw data into source detections).

(3) Acquisition and storage of detected sources.

(4) Multi-sensor/multi-temporal data mining of the products of steps (1), (2) and (3).

The highly distributed nature of the NVO places new twists on all of these steps. Future NVO research is likely to focus on developing the software tools necessary for Step (4) as well as the methods for "federating" data from Steps (1) and (3). However, past experience with individual surveys indicates that Step (2) has dominated computer software and hardware costs and may have a large impact on the NVO. [NOTE: It is quite possible that NASA will spend $0.5B on data reduction software across a dozen missions in the coming decade.]

Federation of data sets from multiple institutions, which is a primary NVO goal, will be made significantly easier if improvement of the data reduction pipeline software is also undertaken. Addressing the challenges of Step (2) for the NVO can be accomplished significantly improving the software environment for data reduction pipelines. Although the NVO can and should influence this effort it may be outside of the NVOs core activities.

The rest of this paper presents a further analysis of the computing and networking requirements of the NVO and provides a discussion of some of the challenges and solutions for addressing data reduction for these massive NVO data sets.

2. Large Survey Requirements

Large area imaging surveys are generating data at an exponentially increasing rate. Reducing this data is a significant hardware and software challenge that is likely to push the limits of computing for some time to come. The computing requirements of these surveys are best understood in terms of the number of operations required to process a single pixel or a single detection. By looking

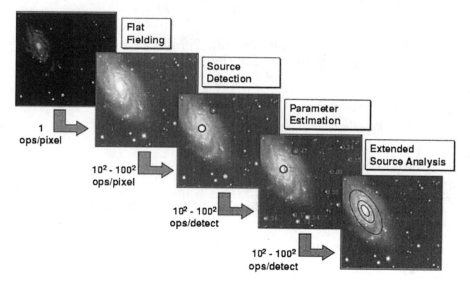

Figure 1. Standard steps in a data reduction pipeline. The computing requirements to fulfill the operation are shown below each step.

at the data reduction pipeline in these terms it is easy to determine the overall computing needs of a given data stream by multiplying the data rate by the per pixel processing requirement.

2.1. Computing Requirements

Figure 1 shows the steps that are common to most data reduction pipelines. The primary driver is the matched filtering step which involves 2D convolutions of the entire image. Such convolutions can easily result in as many as 10^4 operations per pixel and requires a sizable amount of computing to keep up with the data rate. Detection processing also involves 2D convolutions but of a sparser nature. Although the real-time processing requirements of a survey may be readily satisfied by a few tens of computers, it is quite common to re-run the data processing pipeline on several years worth of data after the pipeline software has been upgraded. In such cases, hundreds or even thousands of computers will be needed to re-process the data in a reasonable time.

For example, to support a camera with a real time data rate of 100 million pixels per second (*e.g.*, the LSST) requires a computer system that can *deliver* 100 Gigaflops, which is approximately 1000 of todays state of the art workstations. To re-process such a data stream could require millions of such computers.

2.2. Software Requirements

The primary difficulty in developing high performance pipeline software is the large number of systems architecture issues (number of processors, total memory, network bandwidth, disk bandwidth, ...) that need to be considered in order to keep up with the data rate (see Figure 2). In other words, the software pipeline

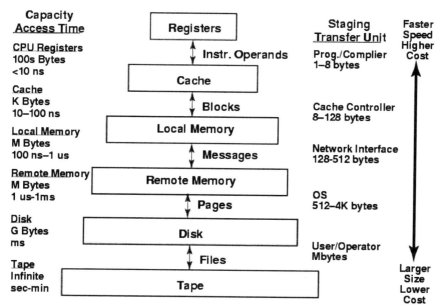

Figure 2. The complex memory hierarchy of a modern computer system. Typically, a data reduction pipeline incorporates the details of this memory hierarchy into the program, which limits portability and re-use of pipeline software.

becomes highly tuned to the system, which in turn increases the size of the code and the expertise necessary to maintain the code. In addition, upgrading to a new system can require a significant re-write. All of these issues mean that there is very little code re-use across different data pipelines.

Unfortunately, the need for highly tuned software will only grow in the future as next generation computers incorporate more complex features (e.g., multi-processor chips, on chip vector units, multi-threading, ...). While it is now possible to get 1% or 10% of a computers peak performance with little or no optimization, in the future an un-optimized program can expected to give less than 0.1% of peak performance. In other words, the price of not being optimized to the hardware will increase.

3. Enabling Technologies

The primary goals of a computer system architect are to manage complexity and to create systems that avoid the need for heroic programming efforts to meet the program milestones. In the context of a real-time data reduction pipeline this means developing simplified abstractions of the hardware and software so as to isolate the specifics of the application from the specifics of the hardware.

In the previous section, the need for complex, highly parallel computing systems was made apparent. Without such systems it will require months to

	One-way link peak bandwidth (MB/s)	Bisection width (links) vs processors (P)
Cray T3E-900	480	$\sim 4P^{2/3}$, 256 max
CSPI	160	$\sim P/2$, 64 max
Mercury Race++	267	8 max (Note 1)
SKY	320	$\sim P/32$, 20 max (Note 2)
VLSI Photonics	1000	P (Note 3)

(1) Race++ ILK16P Interconnect. Multiport configurations may have more links.
(2) Assumes 6U boards, 3-level crossbar tree. 9U would give fewer links per processor, same max.
(3) Projected performance; assumes fully connected, single-level crossbar switch

Figure 3. Current and future capabilities of various interconnect technologies.

analyze a single nights worth of observations, and decades to re-process a significant portion of a survey.

3.1. High Speed Interconnects

One way to create a less complex parallel computing systems is to invest in extremely capable networks. Minimizing network usage is the single most common optimization step performed in writing parallel software. High performance networks greatly alleviate the pressure on the programmer to implement to the specific hardware. In other words, it is incumbent the system architect to assemble a "balanced" system (*i.e.*, computation *vs.* communication). Fortunately, the rapid rise of the Internet and cluster computing has driven the need for ever more capable interconnects. As a result a variety of technologies will be available for producing systems that are well balanced for data processing pipelines (see Figure 3).

3.2. High Performance Parallel Software Libraries

A fast method for implementing high performance data pipelines is to re-use already optimized code. The best way is to leverage existing libraries (*e.g.*, Lapack, ScaLapack, FFTW, VSIPL, ...) developed by other communities (see Figure 4). These software packages remove the majority of the effort required to achieve optimal performance on a given computer. In addition, it is important for the community to increase the capability of the data pipeline applications it has developed (*e.g.*, IRAF, IDL, ...). Currently, these tools provide a variety of application specific functions. Unfortunately, they are not designed for real-time parallel data pipelines. It would be highly beneficial to upgrade (and add to) these tools so that they can exploit the hardware technology that will be required to effectively process large surveys.

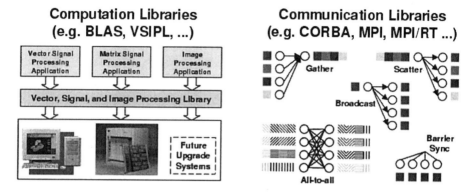

Figure 4. Summary of various existing software libraries and their capabilities.

4. Summary

The NVOs core data mining and archive federation activities are heavily dependent on the underlying data pipeline software necessary to translate the raw data into scientifically relevant source detections. The data pipeline software dictates: the raw data storage and retrieval mechanisms, the meaning and format of the fields in the source catalogs, and the ability of the NVO users to re-analyze raw data for their own purposes. Increasing the performance of the core data pipeline software so that it can address the needs of current and future high data rate surveys is an important activity that should be addressed in concert with the development of the NVO.

Acknowledgments. This work is sponsored by DARPA, under Air Force Contract F19628-95-C-0002. Opinions, interpretations, conclusions and recommendations are those of the author and are not necessarily endorsed by the United States Air Force.

Approaches to Federation of Astronomical Data

Roy Williams

Center for Advanced Computing Research, California Institute of Technology, Pasadena, CA, 91125

Abstract. We discuss some approaches to rich semantic interoperation of web-based services, so that a user of the Virtual Observatory can create a distributed network of services to read data from archive, calibrate it, compute with it, fuse it to their data, estimate query execution cost, and many others. Each of these services can be maintained by different people and connected by standard protocols. We point out the need for interoperating the FITS and XML ways of representing structured information, and the need for standard semantics and representation of generic data objects. We consider Capability documents as a way to interoperate remote services, as well as concrete implementations of these from the the geography and business communities.

1. Remote Data

This paper is about working with remote data. First let us define the idea of a *service*, by which we mean a program (perhaps a web server), that is listening on a socket for requests, then responding. The Yahoo! search engine or an FTP server are both services. We will define a *remote service* to be one where the (human) data client has never met the curator of the archive, and they may speak no language in common. Remoteness is not about geographical distance, but about relying on standards: both client and server must adhere to standard protocols for request and response before the remote data service can be used.

Suppose an astronomer has been told about a catalog of astronomical objects that she finds interesting, and wishes to compare it with a catalog she already has. Let us consider three ways in which this can happen: the past, the present, and the (Virtual Observatory enabled) future.

Past: In the old days, this would mean a trip to the library to find out who has the data, writing a letter to him requesting a copy of a tape, followed by a wait of weeks, then hours of software installation, extraction of the required data, custom code for changing the coordinate system, and more custom code to convert to the required file format, then creating the "payload" code that does the client's actual scientific analysis.

Present: These days we imagine these difficulties to be almost gone. Those who own data are embarrassed not to have put it on the web; also we have email and 100 Megabit/second connections, so there is no tedious waiting for the mailman to bring the tape. But in fact, most of the stumbling blocks to data federation are still present. The data owner has put a big text file at an FTP site and told his friends where it is. The client uses a web search engine or email to find the FTP site, then downloads the text file. She looks at the top of the big table, and sees columns called "RA1" and "RA2". An hour later, she realizes that the 1 and 2 are footnotes, and at the bottom of the table is an explanation that they are different Equinoxes. Still there is lots of custom code, or the use of tools like Matlab or Excel or IDL to bring about the cross-comparison.

Virtual Observatory Future: Let us suppose that the data owner has put the data not just "on the web", but done so with *Virtual Observatory Compliance*. This means that the semantic meaning of the data is exposed, and not only to the sharp intelligence of a human, but also to the dim wit of a computer. If a pair of numbers represents a position in the sky, the computer knows that they can be converted to other coordinate systems, and can do this silently. VO Compliant data services will be registered with an information service — like Napster does for music files — although we would expect it to be distributed, more like the Gnutella music service. Thus it will be much easier to find relevant data. When a VO-compliant client connects to a VO-compliant server, there is a conversation about their capabilities that the humans need not worry about. Our astronomer client will use a shrink-wrapped catalog-comparison service, connecting the remote catalog and her own catalog, receiving a data object which is the result, complete with provenance data and a way to cite the result in publications.

Thus the VO will be an example of a "semantic web" (Berners-Lee, 1998), a web of not just data, but semantically meaningful content. Without being able to do this, much astronomy data will remain effectively inaccessible for meaningful research just because the science community will not be able to access, manage and manipulate all the available data.

Services could be connected together by a user with a "modules and pipes" model of component computing, each module representing a service that is remote, where the user has not met the curator. As soon as the module is brought on to the desktop, its capability document would be fetched, so that decisions can be made about how to connect it to other modules/services. The user could decide that an output from one service (for example "*g_magnitude*") should be used as the input to another service (perhaps "*mag*"). The computer checks that the data-types are compatible, and makes any necessary conversion of physical units, then the human decides that these quantities are semantically equivalent.

1.1. Data Federation

Data federation (Williams *et al.* 1999) is simply the use of multiple data sources to create knowledge, for example visual identification of radio objects is federation of different wavelengths; identifying variable stars is federation over different times. Given two catalogs, it is often interesting to find the set-wise intersection (find the same physical objects represented in both catalogs). In this new joined catalog, there is more data with each object, more discrimination from others, a better chance to find the rare objects and see the trends and clusters.

Federation of data is also a nonlinear effect: new knowledge can be created from the fusion of datasets that could not be seen in the isolated data. This new knowledge does not require a rocket launch or a telescope, indeed it is at very little cost. When the semantic web makes it easy to federate data sources, it can be done easily, checking an unlikely intuitive hunch, or just looking for interesting things.

2. Structured Information

2.1. XML for Structured Information

XML is a "file format for creating file formats", and is rapidly becoming an unassailable standard across the web. There is no doubt that it will become one of the cornerstone technologies of the Virtual Observatory. For a general introduction, go to any bookstore or visit xml.com or xml.org on the web.

XML looks superficially like HTML, in that it has both control elements and text. A date in the recent past might be represented in HTML as <i>April 12, 1997</i>, where the surrounding tag <i> means that the text should be in italic. An English-speaking human recognizes this as a date, but computers and non-English speakers may not. In an XML version of the same data, the date might be written:

```
<date>
<day>12</day>
<month>4</month>
<year>1997</year>
</date>
```

Now we have structured information. There are many tools that can display and edit such data. Such tools can be used to automatically check the schema of the document — for example, a memo can only have one sender, but can have many receivers, or that in a date, the day, month and year cannot be negative.

An XML representation of a date is more flexible than just a text string. A suitably informed computer can read and understand this date, doing such useful things as sorting documents in order, making histograms of the number of documents received in each of the last 12 months. In rendering the date for human consumption, it could write Month/Day/Year for Americans, and Day/Month/Year for the rest of the world, or substitute locale-specific month names for the numbers.

XML is an excellent vehicle for expressing documents and metadata, and is now a powerful and universal web standard. However, it is not good at expressing bulk binary data; this is usually done with a link to a file or URL, or it can be converted to text with Base64 or similar.

Obviously if documents are exchanged, both sender and receiver should agree to the same standard. There are several examples of astronomical XML standards emerging, including Astores (Accomazzi et al. 1999) for catalog data and AML (Astronomical Markup Language; see, Guillaume 1998) for bibliographic and other information.

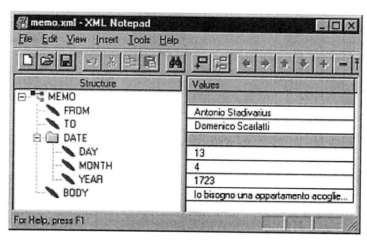

Figure 1. A memo in XML, rendered with the XML notepad application. The tree structure of elements and text is clear.

2.2. FITS and XML

The bulk of this paper is about ways of exchanging structured, semantically meaningful information, about how publishers of such information can advertise themselves, how clients can automatically configure an information transfer or initiate a remote procedure call. Such mechanisms will go a long way to creating the semantic web with will empower the Virtual Observatory.

But first we can discuss another, more human side to the story, a side that is less well-developed. Astronomers are further along than many fields of science in that they have a universal standard for structured information: FITS. This is a way to attach keyword-value pairs to binary information, originally for images, now extended to tables and other types of data. The development of XML followed a similar path: first it was for human-readable documents, the emphasis being on the separation of style from content in displaying such documents. However, XML has now become a generic way to represent structured information of many abstract kinds, and is becoming the universal language for everyone — with the possible exception of astronomers, who invented structured information first!

In many ways FITS and XML are equivalent. Users quickly realize that the challenge in using these for data exchange is not one of finding processing software, but in agreeing on the meaning of the data structure (the "FITS headers"). There has been a lot of progress in creating standards using FITS as a vehicle, and those who are mandarins of the Virtual Observatory should be very careful not to tyrannize or stifle such grass-roots efforts, but at the same time to choose and discriminate between competing emerging standards.

One project that will advance the VO considerably is a software toolbox to convert back and forth between FITS and XML. This will encourage cross-fertilization between the worlds of business and astronomy. In business, they are considering how to add bulk binary data to XML, and in astronomy, they are trying to get beyond the sometimes unintelligible 8-character names for FITS keywords.

2.3. XSIL: Extensible Scientific Interchange Language

XSIL (Williams 2000) is an XML dialect for common scientific datatypes. It defines a set of basic data objects — Parameter, Array, Table, data Stream, Time, and so on, and is designed for extensibility.

There are extension mechanisms so that people can build their own specialized data objects. For example, XSIL provides base objects Param (parameter) and Array. These might be combined to make TimeSeries (a one-dimensional array plus a parameter StartTime and EndTime). That object could then be extended in turn to make ObservationTimeSeries (*e.g.*, by adding more data about what instrument made the data).

XSIL comes with a Java parser and a browser to read it. If there is an element in the XML side, the Java code looks for certain Java classes of the same name to handle it, view it and edit it. Thus the browser is extensible in parallel with the XML.

XSIL can be used for a complete dataset (all data in XML), or it can serve as metadata, pointing to local or remote data, which can be URL or file, encoded or endian or plain text. Remote data is only read on demand, and stored in a memory-efficient way. The browser uses Java Swing and the graphing and table components from KL Group[1].

2.4. More Standards Needed

In addition to the standards mentioned above, one of the tasks of the Virtual Observatory is to agree on formats (XML or FITS) to represent a broader range of semantic objects. Many of these can come from other fields, the computer scientists and business are building these now.

- Document, Published article, Preprint, Person: we must borrow from the Digital Library community, as well as using existing standards such as AML.

- Table, Link, Parameter, Array, Image: if these basic objects are well-defined and implemented, we can build with them and reuse the software.

- Message, Exception report, Service capability, Program: we need to think sharply about the meaning of these things and the contexts in which they might be used, so that we can exchange them in a meaningful way.

- An astronomical object (*e.g.*, star) is distinct from an Observation of that object. Does every object have a position in the sky and a magnitude? What about large objects such as molecular clouds, and moving objects?

- Groups of objects, Extensions of object, Object handler. These "meta-objects" are the natural next thoughts. Once I have a table, I want a set of tables, the code to deal with tables, an so on.

---

[1] www.klgroup.com

The ISAIA project (Hanisch et al. 2000) is a wide collaboration to make a hierarchy of such semantic standards for astronomy.

3. The Semantic Web

There are many astronomical data services available today, but most of them assume that a human is using the service, not a computer. There is an idiosyncratic form to fill in, and the results come back in a nicely-colored HTML table. Such data cannot be read in any meaningful sense by a computer, making it difficult to federate data services.

What we envision in the Virtual Observatory is a network of services, each feeding data to another, perhaps with very large quantities of data. The services may be independently curated and managed, and only when a small, valuable piece of knowledge comes out is it presented to a human. Alternatively, bulk data is delivered to a user's program, but it may be highly processed already, and fused with other data products.

For example, in the future VO we could combine several services like this:

- A yellow-pages service provides locations of necessary data and computing services.

- A monitoring agent can keep track of a long-running computation, allowing the human client to periodically check in, monitor, and steer the computations, while retaining diagnostic and logging information from the services being coordinated.

- A storage service can handle many Gigabytes for a long time for those with appropriate authentication.

- A crossmatch service can take multiple input catalogs, possibly unordered, and matching criteria, then create a "fuzzy join" on the catalogs, based on the criteria.

- Query estimation services can consider the bulk data transfers and computation that is suggested, and give estimates of resource consumption, both before and during the computation.

- A compute service can take a stream of data objects (*e.g.*, catalog entries), and route the stream to multiple slave processors that do pattern-matching, then collect back the results as an output stream.

- A raw data archive may have some data on tape, some on disk, some at remote locations, but it can respond to queries with a stream of data objects delivered at uniform fast rate.

- A compute service may calibrate the raw data on-the-fly, using a subsidiary database to get the calibration coefficients.

Services could then be connected together by a user with a "modules and pipes" model of component computing, each module representing a service that is remote, where the user has not met the curator. As soon as the module is brought on to the desktop, its capability document would be fetched, so that decisions can be made about how to connect it to other module/services. The user could decide that an output from one service (for example "$g\_magnitude$") should be used as the input to another service (perhaps "mag"). The computer checks that the data-types are compatible, the human decides that these quantities are semantically equivalent.

In the following sections, we describe some ideas that will be used in the geographic community, and in the business world, to implement the semantic web in the next few years.

4. Capability Documents

There is a great deal of astronomical data already available on the web. Let us consider an imaginary catalog of some kind of interesting stars, together with a web-based service to find out what catalog members are close to a given point in the sky. The builder of the service might have decided to use HTTP GET protocol, so that a request might look like this:

http://www.blahblah.edu/getdata?request=table&RA=185.0&Dec=23.0&Radius=0.5

and it produces a response like this:

```
183.22   22.6   17.1   16.8   17.3
186.13   22.9   16.3   15.9   16.4
```

A human could probably figure out that RA is right ascension and Dec is declination, and that $radius$ is a search area. The human would be especially helped by seeing the form that came with the website, and reading the attached documentation. The response shown here is two objects from the catalog, each with RA and Dec and three magnitudes in different filters. However, none of this is clear without a human to read and understand. It would be tedious to manually configure the connection of a dozen or so services by examining the request and response semantics of each one.

In this section we discuss the idea of a **capability document**, where a service responds to a standard request with a document describing what the service can do. For example, to get the capabilities for the service above, we would submit this request:

http://www.blahblah.edu/getdata?request=capabilities

The capability document may contain these fields:

- The name of this service, with a title, abstract, contact information, and a link to a web page that describes it.

- The version number of the archive server software. When a request comes in with a different version number from that which the archive server supports, a negotiation could take place to determine a mutually acceptable version of the communication protocol.

- Acceptable distributed-computing protocols for requests and responses (*e.g.*, SOAP, Nexus, HTTP, RMI, Corba, *etc.*), and specific information about servers, port numbers, *etc.* While we assume that the capability document itself is obtained by HTTP, there is no reason why the request and response should be transmitted this way.

- Available output formats, so that for example the keyword "format" can be used by the client. Then, for example, the request can contain "format=csv" if comma-separated values are wanted. A binary stream might be requested as "encoding=bigendian&timeout=900" for data stored on a slow medium like a tape robot. Ways of specifying binary streams are specified in the XSIL language (Williams 2000).

- Disposition of error, diagnostic and debugging information may be returned. The default is a plain text message in place of the expected response, but more sophisticated mechanisms may also be available.

- It may be that a service is only available to certain authenticated users, or that payment must be made. A section of the capability document defines the ways in which the server expects to be given this information.

Part of the capability document could be a way of building a user-interface, so that a human can frame a request. This would be an HTML form that can be used as part of another page. (Unfortunately, HTML is a much looser language than XML, so that its strict version, XHTML should be used to ensure that the capability document itself is well formed[2].

4.1. Hierarchies of Capability: Subject Index

Once we have a way to describe a single data service, there is the possibility to describe collections of services and links between them. A simple way to do this is by presenting HTML web pages to a human user, who can navigate to a service that she wants to use. But there are advantages to having a machine parse a list of services. Therefore we have chosen to allow a capability document to describe lists of other capabilities documents as well as a data service. In this way we can organize and crosslink data services.

When a link is included in a capability document, it includes a "subject" element. Once we have the name of a subject we can append it to the name of the previous service, then ask for subject-specific information. For example, if we have been informed that subject-specific information is available on gamma ray astronomy, with the subject name *gamma_ray_astronomy*, then we can ask for specific capabilities as follows:

http://www.blahblah.edu/gamma_ray_astronomy?request=capabilities

---

[2]See http://www.w3.org/TR/xhtml1/ for more details

Thus we can reflect a directory hierarchy of services on the server with a service hierarchy that is visible to the client.

4.2. Capability for the Request

In the example above, the capability document explains to a human that "RA" means "right ascension", a coordinate value on the celestial sphere. To a computer, an explanation has also been given, that this is a floating-point number between 0 and 360, and the text "degrees" should be written next to it. Similarly for "Dec".

The capability document might also explain that other coordinate systems are available, and that this server can respond to galactic coordinates as well as equatorial, that it can understand different notations. Thus there are alternate versions of the request, so that this would also be acceptable:

http://www.blahblah.edu/getdata?request=table&Glon=124.3&Glat=19.3

Each service has associated a list of keywords that may be used in a request. Each keyword may have other information associated with it, including:

- Information about the allowed values for this keyword (mathematically, it's domain),
- Default values if it is omitted,
- Short and long descriptions of the semantic meaning,
- Units (if any) implied for the value,

For the example above, the keyword "RA" has range 0 to 360 degrees, the title is "Right Ascension", and the explanation is "Right ascension in the J2000 coordinate system, see http://... for further details".

One further type of interaction we might imagine is making SQL queries to the service, or perhaps by sending some other script or code to be executed. Small pieces of such script can be sent as strings as part of the request, and the capability document can indicate that this is allowed.

Keywords may be arbitrarily grouped, so that we can specify which combinations of keywords can be used for a request. For example, "RA" and "Dec" might form a group, and "Glon" and "Glat" another. Then we can use either pair in a request, but we cannot mix keywords from different groups. Keyword groups may also have a domain specification: if a catalog covers a small domain of the sky, then this domain is naturally expressed on the joint object (RA, Dec) rather than on each coordinate independently.

4.3. Capability for the Response

We have explained how the capability document can define a service and how to make a request. Now we shall consider the explanation of what will be in the response, and what data formats are available. Let us think of the response as a table of data, with each row a data record, and here we consider how the capability document defines the headers for the table. Let us call each part of the data record a "column", to distinguish it from "keyword" that we have used while describing the service request.

Each column is defined by:

- The name of this column, its title, it abstract, its "further info" link,

- The data type to be used for storing it (int, float, complex, string, *etc.*),

- Default value to be used in case a value could not be read by the client,

- The minimum or maximum of the values in this column,

- Semantic meaning of the column: name, title, abstract, URL link.

For the column in the example above that is the chart number, it might be defined as an integer between 1 and 20 (the number of charts), and it has no units.

We would like to have the possibility of embedding links into the response data, so that we can draw catalog information on image data, then hyperlink to more information about that object. For example the response to a request might be a list of galaxies, each with some numerical fields (brightness, color, *etc.*), but also a link to an image, or to bibliographic records. There could be an individual link for each galaxy, or there could be a "template" link that came with the metadata, that is to be combined with some ID number of the galaxy. For example, the template might be

http://www.blahblah.edu/further_info?ID=$ID

so that the response column whose name is "ID" is to be substituted in the relevant place.

4.4. OpenGIS Web Mapping Testbed

The Geographic Information Systems community (GIS) has already defined much of the capability document infrastructure described above, and implemented many services that use it. In the figure below is an example of a web browser showing a composite map of the English Channel, with different map layers coming from different servers. Each server in the list (top of three pull-down menus) has provided a capability document, describing the nature of the different map layers that it can provide. The user has selected several of these (key, to the right), and chosen an order for the layer stack. Here we see evaporation, built areas, railways, population centers, and coastline.

The OpenGIS initiative[3] has already defined the XML format of the capability document for the case of map layers being returned by a server.

---

[3]http://www.opengis.org

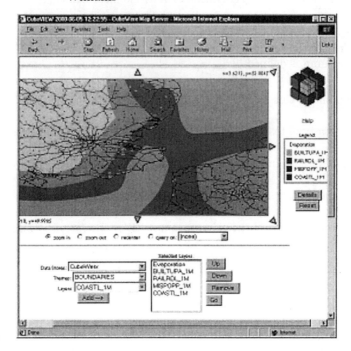

Figure 2. An example client in a web browser, showing London and France. Multiple layers are fused from multiple Open-GIS-compliant servers.

5. Business Initiatives

Business is also realizing the utility of interconnected web-based services. If a retailer has dispatched goods through a package service, he would like to show the customer the status of the package from his own web site, not to simply say "here is the tracking number, go look it up with the package company." Businesses that broker information have an obvious interest in a standard way for their computers to talk with those of their wholesalers. This "B2B" (Business to Business) market sector is growing strongly.

It is becoming clear that the universal vehicle for such communication is by XML-formatted messages delivered by the HTTP protocol. In this section, we summarize some of the initiatives from various companies and consortia that can create a business version of the semantic web.

Table 1 summarizes the protocol stack. We are becoming familiar with the idea of XML-formatted data objects carried on an HTTP protocol, itself carried reliably by TCP/IP.

In the following, we discuss the upper layers of the stack: SOAP to allow rich messages between web-client and web-server; WSDL to provide capability information for web services; and UDDI for publishing the existence of services. We should point out that the use of these protocols does not restrict data transfer to the HTTP protocol; however, we can use the flexibility of these protocols to decide how high-bandwidth communication can occur by some other mechanism, for example parallel FTP.

UDDI	Registration and discovery
WSDL	Service description
SOAP	Remote objects and computing, who does what
XML	Structured information
HTTP	Identified file formats, e.g., Text, JPEG
TCP/IP	Reliable transfer of byte streams

Table 1. A summary of the protocol stack.

5.1. SOAP: Open Distributed Computing

SOAP (Simple Object Access Protocol; see, e.g., Box et al. 2000) is an XML format that carries commands and objects between clients and servers, including who is being commanded by this message, and what reply is expected. While many web services today use a keyword-value combination as a request (such as the example above, with the RA, Dec, radius keywords), SOAP allows the use of a complex object as a request and response, both serialized in XML. There is also capacity for remote-procedure calls, leading to simple distributed computing, but without the complexity of CORBA or the language specificity of Java RMI.

Several prototype SOAP-based services are described at the Xmethods web site.

5.2. WSDL: Service Description

WSDL (Web Services Description Language) uses XML to describe network services or endpoints. It can describe services that use the HTTP GET and POST protocols (keyword-value set), and also services mediated by SOAP.

Requests and responses to a service are known collectively as messages. Each of these is a collection of types parts, for example "the variable alpha in the message is a floating-point number". An operation takes a message as input and produces one as output.

5.3. UDDI: Registration and Discovery

The UDDI initiative (Universal Description, Discovery and Integration of Business for the Web[4], was started by Microsoft, IBM, and Ariba, and many others. It provides a framework for the description and discovery of business services on the web. It does this by using distributed registries of services, and conventions for accessing that registry using SOAP.

There are three components to the UDDI specifications. White pages shows address and contact information for the service, and yellow pages categorizes the service within a hierarchy so it can be found easily. The green pages component defines the technical information about the service capabilities.

---

[4]http://www.uddi.org

6. Conclusions

In addition to high-performance computing and high-performance networking, the Virtual Observatory will require a leap in semantic interoperability. While the web already provides interoperability between humans (mediated by computers), the VO will consist of services that interactively perform multiple steps on the user's behalf. These tasks may require one web service to call on other web services, coordinationg the steps like a traditional software program executes commands. The problem today is that integrating with other services remains difficult, because tools and common conventions for interoperation are lacking.

The astronomical community is well-versed in the art of making standard semantic data objects using FITS files, and they will benefit from further such standardization under the auspices of the VO. However, the range of information objects must be wider, taking such standards from the computer science community. Furthermore, astronomers must embrace XML in addition to FITS as a vehicle for structured information, thereby getting the best of both.

We are all expecting archive-based research to be a fourth arm of the scientific method, in addition to observation, theory, and simulation. For this to happen, it must be possible to connect apparently disparate ideas into a hypothesis and then make appropriate tests. Therefore, data services must be interoperable even when the the people involved have never met and nobody has ever thought of connecting these services. This is why standards are necessary. We do not know in advance what kinds of data will interoperate with what.

High bandwidth data exchange and high-performance computing come after the semantic web has been established. Once it is established that multiple servers can effectively interoperate and they have the necessary data, then the experimenter will be able to connect the services with high-performance data services. In the words of Kernighan, "First make it work, then make it fast".

References

Berners-Lee, T. 1998, Semantic Web Road Map[5]

Williams, R., Messina, P., Gagliardi, F., Darlington, J., Aloisio, G. 1999, Report of the European-United States Joint Workshop on Large Scientific Databases[6]

Accomazzi, A., et al. 1999, Describing Astronomical Catalogues and Query Results with XML[7]

Guillaume, D. 1998, Astronomical Markup Language[8]

---

[5]http://www.w3.org/DesignIssues/Semantic.html

[6]http://www.cacr.caltech.edu/euus/

[7]http://vizier.u-strasbg.fr/doc/astrores.htx

[8]http://monet.astro.uiuc.edu/ dguillau/these/

Williams, R. 2000, Extensible Scientific Interchange Language (XSIL)[9]
Hanisch, R., McGlynn, T., Plante, R., King, J., White, R., & Mazzarella, J. 2000, ISAIA: Interoperable Systems for Archival Information Access[10]
Box, D., et al. 2000, SOAP, Simple Object Access Protocol[11,12,13]

---

[9]http://www.cacr.caltech.edu/XSIL

[10]http://heasarc.gsfc.nasa.gov/isaia/

[11]http://www.oasis-open.org/cover/soap.html

[12]http://xml.apache.org/soap

[13]http://msdn.microsoft.com/xml/general/soap_webserv.asp

Distributed Archives Interoperability

Cynthia Y. Cheung

NASA Goddard Space Flight Center, Astrophysics Data Facility, Greenbelt, MD 20771

Abstract. Interoperability is an inherent goal of the Virtual Observatory of the Future. It describes the state when the distributed, diverse and multispectral astrophysical data and computational resources are linked together into a cohesive functional unit, enabling systematic data exploration and discoveries. This paper summarizes the technical issues and challenges to be addressed by both the astronomy and computer science communities before this goal can be realized.

1. Introduction

The astrophysics resources[1] currently available on the Internet are diverse and distributed. They include observational data archives, topical data repositories, catalog servers, image servers, data analysis and visualization software packages, discipline-specific knowledge bases and bibliographical services. But they are only loosely connected and few services can exchange data and information easily. In particular, it is necessary for the researcher to learn the specific structure of each data service and user interface before one can determine whether the relevant data or tools exist and then access them. This problem is exacerbated by the on-going multi-spectral sky surveys that generate data on the order of terabytes in volume. The technological challenge is to evolve the design and management of these distributed systems so that researchers can explore, combine, analyze, and cross-correlate the data in an efficient manner. Astronomers not only wish to retrieve the diverse data types and software as a package, but also wish to be able to integrate data of similar nature (e.g., images or spectra) that are retrieved from the distributed sources. Therefore, the challenge is to integrate both *vertically* across all levels of data and complexities and *horizontally* within each data category.

The computer and information technology community has been working on the fundamental issues related to interoperability. These include the next generation Internet, transfer protocols, data search and access techniques, and the semantic web. Astronomers can leverage the latest development in physical connectivity provided by the commercial sector, such as the computer-to-computer communication protocols and high-speed network links. However, the discipline-specific implementations of the data search and access techniques need

---

[1]http://www.stsci.edu/astroweb/astronomy.html

to be worked out by the community. Often the requirements for science research support are quite different from those in the commercial sector. The community also needs to establish standards for both the syntax and semantics of the astronomical data transfer protocols. These standards are essential to enable data discovery and interoperability.

2. Levels of Interoperability

Interoperability is an inherent requirement for realizing the vision of the Virtual Observatory (VO). Due to the sheer volume of the on-going sky surveys, very few data centers or users would have the resource to transfer a sizable subset of the multi-spectral data to local storage before initiating data analysis and cross correlation. This implies that both the data archives and compute services will remain distributed and the resources that are required to fulfill specific queries must be harnessed dynamically. The selection, visualization, discovery and even initial analysis will be done most efficiently at the data site, or at certain terascale supercomputing facility to minimize large data transfers. It is in fact advantageous for data to be resident and maintained by the originating data centers, because scientific expertise are available to ensure proper quality control and continuous improvement (CODMAC 1983). Users should be able to enter the VO from multiple gateways, possibly from a computer at their own home institutions, and submit their queries to all the components federated within the VO.

Figure 1 illustrates the successively deeper levels of interoperability that we strive to achieve in building the VO. Currently, an astronomer is able to access data services one at a time using an interface that is unique to that service. Though most services provide hyperlinks to other related sites, the interaction is strictly *one-to-one*, between one user and one service.

In the near future, users would be able to access multiple data sites simultaneously through a uniform Data access layer (DAL). Projects such as *AMASE* (Cheung *et al.* 1998) and *AstroBrowse* (Heikkila *et al.* 1998) represent initial prototype efforts to build a DAL. These implementations encapsulate the structure and semantics of the different data centers in middleware systems, by creating a federated view or a shared schema for the heterogeneous sources. The DAL accepts queries from the user, then transforms and redistributes the queries to the appropriate data centers. The results are returned to the DAL for presentation to the user. The middleware mediates between the user and the disparate data sources and provides the capability for queries to be sent to many sites simultaneously (*one-to-many* connectivity). This is in fact a modified *one-to-one* interaction. There is no data exchange between the data centers. Although this scheme provides good support for data access, it does not scale-up efficiently because of the complexities involved in creating and maintaining a shared schema for a large number of sources.

The vision for interoperability in the VO is for users and data centers to have multiple connectivity to one another (*many-to-many* connectivity). Intelligent software agents will consult user-defined profiles and preferences to mediate the interactions between the different elements of the VO, ensuring that the communication is conducted in the proper context. Standard protocols and metadata

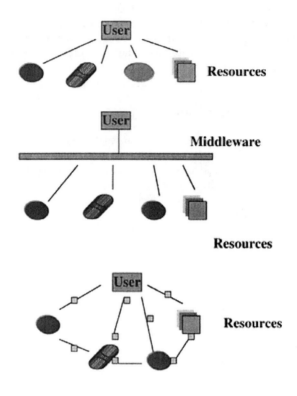

Figure 1.   Levels of Interoperability.

play very important roles in this multiple connectivity environment. To achieve this vision, we need to gradually evolve the astrophysics infrastructure towards *semantic interoperability*.

3. Semantic Interoperability

Many technical challenges must be overcome to reach semantic interoperability in the VO. Foremost of these is the derivation of metadata standards. Metadata defines the semantic context of a data set and is fundamental to its proper understanding and correct usage, especially in large scale cross-correlation that requires automated transformation. Metadata is more than the 'classical' syntactic description of a data set in terms of volume, format, physical location, mission/observatory, instrument, and date of acquisition. Metadata should provide the *conceptual data model*, *i.e.*, how the data is structured in terms of scientific contents: the measured parameters, units, accuracy, algorithms and theoretical models used in data reduction and calibration, caveats, and range of validity. Much of this form of metadata do not exist for the legacy data archives and must be generated according to a common VO data model or schema. It is worth mentioning that the FITS headers for much of astronomical data contain a mixture of both syntactic and semantic metadata. The current use of FITS keywords across the different wavelength regimes is also not uniform.

Another useful form of metadata already exists in astronomy, though often not considered as such, is the set of classical astronomical catalogs. These are aggregates of pre-computed or derived results of astronomical measurements. The authors had often performed some smoothing of the semantics when compiling the catalogs and attempted cross-correlation with other publications. The ADC[2] has started several efforts to utilize these catalogs as the metadata gateway to the original observation data and other online resources. It is quite challenging to define standards for both the syntax and semantics across the entire catalog collection, which encompass virtually all sub-disciplines in astronomy. This work cannot be accomplished without collaboration from the international astronomical community.

The metadata should also be cast in a computer understandable and human understandable language like the eXtensible Markup Language (XML) to facilitate automated exchange and updates. Metadata standards defined in XML is *extensible* so differences between disciplines can be accommodated. Fortunately, computer scientists and many other disciplines are also working on the issues of semantic interoperability, so the astronomical community would be able to leverage their development of tools and methodologies. But the community needs to work on the standardization of keywords, terminology and astronomical naming conventions. When the metadata (or data model) for each VO site is cast in well-defined standards, tools can be developed to transform data from one schema to another. Intelligent software agents can then be built to mediate the interactions between data centers.

4. Interoperability Functions in the VO

Several basic functional components are required in the VO framework to enable interoperable capabilities for scientific research. In the following subsections, we shall describe them briefly.

4.1. Integrated Search and Discovery

Users will have the capability in the VO to issue a single query against all the available resources and receive the integrated search results. This ranges from simple queries to locate data to rather complex queries in data mining. The basic components required to enable integrated search and discovery include a VO registry, a query processor, and tools to support data selection and integration of results. Prototypes of these basic components are already in use, but they must be further expanded and refined, and also updated with the latest technological innovation.

- A resource directory must be created to register the data contents and computational resources at each federated VO site. An example of such a

---

[2]http://adc.gsfc.nasa.gov

registry is the implementation of the CDS GLU^3 system in $Astrobrowse^4$ for various astrophysics web sites. The directory also contains metadata that describe the syntax and semantics of queries acceptable to a specific site.

- A query processor will take the user query and translate its contents into multiple queries, then direct them to the available resources listed in the directory. $AMASE^5$ is a prototype of a query processor with a metadata directory.

- Browsing and visualization tools are needed to allow users to browse data samples, do simple analysis and plotting, and create relevant subsets before initiating a transfer. Because of the huge volume of data in the VO, data transfers must be done judiciously to minimize the network traffic. The ADC Data Viewer[6] is an example of an interactive browsing tools designed for astronomical catalogs.

- Data integration tools must be available to support integration and analysis of query results that are retrieved from the distributed data centers. For example, tools that support the re-normalization and superposition of data from different wavelength regime and enable the visualization of multi-dimensional datasets would be very useful. A batch query capability that enables users to feed the query result from one data service as the query input to another service is also highly desirable.

4.2. Data and Software Exchange

An efficient method to transfer data and software and to propagate updates is an important enabling component of the VO. Most of the current data and software transfers are conducted using *pull* technology such as FTP. Users initiate the updates after they are notified by a newsletter or email. There is only limited use of *push* technology for automated updates. For example, the metadata at the distributed Hubble Space Telescope archive sites are maintained automatically by a replication server. In the VO, the application of hybrid data transfer techniques using a combination of *push* and *pull* will allow users to receive only the relevant and up-to-date information through broadcast from a data cache[7], resulting in reduced storage and communications bandwidth requirements for data sites. The diverse data types and relevant software will also be packaged in a logical and efficient manner for delivery to the user. For example, the proposed eXtensible Data Format[8] will integrate different data types and allow the appropriate software to be included by reference with the data.

---

[3]http://simbad.u-strasbg.fr/glu/glu.html

[4]http://heasarc.gsfc.nasa.gov/ab/

[5]http://www.amase.gsfc.nasa.gov

[6]http://tarantella.gsfc.nasa.gov/viewer/

[7]http://www.cs.umd.edu/projects/opsis/topics/aircache.html

[8]http://tarantella.gsfc.nasa.gov/xml/XDF_home.html

4.3. Integrated Analysis Environment

Analysis packages in general use by the community, *e.g.*, AIPS, IRAF, FTOOLS, etc., need to be expanded to include tools for manipulating large quantity of data and to support statistical analysis in data mining. Each package currently has its specific format requirements. Standards interfaces to both data and software must be defined so that user-defined software functions can be easily deployed and data can be interchanged between the different packages. The development of a standard Astronomical Markup Language, as has been done in the Mathematics and Chemistry disciplines, would be a step in the right direction in helping to bridge data and software in legacy formats.

4.4. Data Organization

Data at each VO site must be organized for efficient access. It is possible that different indexing schemes would need to be derived for the same or different data sets in order to support different scientific objectives.

4.5. Query Optimization

Methodologies must be developed to build efficient query execution plans so that a reasonable VO query performance can be achieved for these distributed and large scale data manipulations.

5. An Illustrative Example

To summarize the above discussions and to gain more insight on semantic interoperability, let us outline the steps required for carrying out one VO science scenario. The objective is to do positional correlation for a region of the sky across multiple wavelengths to support object cross-identification. The query would be submitted through a VO gateway. The query processor would take the sky coordinates and other user-specified parameters and send inquires to the distributed archive sites, according to the resources recorded in the VO registry. Intelligent agents will translate the queries into the proper syntax and semantics for the respective data sites. Software will be initiated at the archive sites to determine the number of sources and their locations. Depending on the efficiency of query execution, the results from one data center will either be shipped to successive data centers for cross-correlation, or the results from all the data centers will be transferred to the query processor or to a compute facility for processing. Intelligent agents will again translate the results into a form that can be manipulated by a suite of data integration and visualization tools. The final results will then be sent to the user for further perusal.

References

Committee on Data Management and Computation 1986, *Issues and Recommendations Associated with Distributed Computation and Data Management Systems for the Space Sciences*, National Research Council (National Academy Press, Washington, D.C.)

Cheung, C.Y., Roussopoulos, N., Kelley, S., & Blackwell, J. 1998 "A Search and Discovery Tool—AMASE," *Proc of the ADASS'98*, Eds. D.M. Mehringer, et al. , *ASP Conf. Series*, vol. 172, p. 213

Heikkila, W., McGlynn, T.A., & White, N.E. 1998 "Astrobrowse: a Web Agent for Querying Astronomical Databases," *Proc of ADASS'98*, Eds. D.M. Mehringer, et al. , *ASP Conference Series*, 172, 221

Technologies for Mining Terabytes of Data

Andreas J. Wicenec
European Southern Observatory, Karl-Schwarzschild-Str.2, D-85748 Garching, Germany

Abstract. This paper focuses on adopted or proposed solutions the ESO Science Archive Facility has implemented or tested to solve the problems imposed by the rather heterogeneous data retrieved from multi mode instruments mounted on telescopes on Paranal and on La Silla (Chile). In view of a virtual observatory (VO hereafter) which would aim to connect a much broader range of data sources the discussions are formulated as generalized guidelines and ideas. From the experiences of the first year of operations of ESO's VLT archive it is concluded that data which should be part of a VO has either to follow pretty strict rules, or the interface layer between the data and the VO has to take care of the "VO compliance".

1. Introduction

It is certainly true that a virtual observatory could give rise to a new approach of scientific evaluation of astronomical data gathered all over the world. However to make something like this reality means a lot of discussions and tedious work. Even though a big part of astronomical raw and also some reduced data are stored in a common format (FITS), everyone who worked on combining two or more raw data sets from different sources, knows that it means a lot of handwork to make any reduction system run smoothly on either of the two, not to speak about a mixture of them. Problems start already at very basic levels, like different units used in FITS keywords or catalogue columns. This seems to be trivial, since one could convert one to the other or both to a common one, but one has to remember that there is no commonly adopted way in FITS to describe the units of keyword values. Really tough problems are the ones which stem from specific properties of instruments and/or instrument modes. In the following sections we will discuss some of the issues related to this root level of a VO, give an impression of the amount of data a VO has to deal with if it wants to include data of a big observatory like ESO and finally we name some of the areas which have to be discussed and solved at the root level to make a VO a success story.

2. Building Blocks for a Virtual Observatory

In very general terms a VO consists of the following building blocks:

- Existing and future data archives which should have internally consistent and homogeneous content
- A well defined and standardized archive interface layer
- A well defined and standardized archive access layer
- VO data reduction tools
- User access layer(s)

In this paper we are mainly concerned with the first point, but for a big multi terabyte archive like ESO's VLT archive it is clear that the other points have to be taken into account, in order to provide something more than just a data dump area, where nobody wants or is even able to retrieve data from. The tricky points are related to the word *standardization*. It is well known that standardization processes take a long time and are tedious work. Nevertheless it is clear, that a VO has to rely on a robust environment, too many different formats, protocols and access mechanisms would make such a system fragile if not impossible. The standardized interface and access layers could still be just a wrapper around an existing archive, but we think that at least the big archives should provide and be responsible for their own wrappers.

2.1. Main Root Level Issues

The main issues to be tackled and solved for a big archive can be listed like:

- (Meta-)data consistency
- Homogeneous access for a variety of different data holdings
- Inter-relation and inter-operability between different data holdings
- Traceability of scientific and calibration data and auxiliary parameters
- Performant access
- Quality control of raw data

Note that all the points given above are *internal* to the archive but are necessary to support a VO kind of activity. The three crucial points in the list above are the Meta data and data consistency, the full control over scientific, calibration and auxiliary data and their interrelation and the quality control of the raw data. Actually only if the full list of requirements is fulfilled we can speak about a scientifically useful archive.

3. ESO's Data Tsunami Wave

The data volume of an archive like the ESO archive, mainly containing raw CCD data is rather big. The date rate is growing exponentially, because new instruments will come on-line during the next years. Currently there are 2 telescopes with 4 instruments in normal operation on Paranal. The ESO archive is storing data from another 3 telescopes with 6 instruments on La Silla. The current data rate is about 0.7 TB/month. It is planned to have all 4 main VLT telescopes, 3 auxiliary telescopes, VST and VISTA on Paranal by the year 2004. The data rate is expected to grow to about 7 TB/month by 2003 already (not including numbers from VISTA nor from the VLTI). VST alone will produce of the order of 75 GB/night. The total data holdings will grow from about 5 TB today to about 103 TB in 2003. The following subsection will give some idea of the technology we are currently using and/or testing in order to be able to deal with this volume of data.

3.1. Hardware Solutions

Currently we are using DVD-R technology for the archive, *i.e.*, data from the mountain is sent to ESO Headquarters in Garching/Germany on DVDs, usually once a week. The media are then inserted in DVD juke boxes and are then available on-line. Processing and quality control of the data is done on normal workstations and/or bigger server machines. All meta information of the data is stored in a RDBMS (Sybase). In addition to the instrument data there are several data streams coming in, which contain meteo and ambient data as well as engineering data and log-files. All this data is stored directly in a data base.

Since we are facing a factor of 10 increase in data rate during the next 2.5 years there is a vital need to think carefully about the concepts and requirements and new technology to be used in the coming month and years. We are currently testing different technologies like Linux clusters (Beowulf), new optical media generations and pure magnetic disk solutions. CD-R/DVD-R robots for the media production on the mountain are about to be installed.

3.2. Software Solutions

The current quality control pipeline systems used at ESO are tailored for the instruments we have right now and they would probably work for some of the future instruments as well. Problems arise when talking about "monsters" like VST and VISTA and huge survey projects. For this kind of challenge we need to redesign the current infrastructure and go for a system which is able to use parallel computing facilities like a Beowulf system. The data transfer layer of the data flow system has to be adjusted, tested and upgraded to be able to deal with the expected data rate. In order to support the big surveys to be carried out at VST and VISTA the supporting data base has to be adjusted as well. We are looking into OODBMS (Objectivity) and data warehouse systems (Sybase IQ). The Objectivity data base is already in use at several places (GSC-II, SDSS) in the astronomical community and elsewhere in science (CERN) and has proven to be a valuable tool.

4. Standardization

One key issue to be tackled for an VO to work is to use standards at several different levels. It is obvious that a standardized access mechanism to meta data as well as the data itself is a must. Other areas of standards are the query language itself, messaging and the inter-operability layer in general. There are some initiatives already under way or in use, like the Astronomical Server URL (ASU), and some XML document type definitions (DTD) such as Astrores, AML, AIML, unit DTD). The XML discussion mailing list[1] is a forum where XML activities are bundled. At ESO we are working on DTDs which should be used for log data, ancillary data and data dictionaries. While some of these initiatives are well under way, both the scope and the group of involved people is too small to cover the areas for a VO.

5. The Archive as an Instrument

Even more work is necessary in order to make a single archive useful as an astronomical instrument. To make a collection of archives work as an instrument is a real challenge. Starting with problems like the pixelization of the sky (HTM, Healpix) down to purely data base issues, this problem first needs to be sized. One big technical field is also the implementation of multi-dimensional index structures. From the discussions in the splinter groups during the conference it became very clear, that in a lot of the technical areas we could learn a lot from other scientific areas and build up completely new collaborations with computer scientists, statisticians and especially designers of huge data base systems in biological, meteo and earth sciences.

Also in these areas a VO would gain a lot in defining standards, *e.g.*, for the sky pixelization or for the definition of standardized content descriptors of catalogues. For imaging data it would also make sense to think about storing them in a multi-resolution format, where the full resolution is a lossless compression. There are already very fast algorithms around, which could be useful.

6. Experience with Commercial Data Warehouse Systems

Tests carried out *by* two of the leading data base companies unveiled some problems with existing commercial systems. In principle data warehouse technologies could be bought off the shelve and used for some of the applications we have in mind. The tests have been carried out in the following way:

We provided astronomical data in form of the USNO-A2.0 catalogue and asked some very specific questions:

- How does the test environment look like (Hardware, Software)
- How long does it take to load the data?
- What are the overheads in terms of tuning the data base?
- What volume has the complete data base after loading?
- How long does a simple coordinate box query take?
- How long does a multidimensional query take?

---

[1] xml@pioneer.gsfc.nasa.gov

The simple queries have been randomly selected fields all over the sky. The same queries have been sent to ESO's USNO-A2.0 server.

The multi-dimensional query (just one) was of the following type:

- Find all stars with R magnitudes between 14 and 16 closer than 10 arcmin around a star with B magnitude between 8 and 10

This query is quite tricky to perform in a normal environment, nevertheless we did a test with about half of the USNO-A2.0 catalogue on a 400Mhz Celeron PC running Linux. The test has been carried out using a brut force method, *i.e.*, simply scanning the whole catalogue twice. It took about half a day to write a couple of small C programs for the scanning. Copying the data from the CDs to the hard disk has been carried out in parallel. Thus all together it took about half a day to prepare the data and the specialized software. Counting for the parts of the software we stole from the currently existing server we can safely say, that it is less than 2 days of work. Running and combining the query results took 4 hours. Since it is a pure scan this can be extrapolated to 8 hours for the full catalogue. The catalogue has been used in it's native compressed format together with a very small index of about 30 kB.

For the data warehouse systems it took in both cases several attempts for the specialists to load the data in a way that the queries performed fast afterwards. The data base size was up to a factor 30 bigger than the original (compressed) size of the catalogue.

As a conclusion it seems that data warehouse systems were not able to deliver good results, but there is a huge variety of other systems on the market and customized software is certainly not the way to go if thousands of different data sources are involved.

Especially in this area it would make a lot of sense to start new collaborations with data base scientists and data warehouse specialists. We are trying to get in contact with other DB/DataWarehouse vendors who are willing to carry out similar benchmarks using their products.

7. Conclusion

Apart from other difficulties it is clear that standardization and data base issues in general are two of the main topics which have to be tackled in a global manner to make a VO possible. There is a relatively small group of institutions and projects in the astronomical area which are working since some time already in the field, but up to now collaborations are rather limited in scope. The experience of the existing small collaborations should be distilled and refined with some additional ingredients from other areas. In this sense a VO not only is a technical challenge, but also a political and social one. Even more so if one takes data and individual rights into account.

References

Since this is a paper on *virtual* observatories all the references given here are URLs. Further information about the listed topics as well as pointers to refereed/printed papers usually can be found on the respective web sites.

ESO's VLT project: http://www.eso.org/projects/vlt/
ESO's Science Archive Facility (SAF): http://archive.eso.org/
Flexible Image Transport System (FITS): FITS http://fits.gsfc.nasa.gov/
Astronomical Server URL (ASU): http://vizier.u-strasbg.fr/doc/asu.html
Astronomical Markup Language (AML): http://monet.astro.uiuc.edu/~dguillau/these/
Astronomical Instrument Markup Language (AIM): http://pioneer.gsfc.nasa.gov/public/aim
Astrores: http://vizier.u-strasbg.fr/doc/astrores.htx
Hierarchical Triangular Mesh (HTM): http://www.sdss.jhu.edu/htm/
Hierarchical Triangular Mesh (HTM): http://www.sdss.jhu.edu/htm/doc/adass99.ps
HealPix: http://www.eso.org/~kgorski/healpix/
GSC2 COMPASS data base: http://www-gsss.stsci.edu/gsc/compass/compass_home.html
SDSS: http://www.sdss.org/
Objectivity at CERN: http://wwwinfo.cern.ch/db/objectivity/
USNO-A V2.0: http://www.nofs.navy.mil/projects/pmm/USNOSA2doc.html

Computer Science Issues in the National Virtual Observatory

Michael T. Goodrich
Deptartment of Computer Science, Johns Hopkins University, Baltimore, MD 21218

Abstract. This paper summarizes a discussion on general technological issues related to a national virtual observatory in a special session at the 2000 Workshop on Virtual Observatories of the Future, June 15, 2000. The discussion leaders were Michael Goodrich, of Johns Hopkins University, Jim Gray, of Microsoft, Michael Kurtz, of the Harvard-Smithsonian Center for Astrophysics, and Giuseppe Longo, of the Osservatorio Astronomico di Capodimonte.

1. Five Key CS Technologies

The discussion was opened by a presentation of five key technologies that are related to the needs of a National Virtual Observatory (NVO).

1.1. External-Memory Computations

When computations and data sets are too large to fit into the internal memory of a computer, they must necessarily involve external-memory computation. Moreover, because of the huge orders of magnitudes differences in the time to access storage in internal memory versus external memory, large-scale algorithms and data structures must be designed from the start with external memory in mind from the start. Thus, algorithm engineering technology developed for external-memory computations should be ported to the NVO. Relevant references include Agarwal *et al.* (1999), Arge (1995), Arge *et al.* (1998), Arge and Vitter (1996), Chiang *et al.* 1995), Crauser *et al.* (1998), Goodrich *et al.* (1993), Nodine *et al.* (1996), and Vengroff and Vitter (1996).

1.2. Geometric Indexing

NVO data is inherently geometric, in that much of it can be modelled as points in a multi-dimensional space. Decades of research in computer science has shown that such data is best indexed geometrically using structures such as the hierarchical triangular mesh (HTM), k-d tree, or BAR-tree. Relevant references include Bentley (1990), Ding and Weiss (1993), Duncan *et al.* (1998), Duncan *et al.* (1999), Eastman (1981), Overmars and van Leeuwen (1982), Samet (1990a, 1990b, 1990c, 1992, 1995), and Sproull (1988).

1.3. Meta-Computing

An interesting concept for performing computations over the Internet involves using a loosely-configured coalition of computers to perform that computation. Such coalitions are known as meta-computers or computational grids, and they offer the potential for the efficient exploitation of significant parallelism for performing large computations. Relevant references include Hernández et al. (2000), Foster et al. (1998), Haupt et al. (1999), and Patten et al. (1999).

1.4. Information Security

Computations that are to be performed over the Internet and data resources that are to be housed on Internet computers must unfortunately content with computer security issues. Data must be kept so as to achieve certain security properties, which including propriety rights, integrity, availability, and (in some cases) confidentiality. Thus, modern cryptographic techniques may need to be incorporated into the architecture for an NVO. Relevant references include Foster et al. (1998), Hughes and Hughes (1998), Knudsen (1998), and Schneier (1996).

1.5. Information Visualization

Information that extracted from an NVO must be visualized. Several techniques exist for visualizing multi-dimensional and even relational data. These should be exploited. Relevant references include Alberts et al. (1997), Di Battistta et al. (1994), Di Battistta et al. (1999), Dodson, (1996), and Johnson (1993).

2. Discussion

In addition to these topics, a discussion leader raised the point that the community will need to decide if the NVO should be used to *perform* science or to *facilitate* science. Many in the audience commented that it should do both.

A discussion leader asked how computational resources will be performed, namely, whether or not sophisticated computations should be performed at client computers or at server computers. An audience member commented that he felt that simple computations could be performed without fees for all users, but that he felt that NVO time for complex computations should be requested in the same manner that instrument time is requested for physical observatories. Likewise, a small panel should allocated time for complex queries on the NVO in the same manner as panels allocate time at physical observatories.

Acknowledgments. The research of this author is supported by the NSF under grants CCR-9732300 and PHY-9980044.

References

Agarwal, P.K., Arge, L., Brodal, G., & Vitter, J.S. 1999, *I/o-efficient dynamic point location in monotone subdivisions*, In *Proc. 10th ACM-SIAM Sympos. Discrete Algorithms*, pp. 11–20

Alberts, D., Gutwenger, C., Mutzel, P., & Näher, S. 1997, *AGD-Library: A library of algorithms for graph drawing*, In Proc. Workshop on Algorithm Engineering, pp. 112–123

Arge, L. 1995, *The Buffer Tree: A new technique for optimal I/O-algorithms*, In Proc. 4th Workshop Algorithms Data Structures, number 955 in Lecture Notes Computer Science, pp. 334–345

Arge, L., Procopiuc, O., Ramaswamy, S., Suel, T., & Vitter, J.S. 1998, *Theory and practice of I/O-efficient algorithms for multidimensional batched searching problems*, In Proc. 9th ACM-SIAM Sympos. Discrete Algorithms, pp. 685–694

Arge, L., & Vitter, J.S. 1996, *Optimal interval management in external memory*, In Proc. 37th Annu. IEEE Sympos. Found. Comput. Sci., pp. 560–569

Bentley, J.L. 1990, *K-d trees for semidynamic point sets*, In Proc. 6th Annu. ACM Sympos. Comput. Geom., pp. 187–197

Chiang, Y.-J., Goodrich, M.T., Grove, E.F., Tamassia, R., Vengroff, D.E. & Vitter, J.S. 1995, *External-memory graph algorithms*, In Proc. 6th ACM-SIAM Sympos. Discrete Algorithms, pp. 139–149

Crauser, A., Ferragina, P., Mehlhorn, K., Meyer, U., & Ramos, E. 1998, *Randomized external-memory algorithms for some geometric problems*, In Proc. 14th Annu. ACM Sympos. Comput. Geom., pp. 269–268

Di Battista, G., Eades, P., Tamassia, R., & Tollis, I.G. 1994, *Algorithms for drawing graphs: an annotated bibliography*, Comput. Geom. Theory Appl., 4, pp.235–282

Di Battista, G., Eades, P., Tamassia, R., & Tollis, I.G. 1999, *Graph Drawing*, Prentice Hall, Upper Saddle River, NJ

Ding, Y., & Weiss, M.A. 1993, *The k-d heap: An efficient multi-dimensional priority queue*, In Proc. 3rd Workshop Algorithms Data Struct., volume 709 of Lecture Notes Comput. Sci., pp. 302–313

Dodson, D. 1996, *COMAIDE: Information visualization using cooperative 3D diagram layout*, In F. J. Brandenburg, editor, Graph Drawing (Proc. GD '95), volume 1027 of Lecture Notes Comput. Sci., pp. 190–201

Duncan, C.A., Goodrich, M.T., & Kobourov, S.G. 1998, *Balanced aspect ratio trees and their use for drawing very large graphs* In Graph Drawing, Lecture Notes in Computer Science, pp. 111–124.

Duncan, C.A., Goodrich, M.T., & Kobourov, S.G. 1999, *Balanced aspect ratio trees: Combining the benefits of k-d trees and octrees*, In 10th ACM-SIAM Symp. on Discrete Algorithms, pp. 300–309

Hernández, A.T.E, Cardinale, Y., Figueira, C. 2000, *SUMA: A scientific metacomputer*, In E. H. D'Hollander, J. R. Joubert, F. J. Peters, and H. Sips, editors, Parallel Computing: Fundamentals & Applications, Proceedings of the International Conference ParCo'99, pp. 566–573

Eastman, C.M. 1981, *Optimal bucket size for nearest neighbor searching in k-d trees*, Inform. Process. Lett., 12, pp. 165–167

Foster, I., Kesselman, C., Tsudik, G., & Tuecke, S. 1998, *A security architecture for computational grids*, In Proceedings of the 5th ACM Conference on Computer and Communications Security (CCS-98), pp. 83–92

Goodrich, M.T., Tsay, J.-J, Vengroff, D.E., & Vitter, J.S. 1993, *External-memory computational geometry*, In *Proc. 34th Annu. IEEE Sympos. Found. Comput. Sci.*, pp. 714–723

Haupt, T., Akarsu, E., & Fox, G. 1999, *WebFlow: A framework for Web based metacomputing*, Lecture Notes in Computer Science, 1593, p. 291

Hughes, M., & Hughes, C. 1998, *Applied Java Cryptography*, Manning Publications, Greenwich, CT

Johnson, B. 1993, *Treeviz: Treemap visualization of hierarchically structured information*, In *Proc. ACM Conf. on Human Factors in Computing Systems*, pp. 369–370

Knudsen, J.B. 1998, *Java Cryptography*, O'Reilly & Associates, Inc., Newton, MA

Nodine, M.H., Goodrich, M.T., & Vitter, J.S. 1996, *Blocking for external graph searching*, Algorithmica, 16(2), pp.181–214

Overmars, M.H., & van Leeuwen, J. 1982, *Dynamic multi-dimensional data structures based on quad- and k-d trees*, Acta Inform., 17, pp.267–285

Patten, C.J., Hawick, K.A., & Hercus, J.F. 1999, *Towards a scalable metacomputing storage service*, Lecture Notes in Computer Science, 1593, p. 350

Samet, H. 1990a, *Applications of Spatial Data Structures: Computer Graphics, Image Processing, and GIS*, Addison-Wesley, Reading, MA

Samet, H. 1990b, *The Design and Analysis of Spatial Data Structures*, Addison-Wesley, Reading, MA

Samet, H. 1990c, *Hierarchical data structures for spatial reasoning*, In L. F. Pau, editor, *Mapping and Spatial Modelling for Navigation*, pp. 41–58

Samet, H. 1992, *Hierarchical data structures for three-dimensional data*, In R. Vinken, editor, *From Geoscientific Map Series to Geo-Information Systems*, volume 122 of Geologisches Jahrbuch A, pp. 45–58

Samet, H. 1995, *Spatial data structures*, In W. Kim, editor, *Modern Database Systems, The Object Model, Interoperability and Beyond*, pp. 361–385

Schneier, B., 1996 *Applied Cryptography (Second Edition)*, John Wiley & Sons

Sproull, R.F 1998, *Refinements to nearest-neighbor searching in k-d trees*, Technical Report SSAPP 184, Southerland, Sproull and Associates

Vengroff, D.E., & Vitter, J.S. 1996, *Efficient 3-d range searching in external memory*, In *Proc. 28th Annu. ACM Sympos. Theory Comput.*

Science User Scenarios for a VO Design Reference Mission

Kirk D. Borne

Raytheon ITSS, NASA-Goddard Space Flight Center, Code 631, Greenbelt, MD 20771

Abstract. The knowledge discovery potential of the new large astronomical databases is vast. When these are used in conjunction with the rich legacy data archives, the opportunities for scientific discovery multiply rapidly. A Virtual Observatory (VO) framework will enable transparent and efficient access, search, retrieval, and visualization of data across multiple data repositories, which are generally heterogeneous and distributed. Aspects of data mining that apply to a variety of science user scenarios with a VO are reviewed. The development of a VO should address the data mining needs of various astronomical research constituencies. By way of example, two user scenarios are presented which invoke applications and linkages of data across the catalog and image domains in order to address specific astrophysics research problems. These illustrate a subset of the desired capabilities and power of the VO, and as such they represent potential components of a VO Design Reference Mission.

1. Science Requirements for Data Mining

One of the major functions of a Virtual Observatory (VO) is to facilitate data mining and knowledge discovery within the very large astronomical databases that are now coming on-line (or soon will be). A similarly important function of the VO is to facilitate linkages and cross-archive investigations utilizing these new data in conjunction with the rich legacy data archives that preceded them. The scientific teams that generate large (multi-Terabyte) databases cannot begin to tap their full scientific potential. Thus a significant portion of the astronomical research community and a comprehensive suite of research tools should be brought to bear on extracting the maximum scientific return for the huge investment in these large astronomical facilities, large surveys, and large scientific data systems. One approach to this problem can be identified as "data mining".

What is data mining and why is applicable to scientific research? Data mining is defined as *an information extraction activity whose goal is to discover hidden facts contained in databases*. Data mining has taken the business community by storm and the phrase has become a bit overworked to describe some fairly routine functions of marketing. Even so, there are consequently now a vast array of resources and research techniques available for exploitation by the scientific communities. It is useful therefore to examine a further categorization of data mining thrusts and their sub-components, since these are likewise applicable to the scientific exploration of large astronomical databases.

In the marketing community, data mining is used to find patterns and relationships in data by using sophisticated techniques to build models — abstract representations of reality. A good model is a useful guide to understanding that reality and to making decisions. There are two main types of data mining models: *descriptive* and *predictive*. *Descriptive* models describe patterns in data and are generally used to create meaningful subgroups or clusters. *Predictive* models are used to forecast explicit values, based upon patterns determined from known results. These models are applicable to scientific inquiry as well.

There is another differentiation of data mining into two categories that we find particularly appropriate to knowledge discovery in large astronomical databases: *event-based mining* and *relationship-based mining*. At the risk of trivializing some fairly sophisticated techniques, we classify event-based mining scenarios into four orthogonal categories:

- Known events / known algorithms — use existing physical models (*descriptive models*) to locate known phenomena of interest either spatially or temporally within a large database.

- Known events / unknown algorithms — use pattern recognition and clustering properties of data to discover new observational (in our case, astrophysical) relationships among known phenomena.

- Unknown events / known algorithms — use expected physical relationships (*predictive models*) among observational parameters of astrophysical phenomena to predict the presence of previously unseen events within a large complex database.

- Unknown events / unknown algorithms — use thresholds to identify transient or otherwise unique ("one-of-a-kind") events and therefore to discover new phenomena.

Similarly, for relationship-based mining, we identify three classes of association-driven scenarios that would find application in astronomical research:

- Spatial associations — identify events (astronomical objects) at the same location in the sky.

- Temporal associations — identify events occurring during the same or related periods of time.

- Coincidence associations — use clustering techniques to identify events that are reasonably co-located within a multi-dimensional parameter space.

Therefore, from this discussion, we can derive a reduced set of science requirements for data mining. These requirements correspond to the following set of exploratory approaches to mining large databases: *Object Cross-Identification*, *Object Cross-Correlation*, *Nearest-Neighbor Identification*, and *Systematic Data Exploration*. (a) "Object cross-identification" refers to the classical problem of connecting the source list in one catalog (or observation database) to the source list in another, in order to derive new astrophysical understanding of the cross-identified objects (*e.g.*, gamma-ray burst counterparts). (b) "Object cross-correlation" refers to the application of "what if" scenarios to the full suite of parameters in a database (*e.g.*, identify distant galaxies as U-band dropouts in a color-color scatter plot from the HDF survey). (c) "Nearest-neighbor identification" refers to the general application of clustering algorithms in multi-dimensional parameter space (*e.g.*, finding a new class of stars, L dwarfs, in the 2MASS database). (d) "Systematic data exploration" refers to the application of the broad range of event-based and relationship-based queries to a database in the hope of making a serendipitous discovery of new objects or a new class of objects (*e.g.*, finding new types of variable stars, such as "bumpers", in the MACHO database).

2. User Scenario #1: Estimating the Galaxy Interaction Rate

It is well established that a significant fraction of all galaxies have been involved in a galaxy-galaxy interaction and perhaps a merger at some time(s) in their past. The rate of these interactions is not yet well determined empirically: either the current rate for galaxies in the nearby Universe, or the cosmologically evolving rate in the distant Universe. Numerical simulations of the galaxy population and of the evolving hierarchical structure within various cosmological scenarios give a handle on the interaction and merger rates, which naturally depend on the choice of cosmological model. In general, the simulations confirm the importance and relatively high frequency of occurrence of interactions and mergers. Given the cosmological significance of interactions to galaxy formation and evolution, it is important to derive a firm value for the galaxy interaction rate observationally, for comparison with the numerical models, which will in turn help to narrow the plausible range of cosmological models, galaxy formation models, and galaxy evolution models.

We attempted an initial exploration of several on-line databases in order to estimate the galaxy interaction rate. We began by exploring an on-line catalog of galaxies (available through NASA's ADC = Astronomical Data Center): the Updated Zwicky Catalog of Falco *et al.* (1999). This catalog identifies multiple-galaxy groupings, which we used to reduce the full list of 19,000 galaxies to the set of 1800 multiples. We then selected a very small sub-sample from this list to conduct a proof-of-concept investigation. We used existing catalog visualization tools and archive linkage tools at the ADC to find all possible NASA mission data and most of the all-sky survey data for these selected objects (Kargatis *et al.* 1998). We then identified characteristics in the optical images or in the IRAS fluxes or in the X-ray emissions to verify that the associated multiple galaxy systems are in fact (to high probability) bound groups (pairs, triples, quartets, *etc.*). The expectation that these small galaxy-galaxy separations and other evidences for physical association do in fact imply an on-going interaction was often confirmed through inspection of the DSS (Digital Sky Survey) imagery, which showed signs of interaction in many cases (*e.g.*, distorted morphologies). Thus, by applying knowledge of astrophysical signatures of interactions, we were able to explore multiple databases (ADC catalogs, NASA mission archives, and ground-based sky surveys) in a coherent organized manner. We estimate that the galaxy interaction rate in the local Universe is approximately 8%.

3. User Scenario #2: In Search of the CIB

Among the exciting results of the COBE mission was the discovery by the DIRBE team of an extragalactic CIB (Cosmic Infrared Background; Hauser et al. 1988). There has been a storm of activity in the research community to identify the sources of the CIB and to understand their power sources (i.e., what powers the strong IR emissions? are they dust-enshrouded quasars? or dusty starbursts? or both?). Some possible counterparts to the CIB include: (a) ultraluminous IR galaxies (ULIRGs; see Sanders & Mirabel 1996 for a review); (b) SCUBA submm sources (Barger et al. 1999; Blain et al. 1999); (c) IR-selected AGN (a new population of AGN identified through the 2MASS survey; Beichman et al. 1998); or (d) Extremely Red Objects (EROs), which have been reported in several recent deep surveys (Smail et al. 1999; Thompson et al. 1999).

We initiated a proof-of-concept search scenario for identifying potential candidate contributors to the CIB. Our approach is similar to that of Haasrma & Partridge (1998), except that we are applying the full power of on-line databases and linkages between these databases, archives, and published literature. Our search scenario involved finding object cross-identifications among the IRAS Faint Source Catalog and FIRST survey catalog, and then attempting to find those commonly identified objects also within other databases, such as the HST observation log. In a very limited sample of targets that we investigated to test our "ADC as a mini-NVO" approach to the problem, we did find one object in common among the HST-IRAS-FIRST databases: a known hyperluminous infrared galaxy (HyLIRG) at $z=0.780$ harboring an AGN, which was specifically imaged by HST because of its known HyLIRG characteristics. In this extremely limited test scenario, we did in fact find what we were searching for: a distant IR-luminous galaxy that is a likely contributor to the CIB, or else similar in characteristics to the more distant objects that likely comprise the CIB.

The preliminary results of the investigations described above for User Scenarios #1 and #2 (including screen shots of the user interfaces employed in the studies) are available on-line[1].

References

Barger, A. J., Cowie, L. L., & Sanders, D. B. 1999, ApJ, 518, L5
Beichman, C. A., et al. 1998, PASP, 110, 367
Blain, A. W., et al. 1999, ApJ, 512, L87
Falco, E. F., et al. 1999, PASP, 111, 438
Haarsma, D. B., & Partridge, R. B. 1998, ApJ, 503, L5
Hauser, M., et al. 1998, ApJ, 508, 25
Kargatis, V. E., et al. 1999, ADASS VIII, 217
Sanders, D. B., & Mirabel, I. F. 1996, ARA&A, 34, 749
Smail, I., et al. 1999, MNRAS, 308, 1061
Thompson, D., et al. 1999, ApJ, 523, 100

---

[1] http://adc.gsfc.nasa.gov/adc/how_to.html

A Processing Automaton for Intensive Data

Jeffrey D. Scargle

NASA Ames Research Center, Moffett Field, CA 94035

James P. Crutchfield

Santa Fe Institute, 1399 Hyde Park Road, Santa Fe, NM 87501

Clark Glymour

Institute for Human and Machine Cognition, University of West Florida, Pensacola, FL 32514, and Carnegie Mellon University, Pittsburgh, PA 15213

Rattikorn Hewett

Institute for Human and Machine Cognition, University of West Florida, Pensacola, FL 32514

Abstract. Automated discovery of new astronomical phenomena and extraction of knowledge about familiar ones will be important for any Virtual Observatory. An autonomous data processor — transforming raw data streams into intelligible structures and then recognizing patterns in the structures — can achieve this grand objective within a restricted but broad scope, namely measurements of the intensity of a physical quantity distributed over a fixed domain.

1. Introduction

The scope of past and future observational projects, and the vast amount of data that will be collected into a Virtual Observatory, mandates the development of automatic, smart analysis engines capable of recognizing meaningful patterns without human intervention. Toward this end we are designing an intelligent, automatic knowledge extraction procedure for scientific data, consisting of the following steps:

1. Pre-processing data reduction (if necessary de-corrupting the raw data)

2. A generic, automatic processing step, yielding informative structures

3. Extraction of patterns from these structures (knowledge discovery and data mining-KDD)

4. Post-processing by humans

The major challenges lie in parts (2) and (3), but of course (1) and (4) are important for the design of the Processing Automaton. It is not known how effective such automatic discovery can be. We believe that by focusing on a restricted class of data (time series, images, spectra, and other *intensive* data — see the next Section) we can take this idea quite far and thereby make a major contribution to the analysis and discovery capabilities of a Virtual Observatory.

The fundamental idea is to transform raw data streams into simplified structures, readily amenable to human understanding, but containing essentially all of the statistically significant information. It is also important that the data itself be allowed, to some extent, to determine the form as well as the content of these structures. By avoiding arbitrary constraints, this adaptivity gives improved representation of intrinsic structure in the underlying physical process. For example, treating event data without binning avoids loss of time resolution and dependence on bin location (Scargle, 1998). In addition, adaptive analysis techniques are likely to be effective in a broader variety of applications.

2. The Data

We consider only intensive data — *i.e.*, measurements of the density of events in a well-defined space S of one or more dimensions. This class includes time series, other 1D sequential data (*e.g.*, spectra and genetic sequences), images, and time sequences of images. What these data types have in common is that they comprise sets of points in some space S of dimension D. For example, $D = 1$ corresponds to time series, $D = 2$ to images, and $D = 3$ to time sequences of images (movies). The quantities actually recorded are often event rates, averaged over bins or pixels. Our test-bed data include several NASA archives (RXTE, the Compton Gamma Ray Observatory data systems, and Mars Orbiter Camera), data from the Naval Research Laboratory-Stanford Linear Accelerator USA X-ray telescope[1], as well as genetic sequence data bases (Scargle, unpublished work on exon-intron boundary identification).

3. Segmentation Yields Structure

Consider this question: *What is the most likely partition[2] of S into sub-volumes within each of which the event rate is constant?* Here the event rates, being nuisance parameters, are marginalized. For Poisson-distributed events the resulting single cell likelihood is an explicit function of just two parameters: the volume of the cell and the number of events in it (Scargle, 1998). The cells are independent, so the total likelihood is the product of the likelihoods of the cells. This simple mathematical structure leads to a powerful algorithm for finding the optimum segmentation of time series data. A key problem, that of determining the number of segmentations, has recently been solved (Scargle, unpublished). Note that all of these results hold in spaces of arbitrary dimension; therefore

---

[1] http://xweb.nrl.navy.mil/www_hertz/usa/usa.html

[2] A *partition* is a division of the whole of S into non-overlapping cells.

generalization to more complex data structures is limited mainly by geometrical complexity.

4. Pattern Discovery

The above formalism yields a robust algorithm, quite successful at identifying a certain class of statistically significant patterns in noisy astronomical time series data. In 1D the patterns — primarily flares or other local outbursts (Scargle et al. 1993, Scargle, Norris and Bonnell 1998) — are straightforward. To deal with more complex patterns we are investigating techniques from knowledge discovery and data mining (KDD; Glymour et al. 1997) and from theoretical physics. Since the data stream has been converted from semi-continuous time, space, or wavelength series into discrete structures, many known KDD methods for cluster analysis, classification, pattern recognition, and the like may be applicable. In addition, we are investigating the following approaches for identification of coherent structures in segmented data.

4.1. A New Approach to Discovering Classification Rules

To avoid limitations of conventional "if-then" rules, typically based on some small subset of the problem variables, we have developed a technique that generates more powerful classification rules (Hewett and Leuchner, 1999, 2000) expressing complex relationships between ranges of variables (not just atomic values), and not constrained to be mutually exclusive.

4.2. Bayesian Networks

Bayesian Networks learned by applying Bayesian methods (e.g., Heckerman, 1997) or by identifying patterns in auto-correlations, or by combinations of the scoring and constraint methods (Glymour and Cooper 1999) are another construct that may be of use in this research, especially in view of our Bayesian approach to the front-end segmentation problem.

4.3. Computational Mechanics

Computational Mechanics (Crutchfield and Young 1989, Shalizi and Crutchfield, 1999) is another applicable technique in which one infers the structure of a stochastic state machine (or ϵ-machine) from time or space-time series data. The ϵ-machine representation is optimal and unique, and possesses other powerful mathematical properties. In addition to their optimality, We find them attractive here for three reasons. First, the algorithm yielding the ϵ-machine does not need to use the time series directly, but instead can use a symbolic dynamical representation, not unlike segmented or blocky structures. Second, the structure of the ϵ-machine represents dynamical equations of motion for the underlying physical process, thus coming very close to the ultimate goal of scientific understanding and theory building. Finally, the approach gives a quantitative way to uncover a process's structural complexity — a useful new invariant.

4.4. Routine Screening

Finally, while it may seem trivial, routinely screening intensive data for generic features does not seem to be in wide use. We incorporate automatic detection of patterns of dynamical significance, including local (occurring at specific times) and global (representing average properties over an interval of observation). Screening targets include periodic signals, quasi-periodic oscillations, "1/f-noise" — plus time delays and other cross-correlation or cross-spectral features in multi-channel data.

5. Summary

We are designing, and plan to implement, an automatic data processing engine that converts time series, images, and spectral data into segmented data structures, followed by discovery of anomalies, transients in time and space, and patterns of dynamical significance. This machinery will be data-adaptive and automatic, but also tunable to specific applications. We are extending the 1-D analysis to data spaces of higher dimension, developing priors suitable for a variety of data types, exercising and tuning the system on existing data bases, and studying synergies with de-noising, polishing, other pre-processing, and — not least — with the human element.

References

Crutchfield, J., and Young, K. 1989, Phys.Rev.Lett, 63, 105

Glymour, C., et al. 1997, in Data Mining and Knowledge Discovery, 1, 11

Glymour, C., and Cooper, G. 1999, Computation, Causation and Discovery, MIT/AAAI.

Heckerman, D. 1997, in Data Mining and Knowledge Discovery 1, 79-119

Hewett, R. and Leuchner, J. 1999, in Proceedings of International Conference on Artificial Intelligence and Soft Computing

Hewett, R. and Leuchner, J. 2000, submitted to Data Mining and Knowledge Discovery: an International Journal

Madigan, D., Raghavan, N., and DuMouchel, W. 2000, "Likelihood-based Data Squashing: A Modeling Approach to Instance Construction," preprint

Scargle, J. 1998, ApJ, 504, 405

Scargle, J., Norris, J., and Bonnell, J. 1998, in Gamma-Ray Bursts: 4th Huntsville Symposium, American Institute of Physics Conference Proceedings #428, 181

Scargle, J., Steiman-Cameron, T., Young, K., Donoho, D., Crutchfield, J., and Imamura, J. 1993, ApJ, 411, 91

Shalizi, C. and Crutchfield, J. 1999, "Computational Mechanics: Pattern and Prediction, Structure and Simplicity,"[3]

---

[3] http://www.santafe.edu/sfi/publications/Abstracts/99-07-044abs.html

Part 4
The US National Virtual Observatory Effort

Part 4

Summary of USNVO Activities

D. S. De Young

National Optical Astronomy Observatory, Tucson, AZ 85719

Abstract. Activities in the U.S. over the past year directed toward the realization of a virtual observatory are described. The genesis and evolution of the Virtual Observatory idea is traced, as is the growing involvement of universities, existing national data centers, national observatories, and federal funding agencies. Planned activities over the forthcoming year are described, together with a brief outlook at possible funding scenarios.

1. The Seminal Steps

The first formal manifestation of the National Virtual Observatory (NVO) in the U.S. appeared in the summer of 1999 in the form of an initial white paper written by Charles Alcock, Tom Prince, and Alex Szalay. This document captured the essence of many informal discussions that had taken place in the previous months among interested participants. The initial white paper laid out in a coherent manner the basic scientific motivation and technical drivers that make a Virtual Observatory both necessary and possible. It summarized the exponential growth in the size of datasets, the coming ubiquity of large scale surveys, the ever increasing cost of new observing facilities that precludes casual repetitions of previous observations, and the explosive growth in both computational capability and high bandwidth connectivity that enable a virtual observatory to come into being.

In addition, the initial white paper laid out some of the essential elements of a viable Virtual Observatory. These include the need for a federation of existing and future archives, a recognition of the essential distributed nature of the NVO, a description of the requirement for new levels of connectivity, and an outline of the requirements for new software tools that an effective NVO will require. Other elements necessary to a complete definition of an NVO were not defined in this first white paper; these awaited further discussions with the community and with the funding agencies. In particular, both a management plan and an implementation plan were not included, and in addition a science reference mission still had yet to be defined. Finally, a detailed budget plan was not included, because the NVO concept had not become well enough defined for such budgetary considerations to be meaningful.

2. Community Engagement

The next major step in the definition of a US NVO took the form of a Workshop, organized by Alex Szalay and Tom Prince, that was held at the Johns Hopkins University on November 15–16, 1999. This workshop was open to all interested participants, and it capitalized on the growing interest in an NVO that had been generated by the circulation of the Alcock, Prince, and Szalay white paper. The workshop attracted 18 participants from eight institutions. Major topics discussed included a communication among participants of the interests they represented with a view to determining common goals, descriptions of the various areas of expertise brought to the meeting, a history of the NVO concept to date, areas of further definition needed for the NVO, and an introduction of technical issues. There were several major outcomes of this workshop. First among these was a discussion that resulted in a preliminary definition of NVO functions, together with an outline of a possible management structure. Further discussions resulted in a very preliminary sketch of an implementation plan, and rough estimates were established for equilibrium staffing levels and budgetary requirements. Thus this workshop made significant headway in addressing those issues of definition, management, and implementation that had not been included in the original white paper. A final very important outcome of this first NVO workshop was the establishment of a path to the future through the formation of working groups to provide further development and to report back to the community. Three working groups were established: Science Justification (Djorgovski, Strom); Management Structure (De Young, Hanisch, Lonsdale); and Technical Issues (Good, Hanisch, Tody).

In the ensuing months from November 1999 to February 2000, while the working groups met to further define the scientific, managerial, and technical issues, other events of significance to the NVO took place. The first of these was a Portfolio Allocation Review held at the National Science Foundation; this review provided an examination of the asset allocation mix within the NSF, and the review endorsed the underlying concepts behind the NVO idea. A second, perhaps more significant, and certainly more concrete, event occurred with the publication of the President's Budget for FY 2001. This budget request contains a 13% increase in the budget of the National Science Foundation, and most significantly, the only new item explicitly called out in this budget request over and above the base budget is for a "National Virtual Observatory". This constituted the first explicit, formal recognition of the NVO concept by any US funding agency, and it represented an essential first step in the realization of the NVO. During the same period, the final form of the AASC Decadal Survey Report to the US National Academy was taking shape. As this process neared completion, it became clear that this very influential report would also highlight and endorse the NVO concept, and in fact the final committee report places the NVO as the highest ranked of the recommended "small" category projects to be funded in the next decade. This endorsement and recommendation provides a key impetus to the NVO implementation in the minds of the Congress, the funding agencies, and the US astronomical community.

A final event of importance in this time period occurred on February 17, 2000, when a teleconference was held among representatives from the national data centers, national astronomy centers, and universities vitally interested in the creation of the NVO. The motivation for this teleconference was to provide a coherent community response to the above events, and a major outcome of this meeting was the creation of an Interim Steering Committee for the NVO initiative. This group was formed in order to more effectively establish liaison, coordination, and communication among all those interested contributing to the efforts to bring the NVO into being. In addition, this meeting produced a consensus that a new version of the NVO White Paper needed to be created in order to provide a document suitable for preliminary consideration by the funding agencies. This new white paper would incorporate the conclusions reached by the working groups on science, management, and technical issues as well as updating some of the arguments made in the first version. A final outcome of this meeting was to establish a rough agenda for a second NVO Workshop to be held in Tucson.

The Second NVO Workshop was hosted by the National Optical Astronomy Observatory and was held in Tucson, February 28–29, 2000. The growing interest in the NVO was evident in the increased participation in this workshop when contrasted with the first NVO workshop. The Tucson meeting drew 28 participants from 12 organizations. The workshop opened with a summary of the recent events described above, together with a discussion of the objectives of this workshop. Major areas of presentation and discussion included several illuminating science scenarios, discussion of the principal elements of NVO infrastructure, and descriptions of the major NVO technical components. The workshop split into two breakout sessions in the afternoon of the first day; these groups were given the tasks of developing an initial design reference mission and of laying out a staged implementation plan for the NVO. The second day included a summary session together with an extensive briefing and discussion with Dan Weedman of the NSF. The principal outcomes of the second workshop were, first and foremost, the definition of a clear path to the completion of the second NVO white paper. This included major structural elements that emerged from the science presentations and from the working groups on implementation plans and the design reference mission. In addition, a specific timetable and writing assignments for sections of the white paper were agreed upon at this meeting. The interaction with the NSF also provided a more clear definition of budget levels that could be expected from this funding agency. Finally, significant planning for the June VO meeting was accomplished during this workshop.

3. Recent Events: March–June 2000

Under the continuing coordination and oversight by the interim NVO Steering Committee, the first meeting between representatives from NASA and NSF to discuss the NVO initiative was held in Washington on April 27, 2000. Three principal points of emphasis emerged from this informal meeting. First, both agencies want to see detailed implementation plans for the NVO. Second, it must be made clear that the NVO is science driven, and that it is not just a technological showpiece. Third, the links to current information technology initiatives must be made very clear in presenting the NVO project to the funding agencies. With this valuable guidance in hand, the Steering Committee continued to refine the final stages of the new NVO white paper. Several draft versions of this document were circulated to the members of the Steering Committee for comments and criticism. This process was essential to obtain widespread community support for the content of the white paper so that it could be seen by both the community and the funding agencies as representing a consensus view.

Presentation of the second white paper was made to a joint meeting of NSF and NASA representatives at NASA Headquarters on 31 May 2000. At this critical meeting, both agencies expressed strong support for the NVO, and both agencies found the white paper very useful in their efforts to advance the cause of the NVO within their respective funding structures. It was clear from the remarks made that NASA will not be able to come forward with significant funding until FY 2002, though some study funds may be available in FY 2001. On the other hand, NSF is prepared to move forward now, but with some restrictions. At this time, any significant NSF funding in FY 2001 must come via a proposal process that flows through the NSF Information Technology initiative. The NSF is currently working on the language of the Announcement of Opportunity for such proposals; once this AO is released, pre-proposals will be due 90 days later. Given these conditions, a major objective of the discussions at this June VO meeting will be to refine our strategies and timetable in order to maximize the success of the Virtual Observatory initiative.

The Design Reference Mission for a Virtual Observatory

Todd Boroson
National Optical Astronomy Observatory, P.O. Box 26732, Tucson, AZ 85726-6732

Abstract. A Design Reference Mission (DRM) must be developed for the National Virtual Observatory (NVO), in order to derive the NVO's functional requirements from its science drivers and to provide metrics by which its success may be measured. A process is proposed for community involvement in this activity. Three possible science projects have been sketched out as a start on the DRM. These include (1) A Panchromatic Census of Active Galactic Nuclei, (2) Formation and Evolution of Large Scale Structure, and (3) The Digital Galaxy.

1. Purpose and Structure of the Design Reference Mission

An important component of the development effort for a large-scale mission or facility such as the National Virtual Observatory (NVO) is the Design Reference Mission (DRM) or Science Reference Mission. This is a series of realistic science projects that could be carried out with the proposed new facility.

The DRM serves a number of purposes, including:

- increasing the involvement of the potential user community, creating support for the facility, and justifying resource requirements;

- establishing a target set of accomplishments with acknowledged scientific value so that the degree of success of the facility can be measured;

- providing a list of functional requirements through a flowdown from the steps necessary to execute the scientific projects;

- predicting the supporting capabilities that might be needed elsewhere in the community to ensure most effective use of the facility.

In order to serve these purposes, the DRM is defined by the following characteristics.

The DRM consists of a set of science projects that are ambitious but credible. These specific projects may not be ones that are actually executed with the NVO, because the community's view of what is interesting or important will change with time. The projects should be ambitious because the functional performance of the NVO will be derived from them.

Each project in the DRM lays out a set of scientific objectives. The background provides context for a discussion of the scientific merit. Why is this project interesting? Is this relevant to our current understanding and interest in this subfield, and is it important in a broader sense? Why would one choose to spend resources solving this particular problem? This section is much like the scientific justification of a proposal to use a telescope or other oversubscribed facility.

In addition, within the section on scientific objectives, there should be a discussion of the relationship of this facility to others in the context of the scientific program. This discussion should lay out why this project could only reasonably be done with the proposed NVO, and how the existence of the NVO makes other facilities more effective with respect to solving this scientific problem.

The next major part of each science project is a description of how it would actually be carried out. This identifies specific data sets, discusses the processing steps and the analysis tools that are needed. It may include decision points at which various outcomes would be treated in different ways, requiring different kinds of follow-up observations or analysis. This section speaks to what sort of supporting or auxiliary data is needed to complete the project.

The last part of the description of each science project is its interpretation into functionality for the NVO. This part takes the sequence of operations in the preceding section and converts them into required capabilities for the NVO. For instance, the need to combine two catalogs at different wavelengths would require the capability to cross-identify objects, understanding what to do when the catalog resolutions or depths differ. The magnitude of a task (how large a dataset one must process) may be used to specify the level of performance that the NVO must achieve.

The DRM concludes with a description of the NVO that includes the capabilities needed to carry out all the science programs. Specific required functions may be generalized in order to accomodate the needs of several projects. Relative priority and sequencing may be derived from the likelihood and timing within the context of the science programs.

The DRM, structured as described above, must receive wide public dissemination if it is to serve its purpose. It has been clear that the NVO is a qualitatively new type of facility and discussions about scientific objectives have been limited by the difficulty of imagining its potential. The DRM provides a context for extending that discussion and making the capabilities real.

2. A Community-Based Process for Development of the DRM

While it has been the hope of the NVO Interim Steering Committee that there will be wide public participation in the development of the DRM, it was necessary to provide some examples of the type of science that the NVO could accomplish for discussion by the community, the technical developers, and the funding agencies. The three example projects laid out in the next sections of this paper are "strawman" science programs included in the NVO white paper. They are only the skeletons of real DRM science projects, and these or other projects will need substantial work to fulfill the role of the DRM.

The first step in the public discussion is this presentation to the Virtual Observatories of the Future meeting participants. This should give this group a chance to begin thinking about the development of the elements of the DRM and its implication for the NVO. Following this meeting an effort will be made to assemble several working groups each with several scientists and technical members to invent, develop, and refine projects for the NVO. This activity will culminate in a small meeting of representatives of these groups to discuss and assemble the DRM out of the working group activities. Following this the DRM will be made available for discussion and further contributions by the broad community.

3. A Strawman DRM for the NVO

Following are three representative science programs for the DRM of the NVO.

3.1. Project 1. A Panchromatic Census of Active Galactic Nuclei

An understanding of the nature and characteristics of Active Galactic Nuclei (AGN) is important both because their luminosities make them visible to large cosmological distances and because they represent a fundamental stage in the evolution of galaxies. Observationally, AGN are distinguishable from stars in that their spectral energy distributions are much broader than black-body functions. However, redshift, variability, (possibly large) obscuration, and a range of intrinsic spectral shapes and characteristics result in a great range of "colors" over wavelengths from x-ray to radio.

Scientific Goals This project aims to construct a complete sample of AGN in order to:

- Test the idea that observable properties are determined by extrinsic factors such as orientation or obscuration (so-called "unification" models).

- Compare the environments of AGN as a function of type, *e.g.*, radio properties *vs.* membership in clusters of galaxies.

- Understand the evolution of the AGN luminosity function, and, in particular, separate density evolution from luminosity evolution.

- Construct the AGN luminosity function for different wavelengths in order to understand the evolution of AGN properties in a statistical sense.

Project Steps

1. Federate a number (N) of surveys covering the same (significant) area on the sky, and together, spanning a large wavelength range (X-ray through radio).

2. Include metadata information so that survey selection effects and constraints can be accounted for. Relevant metadata will include survey area, bandpasses, flux limits, *etc.*

3. Identify distinguishable "clouds" of objects in N-dimensional color space in the resulting joint dataset.

4. Apply a priori astronomical knowledge (*e.g.*, published catalogs, theoretical models) to understand the populations of these clouds.

5. Identify AGN candidates in the joint dataset. Note that confirmation may require new observations or the process might be carried out statistically, resulting in a probability that any particular object is an AGN.

6. Use catalog properties and/or further measurements from image databases to address science questions.

NVO Functionality This project requires the federation of relevant surveys (at a number of wavelengths) including cross-identification of objects in multi-wavelength surveys and interchange/merging of metadata. Since a complete census is desired, the treatment of objects that are detected in some but not all of the surveys will be critical.

The generation of candidate lists requires cluster analysis to identify clouds and sequences. This will include both supervised analysis (in which astronomical knowledge guides the definition and analysis) and unsupervised analysis (in which new patterns are recognized).

A fundamental aspect of the analysis will involve visualization of multi-dimensional datasets.

It is likely that unambiguous identification of all candidate AGNs will not be possible, and tools to perform statistical classification of the populations of defined regions in the multi-dimensional parameter space will be needed.

3.2. Project 2. Formation and Evolution of Large-Scale Structure

Clusters of galaxies represent the largest unambiguous mass concentrations known. Various models of the early evolution of the universe make different predictions for how clusters form and evolve. These models can be tested by comparing their predictions with observed mass and luminosity functions of clusters.

Scientific Goals This project aims to construct an unbiased sample of clusters of galaxies over a cosmologically significant redshift range in order to test various structure formation and evolution models by comparison with evolution of the observed mass and luminosity functions with redshift. In addition, the sample can be used to study the morphology-density relation for galaxies and its evolution over interesting timescales.

Project Steps

1. Create statistical cluster samples using multi-wavelength pixel data in a number of different ways:

    X-ray surveys: identify cluster signature from emission of hot gas in image data.

    Optical/IR surveys: convolution of image data with kernel designed to select clusters

    Millimeter surveys: identify clusters from variation in CMB temperature caused by Sunyaev-Zeldovich effect

    Radio surveys: identify clusters based on presence of radio source morphologies indicative of cluster environment

2. Compare the results of different selection techniques. Quantify the selection effects as functions of cluster mass, density, redshift, *etc.*

3. Use various distance indicators or redshift estimators to supplement measured properties of clusters.

NVO Functionality This project requires the capability to operate on large quantities of imaging data with user-defined algorithms and tools.

Once candidate clusters are identified, the (unknown) effects of the different selection techniques must be evaluated. This would be done by constructing simulated surveys to understand the selection effects in each case; conversely, the user-defined tools may be tested on the simulations.

Given the selection effects, predictions of observed sample properties must then be compared with the observed samples, using statistical tools.

3.3. Project 3. The Digital Galaxy

The Galaxy is composed of a number of structural elements: halo, thin disk, thick disk, bulge, spiral arms. Each of these is characterized by populations of stars that have correlated distributions in age, mass, chemical composition, and orbit (position and kinematics), as well as distributions of non- stellar material such as gas and dust.. These are the fossil tracers of the formation processes. A complete understanding of these, in a global context, has never been possibly because of the difficulty of studying samples of the size needed to disentangle all the variables simultaneously.

Scientific Goals This project aims to construct very large samples of galactic stars together with as much information about the physical properties of each object as can be derived. These datasets, together with maps of non-stellar components of the Galaxy, will be used to:

- Generate a parameterized model of the Galaxy, including positional and kinematic information.

- Confront this model with models for the structure of the Galaxy, based on various formation processes.

- In particular, search for comoving groups that are representative of merger events or tidal debris tails.

Project Steps

1. Federate various optical and IR surveys to generate matched catalogs of stars.

2. Use positions, magnitudes, and colors to construct three-dimensional stellar distributions. This will require using colors to derive luminosity classes and estimate extinction.

3. Quantify dust distribution and obscuration using FIR, H I, and CO images.

4. Iterate with 2) until consistent.

5. Identify bulk flows and sties of star formation using IR and radio images.

6. Use proper motion surveys (and radial velocity information, when available) to deduce motions of subsets of stars.

7. Use multi-epoch imaging to find variables. Use these to provide a distance check.

NVO Functionality This project requires the federation of multi-wavelength and multi-epoch catalog data with consideration of metadata. The image data will be operated on with user-defined tools

Visualization tools will be needed for large multi-dimensional datasets. Components of the galaxy and the discovery of coherent groups of objects will require statistical analysis tools.

Acknowledgments. I wish to thank S. Strom, S.G. Djorgovski, and R.J. Brunner for help in developing the example science projects described above.

Virtual Observatories of the Future
ASP Conference Series, Vol. 225, 2001
R.J. Brunner, S.G. Djorgovski, and A.S. Szalay, eds.

Toward a National Virtual Observatory: Science Goals, Technical Challenges, and Implementation Plan[1]

Abstract. The National Academy of Science Astronomy and Astrophysics Survey Committee, in its new Decadal survey entitled *Astronomy and Astrophysics in the New Millennium*, recommends, as a first priority, the establishment of a National Virtual Observatory. The NVO would link the archival data sets of space- and ground-based observatories, the catalogs of multi-wavelength surveys, and the computational resources necessary to support comparison and cross-correlation among these resources. This White Paper describes the scientific opportunities and technical challenges of an NVO, and lays out an implementation strategy aimed at realizing the goals of the NVO in cost-effective manner. The NVO will depend on inter-agency cooperation, distributed development, and distributed operations. It will challenge the astronomical community, yet provide new opportunities for scientific discovery that were unimaginable just a few years ago.

1. Executive Summary

Technological advances in telescope and instrument design during the last ten years, coupled with the exponential increase in computer and communications capability, have caused a *dramatic and irreversible change in the character of astronomical research*. Large scale surveys of the sky from space and ground are being initiated at wavelengths from radio to X-ray, thereby generating vast amounts of high quality irreplaceable data.

The potential for scientific discovery afforded by these new surveys is enormous. Entirely new and unexpected scientific results of major significance will emerge from the combined use of the resulting datasets, science that would not be possible from such sets used singly. However, their large size and complexity require tools and structures to discover the complex phenomena encoded within them. We propose establishing a **National Virtual Observatory** that can meet these needs through the coordination of diverse efforts already in existence as well as providing focus for the development of capabilities that do not yet exist. The NVO will act as an enabling and coordinating entity to foster the

---

[1]This draft white paper was prepared by the informal NVO interim steering committee, whose members include T. Boroson (NOAO), R. Brissenden (CXC), R.J. Brunner (CIT), T. Cornwell (NRAO), D. De Young (NOAO), S.G. Djorgovski (CIT), R. Hanisch (STScI), G. Helou (IPAC), C. Lonsdale (IPAC), T. Prince (CIT), E. Schreier (STScI), S. Strom (NOAO), A. Szalay (JHU), D. Tody (NOAO), and N.White (HEASARC). This version of the white paper has been modified by the editors to conform to the style guidelines for these proceedings, and is included here by agreement of the NVO interim steering committee. The original version, dated June 8, 2000, is available from the conference web site: http://astro.caltech.edu/nvoconf/.

development of tools, protocols, and collaborations necessary to realize the full scientific potential of astronomical databases in the coming decade. When fully implemented, the NVO will serve as *an engine of discovery for astronomy.*

The new scientific capabilities that will be enabled by the NVO are essential to realize the full value of the terabyte and petabyte datasets that are in hand or soon to be created. Rapid querying of large scale catalogs, establishment of statistical correlations, discovery of new data patterns and temporal variations, and confrontation with sophisticated numerical simulations are all avenues for new science that will be made possible through the NVO. In addition, the NVO and its data archives will require *collaborations with the computer science community,* will provide opportunity for collaboration with other disciplines facing similar challenges, and will be *a venue for educational outreach.* Three examples of scientific programs involving Active Galactic Nuclei, the Large Scale Structure of the Universe, and the structure of our Galaxy illustrate the scientific promise of the NVO. *The NVO will be technology-enabled, but science-driven.*

Implementation of the NVO involves significant technical challenges. These include both the incorporation of existing data archiving efforts in astronomy as well as the development of new capabilities and structures. Major technical components to the NVO include archives, metadata standards, a data access layer, query and computing services, and data mining applications. Development of these capabilities will require close interaction and collaboration with the information technology community.

The implementation plan for the NVO is defined by four phases, beginning with activities initiated prior to the establishment of the NVO and leading to the fully operational phase of the NVO four–five years after its inception. This implementation plan is designed to begin placement of deliverables and capability to the community at the earliest possible time. This early functionality is essential to the success of the NVO.

2. Introduction: Winds of Change

For over two hundred years, the usual mode of carrying out astronomical research has been that of a single astronomer or small group of astronomers performing observations of a small number of objects. In the past, entire careers have been spent in the acquisition of enough data to barely enable statistically significant conclusions to be drawn. Moreover, because observing time with the most powerful facilities is very limited, many astrophysical questions that require a large amount of data for their resolution simply could not be addressed.

This approach is now undergoing a dramatic and very rapid change. The transformation is being driven by technological developments over the last decade that are without precedent. The major areas of change upon which this revolution in astronomy rests are advances in telescope design and fabrication, the development of large scale detector arrays, the exponential growth of computing capability, and the ever expanding coverage and capacity of communications networks.

The advances in telescope design and fabrication have made possible the great space based observatories, opening new vistas in gamma ray, X-ray, optical and infrared astronomy. Advanced technology has also made possible the establishment of a new generation of large aperture ground based optical and IR telescopes as well as the design and construction of single dish and multi element arrays operating at millimeter and centimeter wavelengths. At optical and infrared wavelengths these advances have been coupled with the development of extremely sensitive, high resolution detector arrays of ever increasing size, and the ability to mosaic these arrays has resulted in instruments with fields of view of order 30 arcminutes and with $\sim 10^8$ pixels per image. These technical developments continue to mature, with more sophisticated and larger aperture telescopes being planned in space and on the ground, using ever larger and more capable arrays of detectors incorporated into advanced instrumentation. Just as Moore's Law reflects the exponential increase in computing capability with time, the technological developments in observational astronomy over the last decade have in effect introduced *a Moore's Law for astronomy as well*.

The emergence of more astronomical facilities, on the ground and in space, with larger apertures and more sophisticated instrumentation, will have a critical and inevitable consequence: an enormously increased flow of data. For example, the current data production rate of HST is about 5 Gigabytes per day; but a facility recently recommended for construction by the AASC Decadal survey—the Large-Aperture Synoptic Survey Telescope—could produce up to 10 terabytes per day!

In addition to this increased data rate, *the manner in which observations are being made is also changing*. Although the new observatories in space and on the ground still devote a significant fraction of their time to research in the "single observer/single program" mode where small blocks of time are allocated to many specifically targeted research programs, more time is now being devoted to large scale surveys of the sky, often at multiple wavelengths, that involve large numbers of collaborators.

These large survey programs will produce coherent blocks of data obtained with uniform standards, and with the amount of data often measured in terabytes. This *paradigm shift* has been made possible not only by the increased capabilities of the new facilities that permit much faster acquisition of data, but also by the availability of computational hardware and software that make it possible to acquire, reduce, and archive this data.

A major technological development that will change the character of astronomical research is the advent of high speed information transfer networks with broad coverage. Although the rapid transfer of large amounts of data over common networks is currently unacceptably slow (over 20 *days* to transfer a 1 terabyte data set), future networks will be much faster. The availability of these data rates, together with the high efficiency of data acquisition at both ground and space based facilities, will make possible the efficient transmission of large amounts of data to many different sites. This technology will also enable access to specific subsets of data by an extensive user community that prior to this had no readily available access to these data; the potential scientific yield resulting from this accessibility will be enormous.

It is clear that all of these technological drivers will result in *an unprecedented flow of astronomical data* in the coming years. Moreover, *these data sets will be very different*, in that most of them will be in the form of coherent surveys, often at multiple wavelengths, over significant portions of the sky. Hence they will have *a richness and depth that is unprecedented*, and they will present unique opportunities for application to a variety of scientific programs by a wide range of users. This aspect alone makes the systematic archiving of these data essential. In addition, the data will be obtained through use of costly state of the art facilities that will be highly oversubscribed, and this will essentially preclude repetitions of observations previously made; this also argues for a general archiving of these data.

The existence of such archives, containing multiwavelength data on hundreds of millions of objects, will clearly create a demand within the astronomical community for access to the archives and for the tools necessary to analyze the data they contain. Opportunities for data mining, for sophisticated pattern recognition, for large scale statistical cross correlations, and for the discovery of rare objects and temporal variations all become apparent.

In addition, for the first time in the history of astronomy, such data sets will allow meaningful comparisons to be made between sophisticated numerical simulations and statistically complete multivariate bodies of data. The rapid growth of high speed and widely distributed networks means that all of these scientific endeavors will be made available to the community of astronomers throughout the US and in other countries.

These technological developments have converged in the last few years, and they will *completely alter the manner in which most observational astronomy is carried out*. These changes are inevitable and irreversible, and they will have dramatic effects on the sociology of astronomy itself. Moreover, there is a growing awareness, both in this country and abroad, that the acquisition, organization, analysis and dissemination of scientific data are essential elements to *a continuing robust growth of science and technology*. These factors make it imperative to provide a structure that will enable the most efficient and effective synthesis of these technological capabilities. Hence there is a need now for an entity such as a National Virtual Observatory to oversee the rational disposition of the growing body of astronomical data.

3. The Vision of a National Virtual Observatory

3.1. Structure and Function: Enabling New Science

The NVO is an entity that will enable advances in astronomy and astrophysics previously unattainable. It will be a key ingredient in establishing a new Age of Discovery in astronomy. With its conjugation of terabyte data archives, image libraries of millions of objects at wavelengths from gamma rays to radio frequencies, sophisticated data mining and analysis tools, access to terascale supercomputing facilities with petabyte storage capacities, and very high speed connectivity among major astronomical centers, the NVO will be unique. It will make possible rapid querying of individual terabyte archives by thousands of researchers, enable visualization of multivariate patterns embedded in large catalog and image databases, enhance discovery of complex patterns or rare phenomena, encourage real time collaborations among multiple research groups, and allow large statistical studies that will for the first time permit confrontation between databases and sophisticated numerical simulations. It will facilitate our understanding of many of the astrophysical processes that determine the evolution of the Universe. It will enable new science, better science, and more cost effective science. The NVO will act as a coordinating and enabling entity to foster the development of tools, protocols, and collaborations necessary to realize the full scientific potential of astronomical databases in the coming decade. *It will be an engine of discovery for astronomy.*

To accomplish this, the NVO will first and foremost be built as a science driven, community effort with a major fraction of the funding disbursed via a peer review process. This would be accomplished through regular announcements of opportunity for both software projects that develop NVO infrastructure and for science activities that utilize the NVO. More specifically, NVO activities to fulfill its role would include:

- Establishment of a common systems approach to data pipelining, archiving and retrieval that will ensure easy access by a large and diverse community of users and that will minimize costs and times to completion;

- Enabling the distributed development of a suite of commonly usable new software tools to make possible the querying, correlation, visualization and statistical comparisons described above;

- Coordinating the establishment of high speed data transfer networks that are essential to providing the connectivity among archives, terascale computing facilities, and the widespread community of users;

- Facilitating productive collaborations among astronomy centers and major academic institutions, both national and international, in order to maximize productivity and minimize infrastructure costs;

- Ensuring communication and possible collaborations with scientists in other disciplines facing similar problems, and with the private sector;

- Maintaining a continuing program of public and educational outreach that capitalizes upon the unique resources, in both data and software, of the NVO to provide a unique window into astronomy and scientific methodology.

3.2. Design Philosophies

The NVO will be a unique entity, primarily because its operation will be distributed and will be based upon rapidly developing technologies in communication and computer science. In order to ensure its continuing vitality, the NVO must embrace several major themes.

- The NVO must be **evolutionary**. From its inception, this evolutionary nature and will enable the NVO to respond quickly to changing technical and scientific opportunities and community requirements. Because of the continuing evolution in technical capabilities, this evolutionary nature will be an integral part of the NVO throughout its existence. An immediate consequence of this agility is the need for a management structure that both manages the distributed development efforts of the NVO and that moves quickly to exploit new possibilities. Management and oversight must be effective, efficient, visionary, and accountable to the community, yet minimize overhead and inertia.

- The NVO must be **distributed** in nature. Significant amounts of expertise are already in place at existing centers, and full advantage will be taken of this from the outset. In addition, the most economical and effective progress toward the goals of the NVO may well be realized in its operational phase through a distributed approach. This would entail location at existing centers and at future data centers those key areas of NVO functions that are most effectively carried out at those centers.

- The NVO must be **integrated**. Complementary to its distributed structure will be an enduring theme of integration as the NVO evolves. In order for the NVO to be most effective in facilitating scientific advances, the information technology functions must be integrated over all wavelengths and over space based and ground based facilities. In addition, integration with developments in computer science and information technology will be an essential element of the NVO.

- The NVO must provide **outreach**. The vast datasets and the accompanying analysis tools that will be available through the NVO provide an opportunity for educational and public outreach on a level that has not been possible in the past. An active outreach program that takes full advantage of the educational potential of the NVO resources must be implemented at all stages of the NVO development.

- The NVO must be **globally oriented**. A continuing aspect of the NVO will be its international links to similar efforts in other countries. Though it will not initially be an international collaboration itself, it is clear that the NVO must maintain communication, and collaborations when appropriate, at all levels with these other activities. It seems inevitable that NVO engendered activities will become a worldwide phenomenon.

- The NVO must **provide a path to the future**. The vision of NVO described here is primarily that of a catalytic and enabling entity, with minimal structure and enormous connectivity. A direct product of this will be the enhancement of scientific productivity in astronomy to new levels. However, a larger and perhaps more enduring legacy of the NVO will be its role in the establishment of an **astronomy information infrastructure** within the US and throughout the world. The growth of this infrastructure, expedited by the NVO, will provide unprecedented new vistas and opportunities for astronomical research in the future.

4. The Scientific Case for the NVO

As we look ahead, the astronomical community stands poised to take advantage of the breathtaking advances in computational speed, storage media and detector technology in two ways: (1) by carrying out new generation surveys spanning a wide range of wavelengths and optimized to exploit these advances fully; and (2) by developing the software tools to enable discovery of new patterns in the multi-terabyte (and later petabyte) databases that represent the legacies of these surveys. In combination, new generation surveys and software tools can provide the basis for enabling science of a qualitatively different nature.

Moreover, *the inherent richness of these databases promises scientific returns reaching far beyond the primary objectives of the survey:* for example, repeated imaging surveys aimed at developing a census of Kuiper Belt objects can provide the basis for discovering supernovae at $z > 1$. Indeed, the multiplier effects of survey databases are enormous as exemplified by the world-wide explosion of research ignited by the Hubble Deep Field.

We now have the tools to carry out surveys over nearly the entire electromagnetic spectrum on a variety of spatial scales and over multiple epochs, all with well-defined selection criteria and well-understood limits. The ability to create panchromatic images, and in some cases digital movies of the universe, provide *unprecedented opportunities for discovering new phenomena and patterns that can fundamentally alter our understanding.* In the past, a panchromatic view of the same region of sky at optical and radio wavelengths led to the discovery of quasars. The availability of infrared data led to the discovery of obscured active galactic nuclei and star-forming regions unsuspected from visible images. Repeated images of the sky have led to the discovery of transient phenomena—supernovae, and more recently, micro-lensing events—as well as deeper understanding of synoptic phenomena. The joining together of various large scale digital surveys will make possible new explorations of parameter space, such as the low surface brightness universe at all wavelengths.

The challenges of discovering new patterns and phenomena in huge astronomical databases find parallels in the medical, biological and earth sciences. For example, the size of the human genome is roughly 3 GB, while a digital all sky survey will be about 10 TB in size. The development of tools and techniques to handle astronomical datasets of this size will clearly have to call upon new developments in computer science and will, when developed, have applications to fields outside astronomy. In all cases, the full power of these databases cannot be tapped without the development of new tools and new institutional structures that can consolidate disparate databases and catalogs, enable access to them, and place analysis tools in the hands of a broad community of imaginative scientists. It is this vision that motivates the creation of the NVO.

The major capabilities of the NVO that need to be established in order to enable its scientific goals include the ability to:

- Federate existing large databases at multiple wavelengths and create tools to query them in both the catalog and the pixel domain;

- Develop universal standards for archiving future large databases;

- Provide a framework for incorporating new databases, thus minimizing the cost of new surveys and experiments and maximizing their scientific return;

- Develop analysis tools for discovery in catalog datasets and for statistical analysis of resulting joint datasets;

- Develop tools for object classification in the image datasets;

- Develop tools for visualization in both catalog and image databases;
- Develop new approaches to querying image databases and for image analysis and pattern recognition;
- Incorporate the results of sophisticated numerical simulations and develop a statistical "toolbox" for confronting these simulations with data; and
- Link with existing and future digital libraries and journals.

All of the above are possible, and all are qualitatively different from what we now do because of size, dimensionality, and complexity. Over time, the NVO can evolve to carry out all these functions. However, while enabled by technology, the NVO is not driven by technology. Instead, its structure and evolution are fundamentally driven by science and the needs of the scientific community. A major tool for guiding decisions about developing an NVO capability and the pace of implementing these capabilities will be that of a community-developed "Science Reference Mission" (SRM) for NVO. We envision a structured process to develop the SRM for NVO comprising:

- A broad community discussion at a workshop to be held in Pasadena during 13–16 June, 2000;
- Discussion among multiple community working groups identified at the Pasadena workshop and charged with developing:
  1. The details of a major scientific program enabled by the technical possibilities outlined above;
  2. An explanation of the scientific merit of the program as well as a discussion of the difficulty of accomplishing the same result without the NVO;
  3. An understanding of the flowdown from the needs of the science program to archive, archive access, and software tool requirements; and
  4. Prioritization of requirements.
- A meeting among the chairs of the community working groups and the interim NVO steering committee to combine the input from the working groups into a coherent SRM, complete with a science-to-requirements flowdown for NVO.

In preparation for this process, we have developed a few examples of science programs that give both a sense of the possibilities for discovery enabled by the NVO and of the initial basis for defining a framework and cadence for their implementation. We caution that these examples are as yet incomplete, and that they will require substantial community effort and input to convincingly demonstrate the potential power of the NVO and its required functions. In the material presented here, we emphasize the flowdown from science to capabilities because we believe that this step is necessary to understand how a complete description of the NVO will be obtained.

Example #1: A Panchromatic Census of Active Galactic Nuclei (AGN)

Background: An understanding of the nature and characteristics of AGN is important both because their luminosities make them visible to large cosmological distances, and because they represent a fundamental stage in the evolution of galaxies. Observationally, AGN are distinguishable from stars in that their spectral energy distributions are much broader than black-body functions. However, redshift, variability, (possibly large) obscuration, and a range of intrinsic spectral shapes and characteristics result in a great range of "colors" over wavelength ranges from X-ray to radio.

Scientific Goals: This project aims to construct a complete sample of AGN in order to:

- Test the idea that observable properties are determined by extrinsic factors such as orientation or obscuration (so-called unification models).

- Compare the environments of AGN as a function of type, *e.g.*, radio properties *vs.* membership in clusters of galaxies.

- Understand the evolution of the AGN luminosity function, and, in particular, separate number evolution from luminosity evolution.

- Construct the AGN luminosity function for different wavelengths in order to understand the evolution of AGN properties in a statistical sense.

Outline of the Project:

1. Federate a number (N) of surveys covering the same (significant) area on the sky, and together, spanning a large wavelength range (X-ray through radio).

2. Include metadata information so that survey selection effects and constraints can be accounted for. Relevant metadata will include survey area, bandpasses, flux limits, *etc.*

3. Identify distinguishable "clouds" of objects in N-dimensional color space in the resulting joint dataset.

4. Apply a priori astronomical knowledge (*e.g.*, published catalogs, theoretical models) to understand population of these "clouds".

5. Identify AGN candidates in the joint dataset. Note that confirmation may require new observations the process might be carried out as a statistical one, resulting in a probability that any particular object is an AGN.

6. Use catalog properties and/or further measurements from image databases to address science questions.

NVO Functionality required:

- Federation of relevant surveys including cross-identification of objects in multi-wavelength surveys and interchange/merging of metadata.

- Cluster analysis to identify "clouds" and "sequences". This will include both supervised analysis (in which astronomical knowledge guides the definition and analysis) and unsupervised analysis (in which new patterns are recognized).

- Visualization of multi-dimensional datasets.

- Statistical analysis/classification of the populations of defined regions in the multi-dimensional parameter space.

Example #2: Formation and Evolution of Large-Scale Structure

Background: Clusters of galaxies represent the largest unambiguous mass concentrations known. Various models of the early evolution of the universe make different predictions for how clusters form and evolve. These models can be tested by comparing their predictions with observed mass and luminosity functions of clusters.

Scientific Goals: This project aims to construct an unbiased sample of clusters of galaxies over a cosmologically significant redshift range in order to test various structure formation and evolution models by comparison with evolution of the observed mass and luminosity functions with redshift. In addition, the sample can be used to study the morphology-density relation for galaxies and its evolution over interesting timescales.

Outline of the Project:

1. Create statistical cluster samples using multi-wavelength pixel data in a number of different ways:

    X-ray surveys: identify cluster signature from emission of hot gas in image data.

    Optical/IR surveys: convolution of image data with kernel designed to select clusters.

    Millimeter surveys: identify clusters from variation in CMB temperature caused by the Sunyaev-Zel'dovich effect.

    Radio surveys: identify clusters based on presence of radio source morphologies indicative of cluster environment.

2. Compare the results of different selection techniques. Quantify the selection effects as functions of cluster mass, density, redshift, *etc.*

3. Use various distance indicators or redshift estimators to supplement measured properties of clusters.

Required NVO Functionality:

- Operate on large quantities of imaging data with user-defined algorithms and tools.

- Construct simulated surveys to understand selection effects; test user-defined tools on simulations.

- Generate predictions of observed sample properties based on various theories. Compare with observed samples.

Example #3: The Digital Galaxy

Background: The Galaxy is composed of a number of structural elements: halo, thin disk, thick disk, bulge, spiral arms. Each of these is characterized by populations of stars that have correlated distributions in age, mass, chemical composition, and orbit (position and kinematics), as well as distributions of non- stellar material such as gas and dust. These are the fossil tracers of the formation processes. A complete understanding of these, in a global context, has never been possible because of the difficulty of studying samples of the size needed to disentangle all the variables simultaneously.

Scientific Goals: This project aims to construct very large samples of galactic stars together with as much information about the physical properties of each object as can be derived. These datasets, together with maps of non-stellar components of the Galaxy, will be used to:

- Generate a parameterized model of the Galaxy, including positional and kinematic information.
- Confront this model with models for the structure of the Galaxy, based on various formation processes.
- In particular, search for co-moving groups that are representative of merger events or tidal debris tails.

Outline of the Project:

1. Federate various optical and IR surveys to generate matched catalogs of stars.
2. Use positions, magnitudes, and colors to construct three-dimensional stellar distributions. This will require using colors to derive luminosity classes and estimate extinction.
3. Quantify dust distribution and obscuration using FIR, H I, and CO images.
4. Iterate with 2) until consistent.
5. Identify bulk flows and sites of star formation using IR and radio images.
6. Use proper motion surveys (and radial velocity information, when available) to deduce motions of subsets of stars.
7. Use multi-epoch imaging to find variables. Use these to provide a distance check.

Required NVO Functionality:

- Federation of multi-wavelength and multi-epoch catalog data.

- Operation on large quantities of image data with user-defined tools.

- Visualization tools for large multi-dimensional datasets.

- Statistical tools to analyze components and find coherent groups of objects.

We emphasize again that these projects are meant to be illustrative of the kind of science, previously very difficult, that could be accomplished using capabilities that we foresee for the NVO. As the NVO comes into being, it is clear that more highly defined and diverse sets of projects will be developed as a result of community input and discussion.

5. Technical Issues

5.1. Overview

In assessing the current state of North American astronomy, the following resources are already in place to support the emerging NVO:

- *Data Centers and Supercomputer Centers.* Some tens of Terabytes of data products (catalogs, images, and spectra) already exist for various space missions, public telescopes, and surveys; this will expand to a Petabyte or more of data by the end of the decade. Archive and data analysis capabilities exist at the major NASA centers (STScI, IPAC, HEASARC, and CXC) and at the CADC (Canada); many smaller or more focused archives exist as well. Supercomputer centers such as the SDSC and NCSA are available for addressing large scale computational problems. A high performance national networking infrastructure is already in place.

- *Astronomical Information Services.* Information services such as the ADS, NED, and SIMBAD exist for name resolution and cross-referencing of galactic and extragalactic objects, and are providing increasingly sophisticated levels of interlinking between bibliographic information, the refereed and preprint literature, and the archival data centers.

- *Data Analysis Software.* Various software packages such as AIPS, AIPS++, IRAF, IDL, FTOOLS, SkyView, *etc.* , exist for the general analysis of astronomical data. The development of sophisticated software for large scale data mining is still in its infancy, although new initiatives such as the NPACI-sponsored Digital Sky and the IPAC Infrared Science Archive are showing the potential of such facilities and have prototyped the technology required to correlate and mine such data archives.

Although these resources are significant, anyone who has tried to perform multiwavelength data analysis or large scale statistical studies combining several different catalogs, with the data involved being available from widely distributed and dissimilar archives, will appreciate how far we have to go to implement the vision of the NVO. Ground-based O/IR and radio data need to be pipeline-processed and archived routinely as space-based data are now. Standards and protocols need to be developed to allow widely distributed archives to interoperate and exchange data. Astronomical data analysis software needs to evolve to be able to access data in distributed multiwavelength archives as easily as local datasets are accessed now. New algorithms, applications, and toolkits need to be developed to mine multi-Terabyte data archives. Supercomputer-class computational systems need to be developed to enable large scale statistical studies of massive, multiwavelength distributed data archives. The data, software, and computational resources need to be interconnected at the highest available network bandwidths.

Data Archives Any consideration of the science to be performed by the NVO, or the technical issues involved in implementing the NVO, must start with the data. Although the data from most NASA missions have been routinely archived for over a decade, relatively little data from ground based telescopes is currently available online, other than for a few major surveys. With modern wide-field and multispectra instruments on ground-based telescopes producing ever larger quantities of data, and with ground-based survey projects becoming almost as common as classical observing, *there is an acute need to archive and publish high quality datasets from ground-based instruments and surveys.* The science promised by the NVO will not be possible unless the NVO succeeds in creating true, panchromatic images and catalogs, seamlessly integrating data from both ground- and space-based archives, and thereby enabling exploration of astrophysical phenomena over most of the electromagnetic spectrum.

Experience over the past decade has shown that astronomical archives are complex and diverse, never stop growing, and are best maintained by those close to the data who know it well. In practice this has meant that most data are put online either by individual large survey projects, *e.g.*, the 2-Micron All-Sky Survey (2MASS) or the Sloan Digital Sky Survey (SDSS), or by discipline specific archive centers which serve a given community. To address the need to move to large scale archiving of ground-based astronomical data, archiving facilities will need to be established at the national centers (NOAO, NRAO, NSO, NAIC), and partnerships will need to be formed with the major private and university-operated facilities. The major national data centers for ground- and space-based data will comprise the principle nodes of the distributed NVO data system in the U.S.

Technical Challenges Given archival quality data from all branches of astronomy, physically distributed at 10–20 major archive centers and any number of ancillary datasets together with a distributed community of thousands of scientific users, one can define what new software and services will be required to implement the NVO. Analysis of the data will be complex, due to the heterogeneous nature of datasets from the different branches of astronomy and due to the use of increasingly complex data structures (within the general framework of the FITS data format standard) to accommodate the increasing levels of sophistication of modern astronomical instrumentation.

The sheer scale of the problem is daunting, with catalog sizes approaching the Terabyte range and the total data volume in the Petabyte range. However, an *even more serious challenge comes from the complexity of these datasets*, with tens or hundreds of attributes being measured for each of ten million or more objects. This is *a crucial new aspect to the data mining issue,* and multivariate correlation of such large catalogs is a massive computational problem. If pixel level analysis of candidate objects is required the computational problem can be even more massive. It is important to recognize that current brute-force analysis techniques do not scale to problems of this size! Multidisciplinary research in areas such as metadata representation and handling, large scale statistical analysis and correlations, and distributed parallel computational techniques will be required to address the unprecedented data access and computational problems faced by the NVO.

Fortunately, astronomy is not alone in facing this problem. *The technological challenges for the NVO are similar to those facing other branches of science,* such as high energy physics, computational genomics, global climate studies, and oceanography. Research and development of information systems technology is already underway in areas such as statistical analysis and data mining of large archives, distributed computational grids, data intensive grid computing (data grids), and management of structured digital information (digital libraries). Much of this research is relevant to the problems faced by the NVO. Information technology and data management throughout the sciences will both advance, and be advanced by, the NVO.

The large dataset size and geographic distribution of users and resources also presents major challenges in connectivity. Next generation networking providing cross-continental bandwidths of 100 MB/sec is now available and currently underutilized, but this situation will change rapidly. It will be essential for the major NVO data centers to be interconnected with very high speed networks, and to utilize intelligent server-side software agents in order to make the most efficient use of the network when interacting with end-users.

5.2. Architecture

The technical challenge of implementing the NVO is a study in contrasts. While data will be widely distributed, the large studies at the cutting edge of the science to be enabled by the NVO will need massive computational resources and fast local access to data. While sophisticated metadata standards and access protocols will be required to link together distributed archives and network services, the effort required to interface a small archive to the NVO must be minimized to encourage publication of new data collections by the community. While data collections and compute services will be widely distributed, users will need a straightforward interface to the system which makes the location and storage representation of data and services as transparent as possible.

To meet this wide range of requirements, the NVO needs a distributed system architecture that provides uniform and efficient access to data and services irrespective of location or implementation. *Data archives* are assumed to already exist and will vary considerably in implementation and access policy. *Metadata standards* will be devised to provide a well defined means to describe archives, data collections, and services. A *data access layer* will provide a single uniform interface to all data and services, and will be used both to link archives and services within the framework of NVO, and to allow user applications to access NVO resources. *Query and compute services* will provide the tools for information discovery and large scale correlation and analysis of disparate datasets.

Data mining applications, running on a user workstation at their home institution, as applets within a Web browser, or at a major NVO data center, will provide the main user interface to enable science with the NVO.

5.3. Components

Data Archives Data archives store *datasets* (*e.g.*, catalogs, images, and spectra) organized into logically related *data collections*, as well as *metadata* describing the archive and its data holdings. Access is provided in various ways such as via a structured Web interface, via a standard file-oriented interface such as FTP, or via other access protocols which may vary from archive to archive.

NVO will place no requirements on data archives other than that they be made accessible to NVO via the data access layer (DAL), which serves as the portal by which NVO gains access to the archive. In the simplest cases interfacing an archive to NVO will be little more than a matter of installing the data access layer software and modifying a few configuration files to reflect the data holdings and access permissions of the local archive, much as one would install a Web server. More sophisticated installations may provide expanded support for metadata access and server-side functions, as outlined in the discussion of the data access layer below.

Metadata Standards Metadata (literally, "data about data") is structured information describing some element of the NVO. Metadata will be required to describe archives, the services provided by those archives, the data collections available from an archive, the structure and semantics (meaning) of individual data collections, and the structure and semantics of individual datasets within a collection. Typical astronomical datasets are data objects such as catalogs, images, or spectra. As an example, the semantic metadata for a typical astronomical image is the logical content of the FITS header of the image.

Metadata describing astronomical data is essential to enable *data discovery* and *data interoperability*. Metadata describing archives and services is necessary to allow the components of NVO to interoperate in an automated fashion. *Metadata standards* are desirable to make these problems more tractable. In practice there are limits to what can be done to standardize dataset specific metadata, but mediation techniques such as those being developed by the digital library community provide ways to combine metadata dialects developed by different communities for similar types of data. Current projects such as Astrobrowse and ISAIA (Interoperable Systems for Archival Information Access) represent initial efforts within the astronomical community to establish metadata standards.

Data Access Layer The data access layer (DAL) will provide a uniform interface to all data, metadata, and compute services within NVO. At the lowest level the data access layer is a standard *protocol* defining how the software components of the NVO talk to each other. Reference grade software implementing the protocol will also be provided, which can either be used directly or taken as the basis for further development by the community. This software will include server-side software used to interface archives and compute services to the NVO, and client-side applications programming interfaces (APIs), which can be used to write NVO-aware distributed data mining applications. Since the DAL is fundamentally a protocol, multiple APIs will be possible, *e.g.*, to support legacy software or multiple language environments.

The key aspect of the data access layer is that it provides a uniform interface to *all* data and services within NVO. User applications use the data access layer to access NVO data and services, and archives and compute services *within* the NVO use the data access layer internally to access data or services in other archives, potentially generating a cascade of such references. NVO is thus an inherently hierarchical, distributed system, which nonetheless has a simple structure since all components share the same interface. In addition to such *location transparency*, the data access layer will provide *storage transparency*, hiding the details of how data are stored within an archive. Finally, the data access layer protocol will define standard data models (at the protocol level) for astronomical data objects such as images and spectra. Archive maintainers will provide server-side modules to perform *data model translation* when data objects are accessed, allowing applications to process remote data regardless of its source or how it is stored within a particular archive.

Often a client program using the data access layer will not need an entire dataset, but only a portion. *Server-side functions* will permit subsetting, filtering, and data model translation of individual datasets. In some cases *user defined functions* may be downloaded and applied to the data to compute the result returned to the remote client. This is critical to reduce network loading and distribute computation.

Since the data access layer can be used to read both metadata and actual datasets from a remote archive, *dataset replication* becomes possible, allowing a *local data cache* to be maintained. This is critical to optimizing data access throughout the NVO, and will be necessary to even attempt many large scale statistical studies and correlations. Dataset replication also makes it possible to replicate entire data collections, and to migrate data archives forward in time. Metadata replication and ingest makes it possible for a central site to automatically index entire remote archives.

Query and Compute Services While the data access layer and metadata standards will allow the NVO to link archives and access data, *query and compute services* will be required to support information discovery and provide the statistical correlation and image analysis capabilities required for data mining.

While most archives will provide basic query services for the data collections they support, large scale data mining does not become possible until multiple catalogs are combined (correlated) to search for objects matching some statistical signature. The larger NVO data centers will provide the data and computational resources required to support such *large scale correlations*. While a query or correlation may result in subqueries to remote archives, extensive use of dataset replication and caching will be employed to optimize queries for commonly accessed catalogs or archives. Sophisticated metadata mediation techniques will be required to combine the results from different catalogs.

In some cases pixel-level analysis of the original processed data, using an algorithmic function downloaded by the user, may be required to compute new object parameters to refine a parametric search (in effect this operation is dynamically adding columns to an existing catalog, an extremely powerful technique). Since with NVO candidate object lists may contain several hundred million objects, this is a massively parallel problem such as might require a Terascale supercomputer to address. Even in the case of large scale statistical studies, distributed computing techniques and fast networks may allow the user to work from their home institution, but some form of peer reviewed time allocation may be required to allocate the necessary computational and storage resources. For some larger studies users may need to visit a NVO data center in order to have efficient access to personnel as well as data, software, and computational resources.

Data Mining Applications The field of data mining, including visualization and statistical analysis of large multivariate datasets, is still in its infancy. This will be an area of active research for many years to come. Most current astronomical data analysis software will need to be upgraded to become "NVO-aware", able to be used equally well on both local and remote data. New applications will be developed as part of ongoing research into data mining techniques. While NVO should provide the interfaces and toolkits required to support this development, as well as some initial data mining applications from the major NVO centers, the open-ended nature of the problem suggests the need for a multidisciplinary data mining research grants program once NVO becomes operational.

Information Systems Research In all areas—storage technology, information management, data handling, distributed and parallel computing, high speed networking, data visualization, data mining—NVO will push the limits of current technology. Partnerships with academia and industry will be necessary to research and develop the informations systems technology necessary to implement NVO. Collaborations with other branches of science and with the national supercomputer centers will be required to develop standards for metadata handling, data handling and distributed computing. Data mining is an inherently multidisciplinary problem which will require the partnership of astronomers, computer scientists, mathematicians, and software professionals to address.

A next generation, high speed national research Internet is already in place, but is underutilized at present due to the lack of credible academic applications designed to make use of high performance networking. NVO would be a prime example of a creative new way to use wide area high performance networking for academic research.

Education and Public Outreach Given the wealth of real science data the NVO will make freely available via the Internet, and the keen interest of the public in astronomy, *the NVO will be uniquely suited for education and advancing science literacy*. The intrinsic Internet-based nature of the NVO lends itself to a variety of high-quality science popularization and education methods with *an unprecedented social and geographical outreach*.

We anticipate that education and outreach professionals (educators, staff members of planetaria, science museums, popular science writers and journalists, *etc.*) would become actively involved in utilizing this remarkable set of resources, creating of popular science websites, course materials (from elementary to graduate school), and sophisticated demonstrations. Schools with modest science education resources would be able to find hands-on demonstrations on line. Applet software running in commodity web browsers will permit NVO data to be accessed and visualized by the public, allowing virtual observations to be taken and the resultant data analyzed and interpreted. We expect that *a range of outreach partnerships will be developed with the NVO as a centerpiece and as a catalyst.*

The NVO is especially interesting as a science and technology education focus because *it bridges a physical science (astronomy) and applied computer science.* It thus employs a range of technologies and skills relevant to many aspects of the economy and society as a whole in the 21st century. Real-life examples of the use of such methodologies can be a powerful way to teach material that may otherwise be a very dry or difficult. For example, we note the great popular success of the SETI@home project; one can envision more sophisticated examples where data mining techniques are used by large numbers of people on comparably exciting problems spawned within the NVO.

6. Implementation Plan

Previous sections have described the technological changes that will enable a "new astronomy" and the characteristics of an NVO that can capitalize and build upon those changes to enable new and more cost effective science than would otherwise be possible.

The fundamental basis for the NVO management activities will be to recognize the science driven nature of the NVO and to maximize the community participation in the NVO effort. There will be three levels of activity and funding. These are structured to ensure that there is a usable and well documented infrastructure, that the software projects are science driven, and that bulk of the funding is dispersed to well focused science based proposals that are peer reviewed.

1. The highest priority is to build the archive infrastructure and well documented protocols to access the data. These will be standards for data access that the community can rely on to build higher level tools. These would evolve as the technology advances, but should always be backward compatible. This infrastructure is funded via a base budget and is developed and maintained by the major NVO distributed sites.

2. There will be regular "AOs" for opportunities to build "software tools" that utilize the infrastructure. They would be delivered to the NVO for wider use by the community and would follow standards defined by the NVO. It is important that these tool building opportunities cover a wide range of possibilities and engage a large part of the community. A strong science enabling case for each software tool must be made, but they will be general user facilities that the entire community can use to do research.

3. There are regular "AOs" to use the NVO. These would be more specific research projects with a well defined goal, that might include software development. (This would be similar to the current NASA ADP program). These would be much less structured in the sense of being grants and with the deliverable being a paper to a journal.

In the early phases of the NVO, the emphasis may be on the first two areas, but as the NVO infrastructure develops the balance of the funding between these three areas will evolve.

Implementation of the NVO would occur in several stages, from preliminary preparation to fully operational stages. *A major objective of the implementation plan is to begin providing some levels of functionality as quickly as possible through use of existing tools and services.*

6.1. Phase 0: Prior to NVO Start

Goal: Create the conceptual design of the NVO; begin activities at some centers to provide necessary capability for implementation of the NVO.

- Develop relevant position papers, supporting documents, and a "Science Reference Mission" which identifies the key science goals for the NVO;

- Initiate work within participating organizations to ensure accessibility of data and the establishment of archives;

- Develop essential enabling technologies, such as information exchange protocols and metadata standards;

- Establish catalog search and/or image data retrieval capability for selected data subsets at all major sites;

- Initiate community involvement through meetings and workshops; and

- Open lines of communication with the international community concerning general NVO initiative.

6.2. Phase 1: Months 1–18

Goal: Establish integrated data discovery, data delivery, and data comparison services.

- Expand and formalize the data discovery and data delivery systems, including establishment of metadata standards, transport protocols, and presentation services;

- Continue to work with all sites to improve access to online services;

- Plan for eventual network connectivity and computational requirements;

- Deploy small scale cross-correlation capabilities and visualization tools;

- Prototype large scale cross-correlation facilities;

- Continue community involvement through meetings, workshops, and the establishment of a Users' Committee and a Visiting Committee;

- Establish core technology and management groups and establish reporting and accountability procedures;

- Identify subsets of NVO functionality that can be most effectively developed by existing data centers or other entities;

- Establish an outreach program;
- Move forward in the design and establishment of an international information infrastructure for astronomy; and
- Foster communication and collaborations with efforts to advance information technology in other fields.

6.3. Phase 2: Months 18–36

Goal: Establish initial large scale cross-correlation capabilities; begin full scale operations.

- Begin putting network and associated computing facilities in place;
- Develop and deploy visualization tools for complex datasets;
- Develop and make initial deployment of the data access layer (DAL);
- Ensure that the data discovery and comparison tools are now mature and in routine operation;
- Establish partnerships with international organizations to assure interoperability of US and non-US facilities and services;
- Management structure and advisory committees now in routine operation.

6.4. Phase 3: Months 36–60

Goal: Establish fully operational baseline NVO; enable full scale cross-correlations supported by suitably configured computational and network systems.

- Breadth of data services extended to additional facilities, including international collaborators;
- Data access layer deployed and in routine operation;
- Support user-defined portable processing agents;
- Establish support of higher level data products, such as pre-prepared cross identifications.

This implementation timetable is only approximate and will naturally evolve as the problems become more well defined and as the level of support for the NVO becomes clearer. However, this timetable is "optimal" in that it reflects estimates of an ideal implementation path for the NVO functionality. Services are estimated to be implemented as rapidly as is feasible from a technical point of view; restricted funding levels below the optimal level would clearly slow this process.

Author Index

Andreon, S. 73, 80
Arviset, C. **165**
Abbott, B. 188

Babu, G.J. **272**
Berriman, G.B. 142, **169**
Boller, T. 234
Bonnarel, F. 176
Borne, K.D. **333**
Boroson, T. **347**
Brunner, R.J. **34**, 52, **64**, **135**, 192, 197

Castro, S. 52
Cheung, C.Y. bf 316
Chiu, N. 142, 169
Christian, C. **148**
Connolly, A.J. 265, 279
Crutchfield, J.P. 337

de Carvalho, R.R. 52
De Young, D.S. **343**
Derthick, M. 284
Dickinson, M. 69
Diercks, A.H. **46**
Dimitoglou, G. **173**
Djorgovski, S.G. **52**, 64, 73, 80, 135, 192
Donahue, M. **69**
Dubois, P. 176

Egret, D. **108**, 176
Emmart, C. 188
Englhauser, J. 234

de Filippis, E. 80
Feigelson, E.D. 272
Fernique, P. 176
Fishman, M. 180
Freyberg, M. 234

Gabrielli, A. 90
Gal, R.R. 52, 64, 73, 80, 192
Genova, F. **176**, 205
Genovese, C. 265, **279**
Glymour, C. 337
Good, J. 135, **142**, 169, 197
Goodrich, M.T. **329**
Gray, J. **241**

Grosvenor, S. **180**

Hernandez, J. 165
Handley, T.H. 135, 142, 169, 197
Hanisch, R.J. **97**, **130**
Hewett, R. 337
Hill, F. **184**
Husman, L.E. 291

Jacob, J.C. 192, **291**
Jasniewicz, G. 176
Johnson, A. 142, 169, 197
Jones, J. 180

Kent, S. **40**
Kepner, J. **297**
Kong, M. 142, 169
Koratkar, A. 180
Kunszt, P.Z. 230

Lawrence, A. **114**
Lee, P. 69
Lee, W-P. 142, 169
Lesteven, S. 176
Liu, C.T. **153**, **188**
Longo, G. 73, 80
Lonsdale, C.J. **13**, 135, 142, 169
Luri, X. 201

Paolillo, M. 73
Ma, J. 142, 169, **197**
Mac Low, M.-M. 188
Mack, J. 69
Mahabal, A.A. 52, 64, **192**
McDonald, L.M. 125
McGlynn, T.A. **125**
McLean, B. **103**
McMahon, J. 297
Monkewitz, S. 142, 169
Moore, A.W. 265, 279
Moore, R.W. **257**

Nichol, R.C. **265**, 279
Norton, S.W. 142, 169

O'Mullane, W. **201**
Ochsenbein, F. 176, 205
Odewahn, S.C. 52, 64, 192

Ortiz, P.F. **205**

Paolillo, M. 73
Piranomonte, S. **73**
Prusti, T. 165
Puddu, E. 73, **80**
Postman, M. 69
Pravdo, S.H. **118**
Prince, T. 135, 197

Rosati, P. 69
Rots, A. **209**, **213**
Ruley, L. 180
Rutledge, R. 197

Sánchez Duarte, L. 173
Sanchez-Ibarra, A. **217**
Saucedo-Morales, J. 217
Scaramella, R. 73, 80
Scargle, J.D. 337
Schade, D. **221**
Shara, M. 188
Scharf, C.A. 69, **86**
Schneider, J. 265, 279
Scholl, I. **225**
Schombert, J.M. **28**
Squires, G.K. **21**
Stocke, J. 69
Strazzullo, V. 80
Summers, F.J. 188
Supper, R. 234
Sylos Labini, F. **90**
Szalay, A.S. **3**, 230

Thakar, A.R. **230**
Tyson, J.A. 21
Tyson, N.D. 188

Voges, W. **234**

Wasserman, L. 265, 279
Welling, J. **284**
Wenger, M. 176
White II, J.C. **159**
White, N.E. 125
Wicenec, A.J. 205, **323**
Williams, R. 197, **302**
Wolf, K. 180

Zhang, A. 142, 169

ASTRONOMICAL SOCIETY OF THE PACIFIC
CONFERENCE SERIES
(ASP CS) VOLUMES

and

INTERNATIONAL ASTRONOMICAL UNION
(IAU) VOLUMES

Published
by

The Astronomical Society of the Pacific
(ASP)

ASP CONFERENCE SERIES VOLUMES
Published by the Astronomical Society of the Pacific

PUBLISHED: 1988 (* asterisk means OUT OF STOCK)

Vol. CS-1   PROGRESS AND OPPORTUNITIES IN SOUTHERN HEMISPHERE
OPTICAL ASTRONOMY: CTIO 25TH Anniversary Symposium
eds. V. M. Blanco and M. M. Phillips
ISBN 0-937707-18-X

Vol. CS-2   PROCEEDINGS OF A WORKSHOP ON OPTICAL SURVEYS FOR QUASARS
eds. Patrick S. Osmer, Alain C. Porter, Richard F. Green, and Craig B. Foltz
ISBN 0-937707-19-8

Vol. CS-3   FIBER OPTICS IN ASTRONOMY
ed. Samuel C. Barden
ISBN 0-937707-20-1

Vol. CS-4   THE EXTRAGALACTIC DISTANCE SCALE:
Proceedings of the ASP 100th Anniversary Symposium
eds. Sidney van den Bergh and Christopher J. Pritchet
ISBN 0-937707-21-X

Vol. CS-5   THE MINNESOTA LECTURES ON CLUSTERS OF GALAXIES
AND LARGE-SCALE STRUCTURE
ed. John M. Dickey
ISBN 0-937707-22-8

PUBLISHED: 1989

Vol. CS-6   SYNTHESIS IMAGING IN RADIO ASTRONOMY: A Collection of Lectures
from the Third NRAO Synthesis Imaging Summer School
eds. Richard A. Perley, Frederic R. Schwab, and Alan H. Bridle
ISBN 0-937707-23-6

Vol. CS-7   PROPERTIES OF HOT LUMINOUS STARS: Boulder-Munich Workshop
ed. Catharine D. Garmany
ISBN 0-937707-24-4

PUBLISHED: 1990

Vol. CS-8*   CCDs IN ASTRONOMY
ed. George H. Jacoby
ISBN 0-937707-25-2

Vol. CS-9   COOL STARS, STELLAR SYSTEMS, AND THE SUN: Sixth Cambridge Workshop
ed. George Wallerstein
ISBN 0-937707-27-9

Vol. CS-10*   EVOLUTION OF THE UNIVERSE OF GALAXIES:
Edwin Hubble Centennial Symposium
ed. Richard G. Kron
ISBN 0-937707-28-7

Vol. CS-11   CONFRONTATION BETWEEN STELLAR PULSATION AND EVOLUTION
eds. Carla Cacciari and Gisella Clementini
ISBN 0-937707-30-9

Vol. CS-12   THE EVOLUTION OF THE INTERSTELLAR MEDIUM
ed. Leo Blitz
ISBN 0-937707-31-7

ASP CONFERENCE SERIES VOLUMES
Published by the Astronomical Society of the Pacific

PUBLISHED: 1991 (* asterisk means OUT OF STOCK)

Vol. CS-13 THE FORMATION AND EVOLUTION OF STAR CLUSTERS
ed. Kenneth Janes
ISBN 0-937707-32-5

Vol. CS-14 ASTROPHYSICS WITH INFRARED ARRAYS
ed. Richard Elston
ISBN 0-937707-33-3

Vol. CS-15 LARGE-SCALE STRUCTURES AND PECULIAR MOTIONS IN THE UNIVERSE
eds. David W. Latham and L. A. Nicolaci da Costa
ISBN 0-937707-34-1

Vol. CS-16 Proceedings of the 3rd Haystack Observatory Conference on ATOMS, IONS, AND MOLECULES: NEW RESULTS IN SPECTRAL LINE ASTROPHYSICS
eds. Aubrey D. Haschick and Paul T. P. Ho
ISBN 0-937707-35-X

Vol. CS-17 LIGHT POLLUTION, RADIO INTERFERENCE, AND SPACE DEBRIS
ed. David L. Crawford
ISBN 0-937707-36-8

Vol. CS-18 THE INTERPRETATION OF MODERN SYNTHESIS OBSERVATIONS OF SPIRAL GALAXIES
eds. Nebojsa Duric and Patrick C. Crane
ISBN 0-937707-37-6

Vol. CS-19 RADIO INTERFEROMETRY: THEORY, TECHNIQUES, AND APPLICATIONS, IAU Colloquium 131
eds. T. J. Cornwell and R. A. Perley
ISBN 0-937707-38-4

Vol. CS-20 FRONTIERS OF STELLAR EVOLUTION:
50th Anniversary McDonald Observatory (1939-1989)
ed. David L. Lambert
ISBN 0-937707-39-2

Vol. CS-21 THE SPACE DISTRIBUTION OF QUASARS
ed. David Crampton
ISBN 0-937707-40-6

PUBLISHED: 1992

Vol. CS-22 NONISOTROPIC AND VARIABLE OUTFLOWS FROM STARS
eds. Laurent Drissen, Claus Leitherer, and Antonella Nota
ISBN 0-937707-41-4

Vol CS-23 ASTRONOMICAL CCD OBSERVING AND REDUCTION TECHNIQUES
ed. Steve B. Howell
ISBN 0-937707-42-4

Vol. CS-24 COSMOLOGY AND LARGE-SCALE STRUCTURE IN THE UNIVERSE
ed. Reinaldo R. de Carvalho
ISBN 0-937707-43-0

ASP CONFERENCE SERIES VOLUMES
Published by the Astronomical Society of the Pacific

PUBLISHED: 1992 (asterisk means OUT OF STOCK)

Vol. CS-25 ASTRONOMICAL DATA ANALYSIS, SOFTWARE AND SYSTEMS I - (ADASS I)
eds. Diana M. Worrall, Chris Biemesderfer, and Jeannette Barnes
ISBN 0-937707-44-9

Vol. CS-26 COOL STARS, STELLAR SYSTEMS, AND THE SUN:
Seventh Cambridge Workshop
eds. Mark S. Giampapa and Jay A. Bookbinder
ISBN 0-937707-45-7

Vol. CS-27 THE SOLAR CYCLE: Proceedings of the
National Solar Observatory/Sacramento Peak 12th Summer Workshop
ed. Karen L. Harvey
ISBN 0-937707-46-5

Vol. CS-28 AUTOMATED TELESCOPES FOR PHOTOMETRY AND IMAGING
eds. Saul J. Adelman, Robert J. Dukes, Jr., and Carol J. Adelman
ISBN 0-937707-47-3

Vol. CS-29 Viña del Mar Workshop on CATACLYSMIC VARIABLE STARS
ed. Nikolaus Vogt
ISBN 0-937707-48-1

Vol. CS-30 VARIABLE STARS AND GALAXIES
ed. Brian Warner
ISBN 0-937707-49-X

Vol. CS-31 RELATIONSHIPS BETWEEN ACTIVE GALACTIC NUCLEI
AND STARBURST GALAXIES
ed. Alexei V. Filippenko
ISBN 0-937707-50-3

Vol. CS-32 COMPLEMENTARY APPROACHES TO DOUBLE
AND MULTIPLE STAR RESEARCH, IAU Colloquium 135
eds. Harold A. McAlister and William I. Hartkopf
ISBN 0-937707-51-1

Vol. CS-33 RESEARCH AMATEUR ASTRONOMY
ed. Stephen J. Edberg
ISBN 0-937707-52-X

Vol. CS-34 ROBOTIC TELESCOPES IN THE 1990's
ed. Alexei V. Filippenko
ISBN 0-937707-53-8

PUBLISHED: 1993

Vol. CS-35* MASSIVE STARS: THEIR LIVES IN THE INTERSTELLAR MEDIUM
eds. Joseph P. Cassinelli and Edward B. Churchwell
ISBN 0-937707-54-6

Vol. CS-36 PLANETS AROUND PULSARS
ed. J. A. Phillips, S. E. Thorsett, and S. R. Kulkarni
ISBN 0-937707-55-4

ASP CONFERENCE SERIES VOLUMES
Published by the Astronomical Society of the Pacific

PUBLISHED: 1993 (* asterisk means OUT OF STOCK)

Vol. CS-37	FIBER OPTICS IN ASTRONOMY II ed. Peter M. Gray ISBN 0-937707-56-2
Vol. CS-38	NEW FRONTIERS IN BINARY STAR RESEARCH: Pacific Rim Colloquium eds. K. C. Leung and I.-S. Nha ISBN 0-937707-57-0
Vol. CS-39	THE MINNESOTA LECTURES ON THE STRUCTURE AND DYNAMICS OF THE MILKY WAY ed. Roberta M. Humphreys ISBN 0-937707-58-9
Vol. CS-40	INSIDE THE STARS, IAU Colloquium 137 eds. Werner W. Weiss and Annie Baglin ISBN 0-937707-59-7
Vol. CS-41	ASTRONOMICAL INFRARED SPECTROSCOPY: FUTURE OBSERVATIONAL DIRECTIONS ed. Sun Kwok ISBN 0-937707-60-0
Vol. CS-42	GONG 1992: SEISMIC INVESTIGATION OF THE SUN AND STARS ed. Timothy M. Brown ISBN 0-937707-61-9
Vol. CS-43	SKY SURVEYS: PROTOSTARS TO PROTOGALAXIES ed. B. T. Soifer ISBN 0-937707-62-7
Vol. CS-44	PECULIAR VERSUS NORMAL PHENOMENA IN A-TYPE AND RELATED STARS, IAU Colloquium 138 eds. M. M. Dworetsky, F. Castelli, and R. Faraggiana ISBN 0-937707-63-5
Vol. CS-45	LUMINOUS HIGH-LATITUDE STARS ed. Dimitar D. Sasselov ISBN 0-937707-64-3
Vol. CS-46	THE MAGNETIC AND VELOCITY FIELDS OF SOLAR ACTIVE REGIONS, IAU Colloquium 141 eds. Harold Zirin, Guoxiang Ai, and Haimin Wang ISBN 0-937707-65-1
Vol. CS-47	THIRD DECENNIAL US-USSR CONFERENCE ON SETI -- Santa Cruz, California, USA ed. G. Seth Shostak ISBN 0-937707-66-X
Vol. CS-48	THE GLOBULAR CLUSTER-GALAXY CONNECTION eds. Graeme H. Smith and Jean P. Brodie ISBN 0-937707-67-8
Vol. CS-49	GALAXY EVOLUTION: THE MILKY WAY PERSPECTIVE ed. Steven R. Majewski ISBN 0-937707-68-6

ASP CONFERENCE SERIES VOLUMES
Published by the Astronomical Society of the Pacific

PUBLISHED: 1993 (* asterisk means OUT OF STOCK)

Vol. CS-50   STRUCTURE AND DYNAMICS OF GLOBULAR CLUSTERS
eds. S. G. Djorgovski and G. Meylan
ISBN 0-937707-69-4

Vol. CS-51   OBSERVATIONAL COSMOLOGY
eds. Guido Chincarini, Angela Iovino, Tommaso Maccacaro, and Dario Maccagni
ISBN 0-937707-70-8

Vol. CS-52   ASTRONOMICAL DATA ANALYSIS SOFTWARE AND SYSTEMS II - (ADASS II)
eds. R. J. Hanisch, R. J. V. Brissenden, and Jeannette Barnes
ISBN 0-937707-71-6

Vol. CS-53   BLUE STRAGGLERS
ed. Rex A. Saffer
ISBN 0-937707-72-4

PUBLISHED: 1994

Vol. CS-54*   THE FIRST STROMLO SYMPOSIUM: THE PHYSICS OF ACTIVE GALAXIES
eds. Geoffrey V. Bicknell, Michael A. Dopita, and Peter J. Quinn
ISBN 0-937707-73-2

Vol. CS-55   OPTICAL ASTRONOMY FROM THE EARTH AND MOON
eds. Diane M. Pyper and Ronald J. Angione
ISBN 0-937707-74-0

Vol. CS-56   INTERACTING BINARY STARS
ed. Allen W. Shafter
ISBN 0-937707-75-9

Vol. CS-57   STELLAR AND CIRCUMSTELLAR ASTROPHYSICS
eds. George Wallerstein and Alberto Noriega-Crespo
ISBN 0-937707-76-7

Vol. CS-58*   THE FIRST SYMPOSIUM ON THE INFRARED CIRRUS
AND DIFFUSE INTERSTELLAR CLOUDS
eds. Roc M. Cutri and William B. Latter
ISBN 0-937707-77-5

Vol. CS-59   ASTRONOMY WITH MILLIMETER AND SUBMILLIMETER WAVE
INTERFEROMETRY,
IAU Colloquium 140
eds. M. Ishiguro and Wm. J. Welch
ISBN 0-937707-78-3

Vol. CS-60   THE MK PROCESS AT 50 YEARS: A POWERFUL TOOL FOR ASTROPHYSICAL
INSIGHT, A Workshop of the Vatican Observatory --Tucson, Arizona, USA
eds. C. J. Corbally, R. O. Gray, and R. F. Garrison
ISBN 0-937707-79-1

Vol. CS-61   ASTRONOMICAL DATA ANALYSIS SOFTWARE AND SYSTEMS III - (ADASS III)
eds. Dennis R. Crabtree, R. J. Hanisch, and Jeannette Barnes
ISBN 0-937707-80-5

ASP CONFERENCE SERIES VOLUMES
Published by the Astronomical Society of the Pacific

PUBLISHED: 1994 (* asterisk means OUT OF STOCK)

Vol. CS-62 THE NATURE AND EVOLUTIONARY STATUS OF HERBIG Ae/Be STARS
eds. Pik Sin Thé, Mario R. Pérez, and Ed P. J. van den Heuvel
ISBN 0-9837707-81-3

Vol. CS-63 SEVENTY-FIVE YEARS OF HIRAYAMA ASTEROID FAMILIES:
THE ROLE OF COLLISIONS IN THE SOLAR SYSTEM HISTORY
eds. Yoshihide Kozai, Richard P. Binzel, and Tomohiro Hirayama
ISBN 0-937707-82-1

Vol. CS-64* COOL STARS, STELLAR SYSTEMS, AND THE SUN:
Eighth Cambridge Workshop
ed. Jean-Pierre Caillault
ISBN 0-937707-83-X

Vol. CS-65* CLOUDS, CORES, AND LOW MASS STARS:
The Fourth Haystack Observatory Conference
eds. Dan P. Clemens and Richard Barvainis
ISBN 0-937707-84-8

Vol. CS-66* PHYSICS OF THE GASEOUS AND STELLAR DISKS OF THE GALAXY
ed. Ivan R. King
ISBN 0-937707-85-6

Vol. CS-67 UNVEILING LARGE-SCALE STRUCTURES BEHIND THE MILKY WAY
eds. C. Balkowski and R. C. Kraan-Korteweg
ISBN 0-937707-86-4

Vol. CS-68* SOLAR ACTIVE REGION EVOLUTION:
COMPARING MODELS WITH OBSERVATIONS
eds. K. S. Balasubramaniam and George W. Simon
ISBN 0-937707-87-2

Vol. CS-69 REVERBERATION MAPPING OF THE BROAD-LINE REGION
IN ACTIVE GALACTIC NUCLEI
eds. P. M. Gondhalekar, K. Horne, and B. M. Peterson
ISBN 0-937707-88-0

PUBLISHED: 1995

Vol. CS-70* GROUPS OF GALAXIES
eds. Otto-G. Richter and Kirk Borne
ISBN 0-937707-89-9

Vol. CS-71 TRIDIMENSIONAL OPTICAL SPECTROSCOPIC METHODS IN ASTROPHYSICS,
IAU Colloquium 149
eds. Georges Comte and Michel Marcelin
ISBN 0-937707-90-2

Vol. CS-72 MILLISECOND PULSARS: A DECADE OF SURPRISE
eds. A. S Fruchter, M. Tavani, and D. C. Backer
ISBN 0-937707-91-0

Vol. CS-73 AIRBORNE ASTRONOMY SYMPOSIUM ON THE GALACTIC ECOSYSTEM:
FROM GAS TO STARS TO DUST
eds. Michael R. Haas, Jacqueline A. Davidson, and Edwin F. Erickson
ISBN 0-937707-92-9

ASP CONFERENCE SERIES VOLUMES
Published by the Astronomical Society of the Pacific

PUBLISHED: 1995 (* asterisk means OUT OF STOCK)

Vol. CS-74　PROGRESS IN THE SEARCH FOR EXTRATERRESTRIAL LIFE:
1993 Bioastronomy Symposium
ed. G. Seth Shostak
ISBN 0-937707-93-7

Vol. CS-75　MULTI-FEED SYSTEMS FOR RADIO TELESCOPES
eds. Darrel T. Emerson and John M. Payne
ISBN 0-937707-94-5

Vol. CS-76　GONG '94: HELIO- AND ASTERO-SEISMOLOGY FROM THE EARTH
AND SPACE
eds. Roger K. Ulrich, Edward J. Rhodes, Jr., and Werner Däppen
ISBN 0-937707-95-3

Vol. CS-77　ASTRONOMICAL DATA ANALYSIS SOFTWARE AND SYSTEMS IV - (ADASS IV)
eds. R. A. Shaw, H. E. Payne, and J. J. E. Hayes
ISBN 0-937707-96-1

Vol. CS-78　ASTROPHYSICAL APPLICATIONS OF POWERFUL NEW DATABASES:
Joint Discussion No. 16 of the 22nd General Assembly of the IAU
eds. S. J. Adelman and W. L. Wiese
ISBN 0-937707-97-X

Vol. CS-79*　ROBOTIC TELESCOPES: CURRENT CAPABILITIES, PRESENT
DEVELOPMENTS, AND FUTURE PROSPECTS
FOR AUTOMATED ASTRONOMY
eds. Gregory W. Henry and Joel A. Eaton
ISBN 0-937707-98-8

Vol. CS-80*　THE PHYSICS OF THE INTERSTELLAR MEDIUM
AND INTERGALACTIC MEDIUM
eds. A. Ferrara, C. F. McKee, C. Heiles, and P. R. Shapiro
ISBN 0-937707-99-6

Vol. CS-81　LABORATORY AND ASTRONOMICAL HIGH RESOLUTION SPECTRA
eds. A. J. Sauval, R. Blomme, and N. Grevesse
ISBN 1-886733-01-5

Vol. CS-82*　VERY LONG BASELINE INTERFEROMETRY AND THE VLBA
eds. J. A. Zensus, P. J. Diamond, and P. J. Napier
ISBN 1-886733-02-3

Vol. CS-83*　ASTROPHYSICAL APPLICATIONS OF STELLAR PULSATION,
IAU Colloquium 155
eds. R. S. Stobie and P. A. Whitelock
ISBN 1-886733-03-1

ATLAS　INFRARED ATLAS OF THE ARCTURUS SPECTRUM, 0.9 - 5.3 μm
eds. Kenneth Hinkle, Lloyd Wallace, and William Livingston
ISBN: 1-886733-04-X

Vol. CS-84　THE FUTURE UTILIZATION OF SCHMIDT TELESCOPES, IAU Colloquium 148
eds. Jessica Chapman, Russell Cannon, Sandra Harrison, and Bambang Hidayat
ISBN 1-886733-05-8

ASP CONFERENCE SERIES VOLUMES
Published by the Astronomical Society of the Pacific

PUBLISHED: 1995 (* asterisk means OUT OF STOCK)

Vol. CS-85* CAPE WORKSHOP ON MAGNETIC CATACLYSMIC VARIABLES
eds. D. A. H. Buckley and B. Warner
ISBN 1-886733-06-6

Vol. CS-86 FRESH VIEWS OF ELLIPTICAL GALAXIES
eds. Alberto Buzzoni, Alvio Renzini, and Alfonso Serrano
ISBN 1-886733-07-4

PUBLISHED: 1996

Vol. CS-87 NEW OBSERVING MODES FOR THE NEXT CENTURY
eds. Todd Boroson, John Davies, and Ian Robson
ISBN 1-886733-08-2

Vol. CS-88* CLUSTERS, LENSING, AND THE FUTURE OF THE UNIVERSE
eds. Virginia Trimble and Andreas Reisenegger
ISBN 1-886733-09-0

Vol. CS-89 ASTRONOMY EDUCATION: CURRENT DEVELOPMENTS, FUTURE COORDINATION
ed. John R. Percy
ISBN 1-886733-10-4

Vol. CS-90 THE ORIGINS, EVOLUTION, AND DESTINIES OF BINARY STARS IN CLUSTERS
eds. E. F. Milone and J. -C. Mermilliod
ISBN 1-886733-11-2

Vol. CS-91 BARRED GALAXIES, IAU Colloquium 157
eds. R. Buta, D. A. Crocker, and B. G. Elmegreen
ISBN 1-886733-12-0

Vol. CS-92* FORMATION OF THE GALACTIC HALO .... INSIDE AND OUT
eds. Heather L. Morrison and Ata Sarajedini
ISBN 1-886733-13-9

Vol. CS-93 RADIO EMISSION FROM THE STARS AND THE SUN
eds. A. R. Taylor and J. M. Paredes
ISBN 1-886733-14-7

Vol. CS-94 MAPPING, MEASURING, AND MODELING THE UNIVERSE
eds. Peter Coles, Vicent J. Martinez, and Maria-Jesus Pons-Borderia
ISBN 1-886733-15-5

Vol. CS-95 SOLAR DRIVERS OF INTERPLANETARY AND TERRESTRIAL DISTURBANCES:
Proceedings of 16th International Workshop National Solar Observatory/Sacramento Peak
eds. K. S. Balasubramaniam, Stephen L. Keil, and Raymond N. Smartt
ISBN 1-886733-16-3

Vol. CS-96 HYDROGEN-DEFICIENT STARS
eds. C. S. Jeffery and U. Heber
ISBN 1-886733-17-1

ASP CONFERENCE SERIES VOLUMES
Published by the Astronomical Society of the Pacific

PUBLISHED: 1996 (* asterisk means OUT OF STOCK)

Vol. CS-97	POLARIMETRY OF THE INTERSTELLAR MEDIUM eds. W. G. Roberge and D. C. B. Whittet ISBN 1-886733-18-X
Vol. CS-98	FROM STARS TO GALAXIES: THE IMPACT OF STELLAR PHYSICS ON GALAXY EVOLUTION eds. Claus Leitherer, Uta Fritze-von Alvensleben, and John Huchra ISBN 1-886733-19-8
Vol. CS-99	COSMIC ABUNDANCES: Proceedings of the 6th Annual October Astrophysics Conference eds. Stephen S. Holt and George Sonneborn ISBN 1-886733-20-1
Vol. CS-100	ENERGY TRANSPORT IN RADIO GALAXIES AND QUASARS eds. P. E. Hardee, A. H. Bridle, and J. A. Zensus ISBN 1-886733-21-X
Vol. CS-101	ASTRONOMICAL DATA ANALYSIS SOFTWARE AND SYSTEMS V – (ADASS V) eds. George H. Jacoby and Jeannette Barnes ISBN 1080-7926
Vol. CS-102	THE GALACTIC CENTER, 4th ESO/CTIO Workshop ed. Roland Gredel ISBN 1-886733-22-8
Vol. CS-103	THE PHYSICS OF LINERS IN VIEW OF RECENT OBSERVATIONS eds. M. Eracleous, A. Koratkar, C. Leitherer, and L. Ho ISBN 1-886733-23-6
Vol. CS-104	PHYSICS, CHEMISTRY, AND DYNAMICS OF INTERPLANETARY DUST, IAU Colloquium 150 eds. Bo Å. S. Gustafson and Martha S. Hanner ISBN 1-886733-24-4
Vol. CS-105	PULSARS: PROBLEMS AND PROGRESS, IAU Colloquium 160 ed. S. Johnston, M. A. Walker, and M. Bailes ISBN 1-886733-25-2
Vol. CS-106	THE MINNESOTA LECTURES ON EXTRAGALACTIC NEUTRAL HYDROGEN ed. Evan D. Skillman ISBN 1-886733-26-0
Vol. CS-107	COMPLETING THE INVENTORY OF THE SOLAR SYSTEM: A Symposium held in conjunction with the 106th Annual Meeting of the ASP eds. Terrence W. Rettig and Joseph M. Hahn ISBN 1-886733-27-9
Vol. CS-108	M.A.S.S. -- MODEL ATMOSPHERES AND SPECTRUM SYNTHESIS: 5th Vienna - Workshop eds. Saul J. Adelman, Friedrich Kupka, and Werner W. Weiss ISBN 1-886733-28-7
Vol. CS-109	COOL STARS, STELLAR SYSTEMS, AND THE SUN: Ninth Cambridge Workshop eds. Roberto Pallavicini and Andrea K. Dupree ISBN 1-886733-29-5

ASP CONFERENCE SERIES VOLUMES
Published by the Astronomical Society of the Pacific

PUBLISHED: 1996 (* asterisk means OUT OF STOCK)

Vol. CS-110   BLAZAR CONTINUUM VARIABILITY
eds. H. R. Miller, J. R. Webb, and J. C. Noble
ISBN 1-886733-30-9

Vol. CS-111   MAGNETIC RECONNECTION IN THE SOLAR ATMOSPHERE:
Proceedings of a Yohkoh Conference
eds. R. D. Bentley and J. T. Mariska
ISBN 1-886733-31-7

Vol. CS-112   THE HISTORY OF THE MILKY WAY AND ITS SATELLITE SYSTEM
eds. Andreas Burkert, Dieter H. Hartmann, and Steven R. Majewski
ISBN 1-886733-32-5

PUBLISHED: 1997

Vol. CS-113   EMISSION LINES IN ACTIVE GALAXIES: NEW METHODS AND TECHNIQUES,
IAU Colloquium 159
eds. B. M. Peterson, F.-Z. Cheng, and A. S. Wilson
ISBN 1-886733-33-3

Vol. CS-114   YOUNG GALAXIES AND QSO ABSORPTION-LINE SYSTEMS
eds. Sueli M. Viegas, Ruth Gruenwald, and Reinaldo R. de Carvalho
ISBN 1-886733-34-1

Vol. CS-115   GALACTIC CLUSTER COOLING FLOWS
ed. Noam Soker
ISBN 1-886733-35-X

Vol. CS-116   THE SECOND STROMLO SYMPOSIUM:
THE NATURE OF ELLIPTICAL GALAXIES
eds. M. Arnaboldi, G. S. Da Costa, and P. Saha
ISBN 1-886733-36-8

Vol. CS-117   DARK AND VISIBLE MATTER IN GALAXIES
eds. Massimo Persic and Paolo Salucci
ISBN-1-886733-37-6

Vol. CS-118   FIRST ADVANCES IN SOLAR PHYSICS EUROCONFERENCE:
ADVANCES IN THE PHYSICS OF SUNSPOTS
eds. B. Schmieder. J. C. del Toro Iniesta, and M. Vázquez
ISBN 1-886733-38-4

Vol. CS-119   PLANETS BEYOND THE SOLAR SYSTEM
AND THE NEXT GENERATION OF SPACE MISSIONS
ed. David R. Soderblom
ISBN 1-886733-39-2

Vol. CS-120   LUMINOUS BLUE VARIABLES: MASSIVE STARS IN TRANSITION
eds. Antonella Nota and Henny J. G. L. M. Lamers
ISBN 1-886733-40-6

Vol. CS-121   ACCRETION PHENOMENA AND RELATED OUTFLOWS, IAU Colloquium 163
eds. D. T. Wickramasinghe, G. V. Bicknell, and L. Ferrario
ISBN 1-886733-41-4

ASP CONFERENCE SERIES VOLUMES
Published by the Astronomical Society of the Pacific

PUBLISHED: 1997 (* asterisk means OUT OF STOCK)

Vol. CS-122	FROM STARDUST TO PLANETESIMALS: Symposium held as part of the 108th Annual Meeting of the ASP eds. Yvonne J. Pendleton and A. G. G. M. Tielens ISBN 1-886733-42-2
Vol. CS-123	THE 12th 'KINGSTON MEETING': COMPUTATIONAL ASTROPHYSICS eds. David A. Clarke and Michael J. West ISBN 1-886733-43-0
Vol. CS-124	DIFFUSE INFRARED RADIATION AND THE IRTS eds. Haruyuki Okuda, Toshio Matsumoto, and Thomas Roellig ISBN 1-886733-44-9
Vol. CS-125	ASTRONOMICAL DATA ANALYSIS SOFTWARE AND SYSTEMS VI – (ADASS VI) eds. Gareth Hunt and H. E. Payne ISBN 1-886733-45-7
Vol. CS-126	FROM QUANTUM FLUCTUATIONS TO COSMOLOGICAL STRUCTURES eds. David Valls-Gabaud, Martin A. Hendry, Paolo Molaro, and Khalil Chamcham ISBN 1-886733-46-5
Vol. CS-127	PROPER MOTIONS AND GALACTIC ASTRONOMY ed. Roberta M. Humphreys ISBN 1-886733-47-3
Vol. CS-128	MASS EJECTION FROM AGN (Active Galactic Nuclei) eds. N. Arav, I. Shlosman, and R. J. Weymann ISBN 1-886733-48-1
Vol. CS-129	THE GEORGE GAMOW SYMPOSIUM eds. E. Harper, W. C. Parke, and G. D. Anderson ISBN 1-886733-49-X
Vol. CS-130	THE THIRD PACIFIC RIM CONFERENCE ON RECENT DEVELOPMENT ON BINARY STAR RESEARCH eds. Kam-Ching Leung ISBN 1-886733-50-3

PUBLISHED: 1998

Vol. CS-131	BOULDER-MUNICH II: PROPERTIES OF HOT, LUMINOUS STARS ed. Ian D. Howarth ISBN 1-886733-51-1
Vol. CS-132	STAR FORMATION WITH THE INFRARED SPACE OBSERVATORY (ISO) eds. João L. Yun and René Liseau ISBN 1-886733-52-X
Vol. CS-133	SCIENCE WITH THE NGST (Next Generation Space Telescope) eds. Eric P. Smith and Anuradha Koratkar ISBN 1-886733-53-8
Vol. CS-134	BROWN DWARFS AND EXTRASOLAR PLANETS eds. Rafael Rebolo, Eduardo L. Martin, and Maria Rosa Zapatero Osorio ISBN 1-886733-54-6

ASP CONFERENCE SERIES VOLUMES
Published by the Astronomical Society of the Pacific

PUBLISHED: 1998 (* asterisk means OUT OF STOCK)

Vol. CS-135   A HALF CENTURY OF STELLAR PULSATION INTERPRETATIONS:
A TRIBUTE TO ARTHUR N. COX
eds. P. A. Bradley and J. A. Guzik
ISBN 1-886733-55-4

Vol. CS-136   GALACTIC HALOS: A UC SANTA CRUZ WORKSHOP
ed. Dennis Zaritsky
ISBN 1-886733-56-2

Vol. CS-137   WILD STARS IN THE OLD WEST: PROCEEDINGS OF THE 13th NORTH
AMERICAN WORKSHOP ON CATACLYSMIC VARIABLES
AND RELATED OBJECTS
eds. S. Howell, E. Kuulkers, and C. Woodward
ISBN 1-886733-57-0

Vol. CS-138   1997 PACIFIC RIM CONFERENCE ON STELLAR ASTROPHYSICS
eds. Kwing Lam Chan, K. S. Cheng, and H. P. Singh
ISBN 1-886733-58-9

Vol. CS-139   PRESERVING THE ASTRONOMICAL WINDOWS:
Proceedings of Joint Discussion No. 5 of the 23rd General Assembly of the IAU
eds. Syuzo Isobe and Tomohiro Hirayama
ISBN 1-886733-59-7

Vol. CS-140   SYNOPTIC SOLAR PHYSICS --18th NSO/Sacramento Peak Summer Workshop
eds. K. S. Balasubramaniam, J. W. Harvey, and D. M. Rabin
ISBN 1-886733-60-0

Vol. CS-141   ASTROPHYSICS FROM ANTARCTICA:
A Symposium held as a part of the 109th Annual Meeting of the ASP
eds. Giles Novak and Randall H. Landsberg
ISBN 1-886733-61-9

Vol. CS-142   THE STELLAR INITIAL MASS FUNCTION: 38th Herstmonceux Conference
eds. Gerry Gilmore and Debbie Howell
ISBN 1-886733-62-7

Vol. CS-143*   THE SCIENTIFIC IMPACT OF THE GODDARD HIGH RESOLUTION
SPECTROGRAPH (GHRS)
eds. John C. Brandt, Thomas B. Ake III, and Carolyn Collins Petersen
ISBN 1-886733-63-5

Vol. CS-144   RADIO EMISSION FROM GALACTIC AND EXTRAGALACTIC COMPACT
SOURCES, IAU Colloquium 164
eds. J. Anton Zensus, G. B. Taylor, and J. M. Wrobel
ISBN 1-886733-64-3

Vol. CS-145   ASTRONOMICAL DATA ANALYSIS SOFTWARE AND SYSTEMS VII – (ADASS VII)
eds. Rudolf Albrecht, Richard N. Hook, and Howard A. Bushouse
ISBN 1-886733-65-1

Vol. CS-146   THE YOUNG UNIVERSE GALAXY FORMATION
AND EVOLUTION AT INTERMEDIATE AND HIGH REDSHIFT
eds. S. D'Odorico, A. Fontana, and E. Giallongo
ISBN 1-886733-66-X

ASP CONFERENCE SERIES VOLUMES
Published by the Astronomical Society of the Pacific

PUBLISHED: 1998 (* asterisk means OUT OF STOCK)

Vol. CS-147 ABUNDANCE PROFILES: DIAGNOSTIC TOOLS FOR GALAXY HISTORY
eds. Daniel Friedli, Mike Edmunds, Carmelle Robert, and Laurent Drissen
ISBN 1-886733-67-8

Vol. CS-148 ORIGINS
eds. Charles E. Woodward, J. Michael Shull, and Harley A. Thronson, Jr.
ISBN 1-886733-68-6

Vol. CS-149 SOLAR SYSTEM FORMATION AND EVOLUTION
eds. D. Lazzaro, R. Vieira Martins, S. Ferraz-Mello, J. Fernández, and C. Beaugé
ISBN 1-886733-69-4

Vol. CS-150 NEW PERSPECTIVES ON SOLAR PROMINENCES, IAU Colloquium 167
eds. David Webb, David Rust, and Brigitte Schmieder
ISBN 1-886733-70-8

Vol. CS-151 COSMIC MICROWAVE BACKGROUND
AND LARGE SCALE STRUCTURES OF THE UNIVERSE
eds. Yong-Ik Byun and Kin-Wang Ng
ISBN 1-886733-71-6

Vol. CS-152 FIBER OPTICS IN ASTRONOMY III
eds. S. Arribas, E. Mediavilla, and F. Watson
ISBN 1-886733-72-4

Vol. CS-153 LIBRARY AND INFORMATION SERVICES IN ASTRONOMY III -- (LISA III)
eds. Uta Grothkopf, Heinz Andernach, Sarah Stevens-Rayburn,
and Monique Gomez
ISBN 1-886733-73-2

Vol. CS-154 COOL STARS, STELLAR SYSTEMS AND THE SUN: Tenth Cambridge Workshop
eds. Robert A. Donahue and Jay A. Bookbinder
ISBN 1-886733-74-0

Vol. CS-155 SECOND ADVANCES IN SOLAR PHYSICS EUROCONFERENCE:
THREE-DIMENSIONAL STRUCTURE OF SOLAR ACTIVE REGIONS
eds. Costas E. Alissandrakis and Brigitte Schmieder
ISBN 1-886733-75-9

PUBLISHED: 1999

Vol. CS-156 HIGHLY REDSHIFTED RADIO LINES
eds. C. L. Carilli, S. J. E. Radford, K. M. Menten, and G. I. Langston
ISBN 1-886733-76-7

Vol. CS-157 ANNAPOLIS WORKSHOP ON MAGNETIC CATACLYSMIC VARIABLES
eds. Coel Hellier and Koji Mukai
ISBN 1-886733-77-5

Vol. CS-158 SOLAR AND STELLAR ACTIVITY: SIMILARITIES AND DIFFERENCES
eds. C. J. Butler and J. G. Doyle
ISBN 1-886733-78-3

Vol. CS-159 BL LAC PHENOMENON
eds. Leo O. Takalo and Aimo Sillanpää
ISBN 1-886733-79-1

ASP CONFERENCE SERIES VOLUMES
Published by the Astronomical Society of the Pacific

PUBLISHED: 1999 (* asterisk means OUT OF STOCK)

Vol. CS-160   ASTROPHYSICAL DISCS: An EC Summer School
eds. J. A. Sellwood and Jeremy Goodman
ISBN 1-886733-80-5

Vol. CS-161   HIGH ENERGY PROCESSES IN ACCRETING BLACK HOLES
eds. Juri Poutanen and Roland Svensson
ISBN 1-886733-81-3

Vol. CS-162   QUASARS AND COSMOLOGY
eds. Gary Ferland and Jack Baldwin
ISBN 1-886733-83-X

Vol. CS-163   STAR FORMATION IN EARLY-TYPE GALAXIES
eds. Jordi Cepa and Patricia Carral
ISBN 1-886733-84-8

Vol. CS-164   ULTRAVIOLET–OPTICAL SPACE ASTRONOMY BEYOND HST
eds. Jon A. Morse, J. Michael Shull, and Anne L. Kinney
ISBN 1-886733-85-6

Vol. CS-165   THE THIRD STROMLO SYMPOSIUM: THE GALACTIC HALO
eds. Brad K. Gibson, Tim S. Axelrod, and Mary E. Putman
ISBN 1-886733-86-4

Vol. CS-166   STROMLO WORKSHOP ON HIGH-VELOCITY CLOUDS
eds. Brad K. Gibson and Mary E. Putman
ISBN 1-886733-87-2

Vol. CS-167   HARMONIZING COSMIC DISTANCE SCALES IN A POST-HIPPARCOS ERA
eds. Daniel Egret and André Heck
ISBN 1-886733-88-0

Vol. CS-168   NEW PERSPECTIVES ON THE INTERSTELLAR MEDIUM
eds. A. R. Taylor, T. L. Landecker, and G. Joncas
ISBN 1-886733-89-9

Vol. CS-169   11th EUROPEAN WORKSHOP ON WHITE DWARFS
eds. J.-E. Solheim and E. G. Meištas
ISBN 1-886733-91-0

Vol. CS-170   THE LOW SURFACE BRIGHTNESS UNIVERSE, IAU Colloquium 171
eds. J. I. Davies, C. Impey, and S. Phillipps
ISBN 1-886733-92-9

Vol. CS-171   LiBeB, COSMIC RAYS, AND RELATED X- AND GAMMA-RAYS
eds. Reuven Ramaty, Elisabeth Vangioni-Flam, Michel Cassé, and Keith Olive
ISBN 1-886733-93-7

Vol. CS-172   ASTRONOMICAL DATA ANALYSIS SOFTWARE AND SYSTEMS VIII – (ADASS VIII)
eds. David M. Mehringer, Raymond L. Plante, and Douglas A. Roberts
ISBN 1-886733-94-5

Vol. CS-173   THEORY AND TESTS OF CONVECTION IN STELLAR STRUCTURE:
First Granada Workshop
ed. Álvaro Giménez, Edward F. Guinan, and Benjamín Montesinos
ISBN 1-886733-95-3

ASP CONFERENCE SERIES VOLUMES
Published by the Astronomical Society of the Pacific

PUBLISHED: 1999 (* asterisk means OUT OF STOCK)

Vol. CS-174   CATCHING THE PERFECT WAVE: ADAPTIVE OPTICS AND INTERFEROMETRY IN THE 21st CENTURY,
A Symposium held as a part of the 110th Annual Meeting of the ASP
eds. Sergio R. Restaino, William Junor, and Nebojsa Duric
ISBN 1-886733-96-1

Vol. CS-175   STRUCTURE AND KINEMATICS OF QUASAR BROAD LINE REGIONS
eds. C. M. Gaskell, W. N. Brandt, M. Dietrich, D. Dultzin-Hacyan, and M. Eracleous
ISBN 1-886733-97-X

Vol. CS-176   OBSERVATIONAL COSMOLOGY: THE DEVELOPMENT OF GALAXY SYSTEMS
eds. Giuliano Giuricin, Marino Mezzetti, and Paolo Salucci
ISBN 1-58381-000-5

Vol. CS-177   ASTROPHYSICS WITH INFRARED SURVEYS: A Prelude to SIRTF
eds. Michael D. Bicay, Chas A. Beichman, Roc M. Cutri, and Barry F. Madore
ISBN 1-58381-001-3

Vol. CS-178   STELLAR DYNAMOS: NONLINEARITY AND CHAOTIC FLOWS
eds. Manuel Núñez and Antonio Ferriz-Mas
ISBN 1-58381-002-1

Vol. CS-179   ETA CARINAE AT THE MILLENNIUM
eds. Jon A. Morse, Roberta M. Humphreys, and Augusto Damineli
ISBN 1-58381-003-X

Vol. CS-180   SYNTHESIS IMAGING IN RADIO ASTRONOMY II
eds. G. B. Taylor, C. L. Carilli, and R. A. Perley
ISBN 1-58381-005-6

Vol. CS-181   MICROWAVE FOREGROUNDS
eds. Angelica de Oliveira-Costa and Max Tegmark
ISBN 1-58381-006-4

Vol. CS-182   GALAXY DYNAMICS: A Rutgers Symposium
eds. David Merritt, J. A. Sellwood, and Monica Valluri
ISBN 1-58381-007-2

Vol. CS-183   HIGH RESOLUTION SOLAR PHYSICS: THEORY, OBSERVATIONS, AND TECHNIQUES
eds. T. R. Rimmele, K. S. Balasubramaniam, and R. R. Radick
ISBN 1-58381-009-9

Vol. CS-184   THIRD ADVANCES IN SOLAR PHYSICS EUROCONFERENCE: MAGNETIC FIELDS AND OSCILLATIONS
eds. B. Schmieder, A. Hofmann, and J. Staude
ISBN 1-58381-010-2

Vol. CS-185   PRECISE STELLAR RADIAL VELOCITIES, IAU Colloquium 170
eds. J. B. Hearnshaw and C. D. Scarfe
ISBN 1-58381-011-0

ASP CONFERENCE SERIES VOLUMES
Published by the Astronomical Society of the Pacific

PUBLISHED: 1999 (* asterisk means OUT OF STOCK)

Vol. CS-186   THE CENTRAL PARSECS OF THE GALAXY
eds. Heino Falcke, Angela Cotera, Wolfgang J. Duschl, Fulvio Melia, and Marcia J. Rieke
ISBN 1-58381-012-9

Vol. CS-187   THE EVOLUTION OF GALAXIES ON COSMOLOGICAL TIMESCALES
eds. J. E. Beckman and T. J. Mahoney
ISBN 1-58381-013-7

Vol. CS-188   OPTICAL AND INFRARED SPECTROSCOPY OF CIRCUMSTELLAR MATTER
eds. Eike W. Guenther, Bringfried Stecklum, and Sylvio Klose
ISBN 1-58381-014-5

Vol. CS-189   CCD PRECISION PHOTOMETRY WORKSHOP
eds. Eric R. Craine, Roy A. Tucker, and Jeannette Barnes
ISBN 1-58381-015-3

Vol. CS-190   GAMMA-RAY BURSTS: THE FIRST THREE MINUTES
eds. Juri Poutanen and Roland Svensson
ISBN 1-58381-016-1

Vol. CS-191   PHOTOMETRIC REDSHIFTS AND HIGH REDSHIFT GALAXIES
eds. Ray J. Weymann, Lisa J. Storrie-Lombardi, Marcin Sawicki, and Robert J. Brunner
ISBN 1-58381-017-X

Vol. CS-192   SPECTROPHOTOMETRIC DATING OF STARS AND GALAXIES
ed. I. Hubeny, S. R. Heap, and R. H. Cornett
ISBN 1-58381-018-8

Vol. CS-193   THE HY-REDSHIFT UNIVERSE:
GALAXY FORMATION AND EVOLUTION AT HIGH REDSHIFT
eds. Andrew J. Bunker and Wil J. M. van Breugel
ISBN 1-58381-019-6

Vol. CS-194   WORKING ON THE FRINGE:
OPTICAL AND IR INTERFEROMETRY FROM GROUND AND SPACE
eds. Stephen Unwin and Robert Stachnik
ISBN 1-58381-020-X

PUBLISHED: 2000

Vol. CS-195   IMAGING THE UNIVERSE IN THREE DIMENSIONS:
Astrophysics with Advanced Multi-Wavelength Imaging Devices
eds. W. van Breugel and J. Bland-Hawthorn
ISBN 1-58381-022-6

Vol. CS-196   THERMAL EMISSION SPECTROSCOPY AND ANALYSIS OF DUST, DISKS, AND REGOLITHS
eds. Michael L. Sitko, Ann L. Sprague, and David K. Lynch
ISBN: 1-58381-023-4

ASP CONFERENCE SERIES VOLUMES
Published by the Astronomical Society of the Pacific

PUBLISHED: 2000 (* asterisk means OUT OF STOCK)

Vol. CS-197 XVth IAP MEETING DYNAMICS OF GALAXIES:
FROM THE EARLY UNIVERSE TO THE PRESENT
eds. F. Combes, G. A. Mamon, and V. Charmandaris
ISBN: 1-58381-24-2

Vol. CS-198 EUROCONFERENCE ON "STELLAR CLUSTERS AND ASSOCIATIONS:
CONVECTION, ROTATION, AND DYNAMOS"
eds. R. Pallavicini, G. Micela, and S. Sciortino
ISBN: 1-58381-25-0

Vol. CS-199 ASYMMETRICAL PLANETARY NEBULAE II:
FROM ORIGINS TO MICROSTRUCTURES
eds. J. H. Kastner, N. Soker, and S. Rappaport
ISBN: 1-58381-026-9

Vol. CS-200 CLUSTERING AT HIGH REDSHIFT
eds. A. Mazure, O. Le Fèvre, and V. Le Brun
ISBN: 1-58381-027-7

Vol. CS-201 COSMIC FLOWS 1999: TOWARDS AN UNDERSTANDING
OF LARGE-SCALE STRUCTURES
eds. Stéphane Courteau, Michael A. Strauss, and Jeffrey A. Willick
ISBN: 1-58381-028-5

Vol. CS-202 PULSAR ASTRONOMY – 2000 AND BEYOND, IAU Colloquium 177
eds. M. Kramer, N. Wex, and R. Wielebinski
ISBN: 1-58381-029-3

Vol. CS-203 THE IMPACT OF LARGE-SCALE SURVEYS ON PULSATING STAR RESEARCH,
IAU Colloquium 176
eds. L. Szabados and D. W. Kurtz
ISBN: 1-58381-030-7

Vol. CS-204 THERMAL AND IONIZATION ASPECTS OF FLOWS FROM HOT STARS:
OBSERVATIONS AND THEORY
eds. Henny J. G. L. M. Lamers and Arved Sapar
ISBN: 1-58381-031-5

Vol. CS-205 THE LAST TOTAL SOLAR ECLIPSE OF THE MILLENNIUM IN TURKEY
eds. W. C. Livingston and A. Özgüç
ISBN: 1-58381-032-3

Vol. CS-206 HIGH ENERGY SOLAR PHYSICS – *ANTICIPATING HESSI*
eds. Reuven Ramaty and Natalie Mandzhavidze
ISBN: 1-58381-033-1

Vol. CS-207 NGST SCIENCE AND TECHNOLOGY EXPOSITION
eds. Eric P. Smith and Knox S. Long
ISBN: 1-58381-036-6

ATLAS VISIBLE AND NEAR INFRARED ATLAS OF THE
ARCTURUS SPECTRUM 3727-9300 Å
eds. Kenneth Hinkle, Lloyd Wallace, Jeff Valenti, and Dianne Harmer
ISBN: 1-58381-037-4

ASP CONFERENCE SERIES VOLUMES
Published by the Astronomical Society of the Pacific

PUBLISHED: 2000 (* asterisk means OUT OF STOCK)

Vol. CS-208   POLAR MOTION: HISTORICAL AND SCIENTIFIC PROBLEMS,
IAU Colloquium 178
eds. Steven Dick, Dennis McCarthy, and Brian Luzum
ISBN: 1-58381-039-0

Vol. CS-209   SMALL GALAXY GROUPS, IAU Colloquium 174
eds. Mauri J. Valtonen and Chris Flynn
ISBN: 1-58381-040-4

Vol. CS-210   DELTA SCUTI AND RELATED STARS: Reference Handbook
and Proceedings of the 6th Vienna Workshop in Astrophysics
eds. Michel Breger and Michael Houston Montgomery
ISBN: 1-58381-043-9

Vol. CS-211   MASSIVE STELLAR CLUSTERS
eds. Ariane Lançon and Christian M. Boily
ISBN: 1-58381-042-0

Vol. CS-212   FROM GIANT PLANETS TO COOL STARS
eds. Caitlin A. Griffith and Mark S. Marley
ISBN: 1-58381-041-2

Vol. CS-213   BIOASTRONOMY `99: A NEW ERA IN BIOASTRONOMY
eds. Guillermo A. Lemarchand and Karen J. Meech
ISBN: 1-58381-044-7

Vol. CS-214   THE Be PHENOMENON IN EARLY-TYPE STARS, IAU Colloquium 175
eds. Myron A. Smith, Huib F. Henrichs and Juan Fabregat
ISBN: 1-58381-045-5

Vol. CS-215   COSMIC EVOLUTION AND GALAXY FORMATION:
STRUCTURE, INTERACTIONS AND FEEDBACK
The 3rd Guillermo Haro Astrophysics Conference
eds. José Franco, Elena Terlevich, Omar López-Cruz, and Itziar Aretxaga
ISBN: 1-58381-046-3

Vol. CS-216   ASTRONOMICAL DATA ANALYSIS SOFTWARE AND SYSTEMS IX -- (ADASS IX)
eds. Nadine Manset, Christian Veillet, and Dennis Crabtree
ISBN: 1-58381-047-1   ISSN: 1080-7926

Vol. CS-217   IMAGING AT RADIO THROUGH SUBMILLIMETER WAVELENGTHS
eds. Jeffrey G. Mangum and Simon J. E. Radford
ISBN: 1-58381-049-8

Vol. CS-218   MAPPING THE HIDDEN UNIVERSE: THE UNIVERSE BEHIND THE MILKYWAY
THE UNIVERSE IN HI
eds. Renée C. Kraan-Korteweg, Patricia A. Henning, and Heinz Andernach
ISBN: 1-58381-050-1

Vol. CS-219   DISKS, PLANETESIMALS, AND PLANETS
eds. F. Garzón, C. Eiroa, D. de Winter, and T. J. Mahoney
ISBN: 1-58381-051-X

ASP CONFERENCE SERIES VOLUMES
Published by the Astronomical Society of the Pacific

PUBLISHED: 2000 (* asterisk means OUT OF STOCK)

Vol. CS-220   AMATEUR - PROFESSIONAL PARTNERSHIPS IN ASTRONOMY:
The 111th Annual Meeting of the ASP
eds. John R. Percy and Joseph B. Wilson
ISBN: 1-58381-052-8

Vol. CS-221   STARS, GAS AND DUST IN GALAXIES: EXPLORING THE LINKS
eds. Danielle Alloin, Knut Olsen, and Gaspar Galaz
ISBN: 1-58381-053-6

PUBLISHED: 2001

Vol. CS-222   THE PHYSICS OF GALAXY FORMATION
eds. M. Umemura and H. Susa
ISBN: 1-58381-054-4

Vol. CS-223   COOL STARS, STELLAR SYSTEMS AND THE SUN:
Eleventh Cambridge Workshop
eds. Ramón J. García López, Rafael Rebolo, and María Zapatero Osorio
ISBN: 1-58381-056-0

Vol. CS-224   PROBING THE PHYSICS OF ACTIVE GALACTIC NUCLEI
BY MULTIWAVELENGTH MONITORING
eds. Bradley M. Peterson, Ronald S. Polidan, and Richard W. Pogge
ISBN: 1-58381-055-2

Vol. CS-225   VIRTUAL OBSERVATORIES OF THE FUTURE
eds. Robert J. Brunner, S. George Djorgovski, and Alex S. Szalay
ISBN: 1-58381-057-9

All book orders or inquiries concerning ASP or IAU volumes listed should be directed to the:

The Astronomical Society of the Pacific Conference Series
390 Ashton Avenue
San Francisco CA 94112-1722 USA

Phone: 415-337-2126
Fax: 415-337-5205
E-mail: catalog@aspsky.org
Web Site: http://www.aspsky.org

Complete lists of proceedings of past IAU Meetings are maintained at the
IAU Web site at the URL: http://www.iau.org/publicat.html

Volumes 32 - 189 in the IAU Symposia Series may be ordered from
Kluwer Academic Publishers
P. O. Box 117
NL 3300 AA Dordrecht
The Netherlands

INTERNATIONAL ASTRONOMICAL UNION (IAU) VOLUMES
Published by the Astronomical Society of the Pacific

PUBLISHED: 1999

Vol. No. 190 NEW VIEWS OF THE MAGELLANIC CLOUDS
eds. You-Hua Chu, Nicholas B. Suntzeff, James E. Hesser,
and David A. Bohlender
ISBN: 1-58381-021-8

Vol. No. 191 ASYMPTOTIC GIANT BRANCH STARS
eds. T. Le Bertre, A. Lèbre, and C. Waelkens
ISBN: 1-886733-90-2

Vol. No. 192 THE STELLAR CONTENT OF LOCAL GROUP GALAXIES
eds. Patricia Whitelock and Russell Cannon
ISBN: 1-886733-82-1

Vol. No. 193 WOLF-RAYET PHENOMENA IN MASSIVE STARS AND STARBURST GALAXIES
eds. Karel A. van der Hucht, Gloria Koenigsberger, and Philippe R. J. Eenens
ISBN: 1-58381-004-8

Vol. No. 194 ACTIVE GALACTIC NUCLEI AND RELATED PHENOMENA
eds. Yervant Terzian, Daniel Weedman, and Edward Khachikian
ISBN: 1-58381-008-0

PUBLISHED: 2000

Vol. XXIVA TRANSACTIONS OF THE INTERNATIONAL ASTRONOMICAL UNION
REPORTS ON ASTRONOMY 1996-1999
ed. Johannes Andersen
ISBN: 1-58381-035-8

Vol. No. 195 HIGHLY ENERGETIC PHYSICAL PROCESSES AND MECHANISMS FOR
EMISSION FROM ASTROPHYSICAL PLASMAS
eds. P. C. H. Martens, S. Tsuruta, and M. A. Weber
ISBN: 1-58381-038-2

Vol. No. 197 ASTROCHEMISTRY: FROM MOLECULAR CLOUDS TO PLANETARY SYSTEMS
eds. Y. C. Minh and E. F. van Dishoeck
ISBN: 1-58381-034-X

Vol. No. 198 THE LIGHT ELEMENTS AND THEIR EVOLUTION
eds. L. da Silva, M. Spite, and J. R. de Medeiros
ISBN: 1-58381-048-X

PUBLISHED: 2001

Vol. No. 204 THE EXTRAGALACTIC INFRARED BACKGROUND AND ITS
COSMOLOGICAL IMPLICATIONS
eds. Martin Harwit and Michael G. Hauser
ISBN: 1-58381-062-5

IAU SPS ASTRONOMY FOR DEVELOPING COUNTRIES
Special Session of the XXIV General Assembly of the IAU
ed. Alan H. Batten
ISBN: 1-58381-067-6